CREEP AND RELAXATION OF NONLINEAR VISCOELASTIC MATERIALS

with an
Introduction to Linear Viscoelasticity

BY

WILLIAM N. FINDLEY
Professor Emeritus of Engineering
Brown University

JAMES S. LAI
Professor of Civil Engineering
Georgia Institute of Technology

and

KASIF ONARAN
Professor of Civil Engineering
Istanbul Technical University

DOVER PUBLICATIONS, INC., New York

Published in Canada by General Publishing Company, Ltd., 30 Lesmill Road, Don Mills, Toronto, Ontario.
Published in the United Kingdom by Constable and Company, Ltd., 10 Orange Street, London WC2H 7EG.

This Dover edition, first published in 1989, is an unabridged, corrected republication of the work originally published in 1976 by North-Holland Publishing Company, Amsterdam, as Volume 18 in the "North-Holland Series in Applied Mathematics and Mechanics." A new section of additional literature relating to Chapter 2 has been added in this edition at the end of the bibliography.

Manufactured in the United States of America
Dover Publications, Inc., 31 East 2nd Street, Mineola, N.Y. 11501

Library of Congress Cataloging-in-Publication Data

Findley, William N. (William Nichols)
Creep and relaxation of nonlinear viscoelastic materials : with an introduction to linear viscoelasticity / William N. Findley, James S. Lai, Kasif Onaran.
p. cm.
Corr. and slightly enlarged ed.
Bibliography: p.
Includes index.
ISBN 0-486-66016-8
1. Viscoelasticity. 2. Materials—Creep. 3. Stress relaxation. I. Lai, James S. II. Onaran, Kasif. III. Title.
TA418.2.F48 1989
620.1′1233—dc20 89-1462
CIP

PREFACE

This book was started in 1968 with the realization that nonlinear viscoelasticity was an active, developing subject, so that a book at this time could not be a definitive treatise. However, the need for information on creep and relaxation in the nonlinear range for critical designs suggested that a book on the subject would be useful.

Accordingly the present volume was prepared with the object of presenting the material in such a way as to be readable and useful to designers as well as research workers and students. To this end the mathematical background required has been kept to a minimum and supplemented by explanations where it has been necessary to introduce specialized mathematics. Also, appendices have been included to provide sufficient background in Laplace transforms and in step functions.

Chapters 1 and 2 contain an introduction and a historical review of creep. As an aid to the reader a background on stress, strain and stress analysis is provided in Chapters 3 and 4, an introduction to linear viscoelasticity is found in Chapter 5 and linear viscoelastic stress analysis in Chapter 6. While the main thrust of the book is creep and stress relaxation, the chapter on linear viscoelasticity includes oscillatory stress and strain as well. This was included because it is sometimes desirable to have a broader time scale than can be provided by creep tests alone when describing the time-dependent behavior of a material. To this end creep or relaxation and oscillatory tests can be complementary.

It was found necessary for reasonable brevity to employ Cartesian tensor notation. However, at appropriate points the components of stress and strain have been expressed in conventional notation as well. To provide the greatest usefulness for those interested in specific applications, expressions for general results have often been reduced to several common states of stress or strain.

While more than one approach to nonlinear viscoelasticity was being explored in the literature at the time the book was started, and others have been proposed since, it was decided to treat one approach in depth while only outlining other methods. The approach considered in detail in Chapters 7, 8 and 9 is the multiple integral theory with simplifications to single integrals. Other simplifications considered in some detail are the assumptions of incompressibility and linear compressibility. These topics are developed in Chapters 7, 8 and 9.

The relation between the responses of viscoelastic materials to (a) stress boundary conditions (creep), (b) strain boundary conditions (relaxation) and (c) mixed stress and strain boundary conditions (simultaneous creep and relaxation) are considered in Chapter 10. Chapter 11 treats the problem of the effect of temperature on nonlinear creep—especially variable temperature.

Methods of characterization of kernel functions and examples of experimental results are presented (wherever possible) in Chapters 8, 9 and 11 to assist the reader to obtain a physical feeling about the nonlinear behavior of various materials.

Stress analysis of nonlinear viscoelastic materials presents many difficulties. As a result, the field is not highly developed. In Chapter 12 the stress analysis of nonlinear viscoelastic materials is introduced and several problems of increasing complexity are discussed.

Finally, in deference to the extensive experimental background of the authors, Chapter 13 discusses experimental methods for creep and stress relaxation under combined stress. This chapter considers especially those experimental problems which must be solved properly when reliable experimental results of high precision are required.

In order that the book be as up-to-date as possible at the time of publication a list of "Additional Literature" has been appended to several chapters This list is presented without citation in the text and (with one exception) without evaluation or comment.

The book was started by the senior author while on sabbatical leave from Brown University as a visiting Professor at the University of Auckland, New Zealand. The quiet office provided by the University of Auckland and the helpful discussions with the staff are much appreciated. The interest of the administration of the Division of Engineering of Brown University in this project is gratefully acknowledged.

The authors are especially indepted to Professor E. H. Lee for reading the completed manuscript and offering many helpful suggestions.

We also thank Professor P. R. Paslay for reading Section 7.5, "Incompressible Material Assumption," and for many helpful discussions, and

Professor A. C. Pipkin for also reading Section 7.5. Helpful discussions with Professors J. E. Fitzgerald and M. L. Williams, Jr., are very much appreciated. The assistance of Dr. B. Erman in checking calculations and R. Mark in checking references is gratefully acknowledged.

Thanks are due the several secretaries at different universities, especially Mrs. Madeline C. Gingrich, for their care in the difficult task of typing the manuscript.

July, 1975

William N. Findley
James S. Lai
Kasif Onaran

CONTENTS

CHAPTER 1

INTRODUCTION

Design of modern high performance machines and structures often must take account of the effect of complex states of stress, strain and environment on the mechanical behavior of different classes of materials. Calculation of the mechanical behavior of a structural member under different conditions of stress or strain and environment requires that the different variables involved be related by means of fundamental equations including the following: (a) The equilibrium equations. These state the relationship among the various stress components at any given point required for equilibrium. (b) The kinematic equations. These express the strain components in terms of displacements which in turn describe the deformation of the body. (c) The compatibility equation. This states the relationship which must exist among the several strain components in order that the strain components in a continuous medium not produce discontinuities. (d) The constitutive equation. This must describe the relationship between stress, strain and time in terms of the material constants for a given material. (e) A set of boundary conditions. These describe the stresses and displacements prescribed at the boundaries. If the material behavior is linear in stress and time independent, then Hooke's law describes the constitutive relationship. A detailed description of equations (a), (b), (c), (d) and (e) for linear, time-independent materials may be found in any book on the theory of elasticity.

In this book emphasis will be placed on discussion of the constitutive equations for time-dependent and nonlinear materials, though an introduction to time-dependent linear behavior is also discussed. Actual materials exhibit a great variety of behavior. However, by means of idealization they can be simplified and classified as follows.

1.1 Elastic Behavior

Most materials behave elastically or nearly so under small stresses. As illustrated by the solid curve in Fig. 1.1, an immediate elastic strain response is obtained upon loading. Then the strain stays constant as long

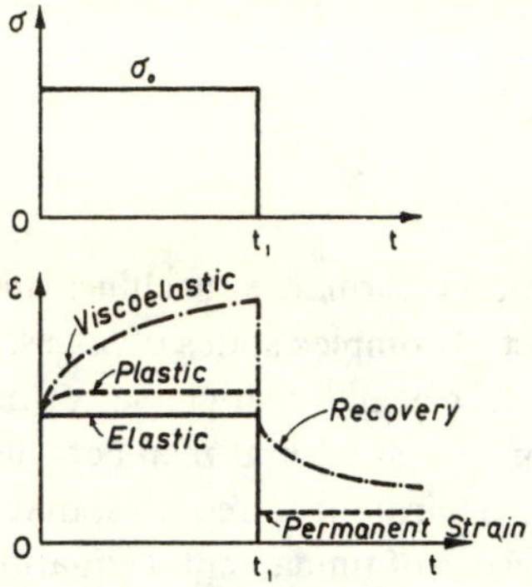

Fig. 1.1. Various Strain Responses to a Constant Load

as the stress is fixed and disappears immediately upon removal of the load. The chief characteristic of elastic strain is reversibility. Most elastic materials are linearly elastic so that doubling the stress in the elastic range doubles the strain.

1.2 Plastic Behavior

If the stress is too high the behavior is no longer elastic. The limiting stress above which the behavior is no longer elastic is called the elastic limit. The strain that does not disappear after removal of the stress is called the inelastic strain. In some materials, the strain continues to increase for a short while after the load is fully applied, and then remains constant under a fixed load, but a permanent strain remains after the stress is removed. This permanent strain is called the plastic strain (dashed curves in Fig. 1.1). Plastic strain is defined as time independent although some time dependent strain is often observed to accompany plastic strain.

1.3 Viscoelastic Behavior

Some materials exhibit elastic action upon loading (if loading is rapid enough), then a slow and continuous increase of strain at a decreasing rate

is observed. When the stress is removed a continuously decreasing strain follows an initial elastic recovery. Such materials are significantly influenced by the rate of straining or stressing; i.e., for example, the longer the time to reach the final value of stress at a constant rate of stressing, the larger is the corresponding strain. These materials are called viscoelastic (dot-dash curve in Fig. 1.1). Among the materials showing viscoelastic behavior are plastics, wood, natural and synthetic fibers, concrete and metals at elevated temperatures. Since time is a very important factor in their behavior, they are also called time-dependent materials. As its name implies, viscoelasticity combines elasticity and viscosity (viscous flow).

The time-dependent behavior of viscoelastic materials must be expressed by a constitutive equation which includes time as a variable in addition to the stress and strain variables. Even under the most simple loading program, as shown in Fig. 1.1, the shape of the strain-time curve, in this case a creep curve, may be rather complicated. Since time cannot be kept constant, reversed or eliminated during an experiment, the experimental study of the mechanical behavior of such materials is much more difficult than the study of time-independent materials.

Recent developments in technology, such as gas turbines, jet engines, nuclear power plants, and space crafts, have placed severe demands on high temperature performance of materials, including plastics. Consequently the time-dependent behavior of materials has become of great importance.

The time-dependent behavior of materials under a quasi-static state may be studied by means of three types of experiments: creep (including recovery following creep), stress relaxation and constant rate stressing (or straining), although other types of experiments are also available.

1.4 Creep

Creep is a slow continuous deformation of a material under constant stress.* However, creep in general may be described in terms of three different stages illustrated in Fig. 1.2. The first stage in which creep occurs at a decreasing rate is called primary creep; the second, called the secondary stage, proceeds at a nearly constant rate; and the third or tertiary stage occurs at an increasing rate and terminates in fracture.

---

* Most creep experiments are performed under a constant load even when the cross-section of the specimen changes significantly with time. For small strains, constant load and constant stress experiments are the same.

Total strain ε at any instant of time t in a creep test of a linear material (linearity will be defined later) is represented as the sum of the instantaneous elastic strain ε^e and the creep strain ε^c,

$$\varepsilon = \varepsilon^e + \varepsilon^c. \tag{1.1}$$

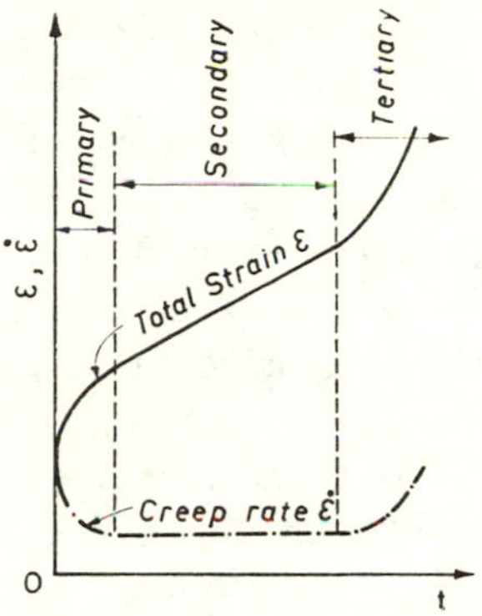

Fig. 1.2. Three Stages of Creep

The strain rate $\dot{\varepsilon}$ is found by differentiating (1.1) and noting that ε^e is a constant:

$$\frac{d\varepsilon}{dt} = \frac{d\varepsilon^c}{dt} = \dot{\varepsilon}. \tag{1.2}$$

1.5 Recovery

If the load is removed, a reverse elastic strain followed by recovery of a portion of the creep strain will occur at a continuously decreasing rate.

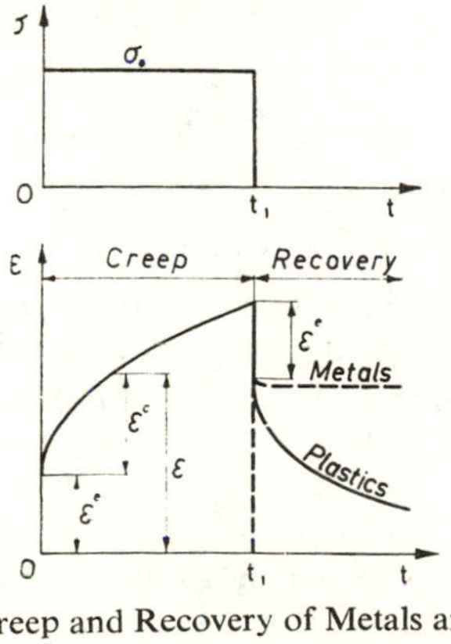

Fig. 1.3. Creep and Recovery of Metals and Plastics

The amount of the time-dependent recoverable strain during recovery is generally a very small part of the time-dependent creep strain for metals, whereas for plastics it may be a large portion of the time-dependent creep strain which occurred (Fig. 1.3). Some plastics may exhibit full recovery if sufficient time is allowed for recovery. The strain recovery is also called delayed elasticity.

1.6 Relaxation

Viscoelastic materials subjected to a constant strain will relax under constant strain so that the stress gradually decreases as shown in Fig. 1.4.

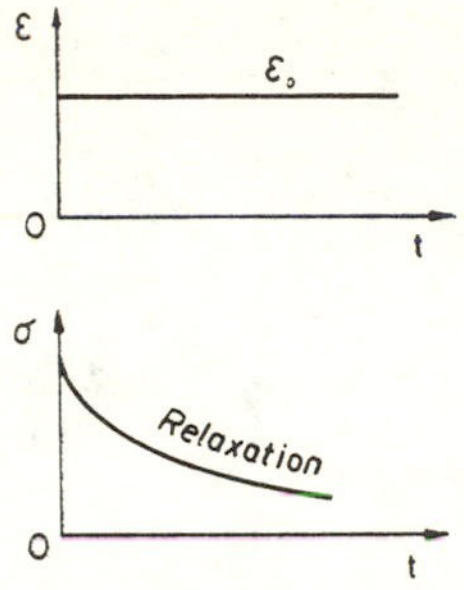

Fig. 1.4. Stress Relaxation at Constant Strain

From a study of the three time-dependent responses of materials explained above, the basic principles governing time-dependent behavior under loading conditions other than those mentioned above, may be established. In actual practice, the stress or strain history may approximate one of those described or a mixture, i.e., creep and relaxation may occur simultaneously under combined loading, or the load or strain history may be a cyclic or random variation.

1.7 Linearity

The material is said to be linearly viscoelastic if stress is proportional to strain at a given time, and the linear superposition principle holds. These linear requirements can be stated mathematically in two equations:

$$\varepsilon[c\sigma(t)] = c\varepsilon[\sigma(t)], \tag{1.3}$$

$$\varepsilon[\sigma_1(t)+\sigma_2(t-t_1)] = \varepsilon[\sigma_1(t)]+\varepsilon[\sigma_2(t-t_1)], \tag{1.4}$$

in which ε and σ are the strain output and stress input, respectively, and c is a constant. Equation (1.3) states that the strain-time output due to a stress input $c\sigma(t)$ equals the scalar c times the strain output due to a stress input $\sigma(t)$. The second requirement, (1.4), states that the strain output due to the combination of two arbitrary but different stress inputs applied at different times, $\varepsilon[\sigma_1(t)+\sigma_2(t-t_1)]$ equals the sum of the strain outputs resulting from $\sigma_1(t)$ and $\sigma_2(t-t_1)$ each acting separately. This second requirement is usually called the Boltzmann superposition principle [1.1].* These two requirements are illustrated in Figure 1.5a and 1.5b.

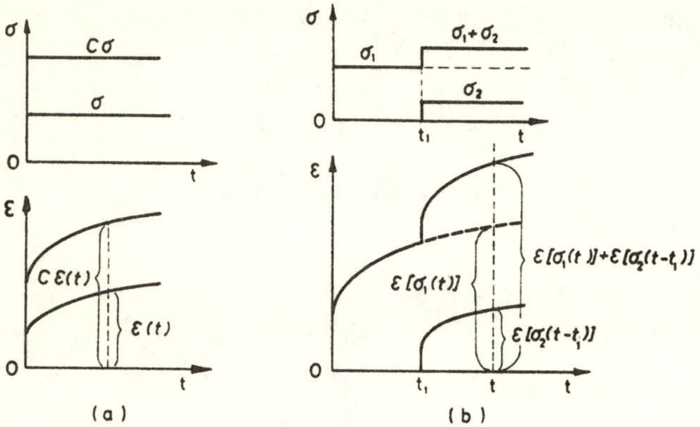

Fig. 1.5. Illustration of Behavior of a Linear Material

Most materials are nearly linear over certain ranges of the variables, stress, strain, time, temperature and nonlinear over larger ranges of some of the variables. Any demarkation of the boundary between nearly linear (where an assumption of linear behavior is acceptable) and nonlinear is arbitrary. The maximum permissible deviation from linear behavior of a material which allows a linear theory to be employed with acceptable accuracy depends on the stress distribution, the type of application and the background of experience. For example, under very short duration of loading many plastics behave linearly even at stresses for which considerable nonlinearity is found if the duration of loading is much longer. In fact, the strain during creep of many plastics can be separated into a time-independent linear part and a time-dependent nonlinear part (see [1.2] for example). Consider another situation. The deflection of a member containing a steep

* Numbers in brackets refer to references at the end of the book (p. 344).

stress gradient, such as a bar in tension containing a hole, may be essentially linear even under forces which cause highly nonlinear behavior in the very localized region of highest stress around the hole. Also in certain applications the accuracy required of deflection calculations is much greater than in some others (gears, compared to pipes, for example).

After all factors are taken into account, a designer must decide on an acceptable range of the variables over which linear theory may be employed. When the requirements of the design exceed these limits the use of linear constitutive equations will yield only a poor approximation of the actual behavior of the material. A more accurate description for viscoelastic materials may be obtained by using nonlinear viscoelastic theory. Such theory will include stress terms of order higher than the first and will therefore be more complex than linear theory.

CHAPTER 2

HISTORICAL SURVEY OF CREEP

2.1 Creep of Metals

The experimental and metallurgical aspects of creep of metals have been treated in a number of books beginning with Norton [2.1] and Tapsell [2.2] and followed by Hoff [2.3], Finnie-Heller [2.4], Kachanov [2.5], Lubahn-Felgar [2.6], Dorn [2.7], Odqvist-Hult [2.8], Kennedy [2.9] and more recently Odqvist [2.10], Hult [2.11], Rabotnov [2.12], and Smith-Nicholson [2.13]. In recent publications on creep of metals the concept of dislocations has been employed to explain the physical phenomena of creep in terms of the structure of metals (see, for example [2.9]). These books have generally concentrated their attention on the primary and secondary stages of creep.

2.2 Creep under Uniaxial Stress

A systematic observation of the creep phenomenon was first reported in 1834 by Vicat [2.14]. Thurston [2.15] seems to be the first to propose three stages of creep as shown in Fig. 1.2 and to study stress relaxation.

Andrade [2.16] made a systematic investigation of creep of lead wires under constant load and in 1910 proposed the first creep law as follows:

$$l = l_0(1+\beta t^{1/3})e^{kt}, \tag{2.1}$$

where l_0 and l are initial and current length of the specimen, respectively, t is the time under load, β and k are material constants which depend on stress. Since then various empirical equations have been proposed for the stress dependence of creep of metals based on experimental observations. The steady state of creep (secondary creep) particularly has been studied in great detail. Bailey [2.17] and Norton [2.1] suggested the empirical equation that yielded good agreement with experimental data for steady creep under low stresses,

$$\dot{\varepsilon} = k\sigma^p, \tag{2.2}$$

where $\dot{\varepsilon}$ is the steady creep rate, σ is the applied stress, k and p are material constants. Values of p have been found to lie between 3 and 7. Equation (2.2) is called the power law or Norton's creep law. The exponential law (2.3), as given below, was proposed by Ludwik [2.18] as a good approximation for the creep rate during steady creep,

$$\dot{\varepsilon} = k e^{\sigma/\sigma^+}, \tag{2.3}$$

where k and σ^+ are material constants. Both the power law (2.2) and the exponential law (2.3) have been used to represent the stress dependence of viscous flow when exhibited by metals or other materials, under constant uniaxial stress σ. If the stress vanishes, (2.3) predicts a finite strain rate, whereas it should be zero as predicted by (2.2). To eliminate this shortcoming Soderberg [2.19] proposed the following empirical equation,

$$\dot{\varepsilon} = c(e^{\sigma/\sigma^+} - 1), \tag{2.4}$$

where c and σ^+ are material constants.

Some writers, Nadai [2.20] for example, employed the hyperbolic sine function originally suggested by Prandtl [2.21] to describe the stress dependence of the steady creep rate.

$$\dot{\varepsilon} = D \sinh \frac{\sigma}{\sigma^+}, \tag{2.5}$$

where D and σ^+ are material constants. Equation (2.5) represents a behavior which is nearly linear for small stresses and nonlinear for large stresses. The above empirical equations have been found adequate to describe the secondary stage (steady-state) creep behavior of certain metals in tension under constant stress at temperatures where creep rates are rather small.

For some metals and applications involving relatively short loading times it is necessary to consider the primary creep range for which the following equation has been widely used (see for example, [2.22]) for the creep strain, ε^c,

$$\varepsilon^c = k\sigma^p t^n, \tag{2.6a}$$

from which the creep rate $\dot{\varepsilon}$ is

$$\dot{\varepsilon} = kn\sigma^p t^{n-1}, \tag{2.6b}$$

where k, p and n are material constants. Equation (2.6a) may be generalized as

$$\varepsilon^c = f(\sigma)\, g(t), \tag{2.7}$$

where $g(t)$ represents any suitable time function. It has been observed [2.23] that a power function of time for $g(t)$, such as t^n is in good agreement with many experiments on materials as different as plastics, metals and concrete. The exponent n is generally smaller than 0.5.

For values of n less than one, as usually encountered, the slope of the strain-time curve starts vertically then decreases continuously, rapidly at first and then more slowly. Thus the latter portion of a linear plot of ε^c vs t (2.6a) for any given time duration will look nearly like a straight line (see Fig. 2.1 for example [2.24]). This means that (2.6a) may be used to approx-

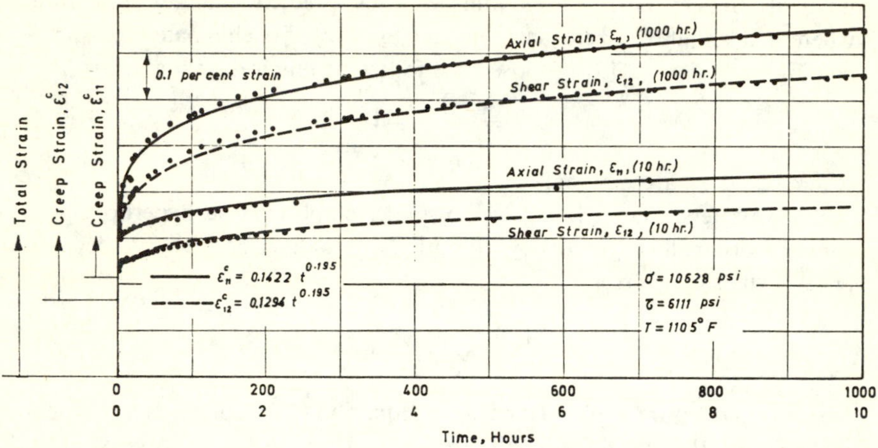

Fig. 2.1. Creep of Prestrained 304 Stainless Steel at 1105°F under Combined Tension $\sigma = 10{,}628$ psi and Torsion $\tau = 6111$ psi together with best representation of 1000 hr. of creep by Eq. 2.6a for each component of strain. Note similarity of the plot of the first 10 hr. compared to 1000 hr. The time independent strains ε^0 were $\varepsilon^0_{11} = 0.2150$, $\varepsilon^0_{12} = 0.1655$ per cent for tension and torsion respectively. From unpublished data by Findley, W. N. and Mark, R. [2.24].

imate both the primary stage of creep and a major portion of the secondary stage without any additional term for the secondary stage [2.23]. A potentially deceiving characteristic of (2.6a) is that if the scale of both ordinate and abscissa in a linear plot are multiplied by the same factor so that the time duration represented is changed by a large amount, the shape of a plot of (2.6a) will remain the same. Thus the apparent linearity of the latter portion of Fig. 2.1 is an illusion. Of course (2.6b) degenerates to (2.2) for second-stage creep by taking $n = 1$.

2.3 Creep under Combined Stresses

Creep experiments under combined stresses are much more difficult to perform than uniaxial creep and mathematical representation is much more involved for combined stresses. In spite of those difficulties, considerable work has been done to study the creep behavior of metals under combined stresses (see for example [2.25]).

Since creep deformation shows some similarities to plastic deformation of metals, the theories of plasticity and creep have been developed together. By transposition of the strain rates for the strains, the constitutive equations of plastic flow have been applied to creep. The elastic (time-independent) part of the strain is obtained from Hooke's law, while the time-dependent strain may be derived from the following assumptions:

a) Incompressibility of the material.

b) Coaxiality of principal directions of stress and strain rate.

It is a common hypothesis that material flow during deformation (both plastic flow and creep) depends on the magnitude of the effective stress σ_e defined as follows:

$$\sigma_e^2 = \sigma_{11}^2+\sigma_{22}^2+\sigma_{33}^3-\sigma_{11}\sigma_{22}-\sigma_{22}\sigma_{33}-\sigma_{11}\sigma_{33}+3(\sigma_{12}^2+\sigma_{23}^2+\sigma_{31}^2), \qquad (2.8)$$

where σ_{11}, σ_{22} and σ_{33} are normal stresses,* σ_{12}, σ_{31} and σ_{23} are shearing stresses.* The function on the right hand side of (2.8) was proposed by Mises [2.26]. It is also proportional to the square of the shearing stress on the octahedral plane, the plane making equal angles with the principal stress directions.

By employing the effective stress defined by (2.8) together with assumptions (a) and (b) stated above, (2.2) may be generalized to represent the strain rate tensor $\dot{\varepsilon}_{ij}$, for three dimensions as described by Odqvist [2.27]:

$$\dot{\varepsilon}_{ij} = \tfrac{3}{2}c\sigma_e^{p-1}\, s_{ij}, \qquad (2.9)$$

where s_{ij} is the tensor* of the stress deviator*, c and p are material constants. If (2.9) is expanded to give the strain rate in one principal* direction the following results

$$\dot{\varepsilon}_{\mathrm{I}} = \frac{c}{2}\,[\sigma_{11}^2+\sigma_{22}^2+\sigma_{33}^2-\sigma_{11}\sigma_{22}-\sigma_{11}\sigma_{33}-\sigma_{22}\sigma_{33}$$

$$+3(\sigma_{12}^2+\sigma_{23}^2+\sigma_{13}^2)]^{\frac{p-1}{2}}\,[2\sigma_{\mathrm{I}}-\sigma_{\mathrm{II}}-\sigma_{\mathrm{III}}], \qquad (2.10)$$

* See Chapter 3 for a discussion of the subscript notation, principal stresses, tensors, stress deviators and invariants.

where $\dot{\varepsilon}_{\mathrm{I}}$ is the strain rate in the direction of the principal stress σ_{I}. The other strain components can be expressed in a similar manner. For uniaxial stress (2.10) reduces to (2.2). Equation (2.10) as given by Odqvist [2.27] includes an effective stress which is a function of the second invariant* J_2 of the stress deviator* tensor* s_{ij}. Later Bailey [2.28] included the third invariant* J_3 of the stress deviation tensor.* This theory was extended to a general tensorial* form by Prager [2.29]. Johnson [2.30] observed that the influence of the third order invariant of the stress deviator* was negligible.

Other equations for multiaxial creep based on similar principles have also been proposed by Soderberg [2.19], Marin [2.31], Kanter [2.32] and Nadai [2.20]. Tapsell and Johnson [2.33] and Marin [2.34] made comparisons of the equations suggested by the preceding authors and observed that they did not vary much and were in reasonable agreement with results of tests of metals provided that the ratios of the stress components remained constant during creep.

In situations where secondary creep dominates the primary creep may be neglected. But under large stresses and especially at high temperatures, primary creep must be taken into account. The primary creep for multiaxial states of stress can be expressed by generalizing (2.6) in a similar manner to that for secondary creep as follows. The strain rate tensor $\dot{\varepsilon}_{ij}$ in the primary stage is,

$$\dot{\varepsilon}_{ij} = \tfrac{3}{2}cn\sigma_e^{p-1}s_{ij}t^{n-1}, \tag{2.11}$$

or for one principal direction,

$$\dot{\varepsilon}_{\mathrm{I}} = \frac{cn}{2}[\sigma_{11}^2+\sigma_{22}^2+\sigma_{33}^2-\sigma_{11}\sigma_{22}-\sigma_{11}\sigma_{33}-\sigma_{22}\sigma_{33} + 3(\sigma_{12}^2+\sigma_{13}^2+\sigma_{23}^2)]^{\frac{p-1}{2}}[2\sigma_{\mathrm{I}}-\sigma_{\mathrm{II}}-\sigma_{\mathrm{III}}]t^{n-1}, \tag{2.12}$$

where the symbols have the same meaning as before. The other strain components can be expressed in a similar way.

Rabotnov [2.35] proposed a theory for metals based on the application of the hereditary elastic theory of Volterra† to plastic deformation. He used a single integral equation in which he introduced a nonlinear function of

* See Chapter 3 for a discussion of the subscript notation, principal stresses, tensors, stress deviators and invariants.

† This theory describes behavior of an elastic material whose elastic moduli change with time.

stress to represent the loading history dependence of the strain. The Rabotnov theory gave better agreement with experiments than the theories described above [2.36].

References [2.10–2.13, 2.37–2.39] include a detailed discussion of recent developments in the theory of creep of metals and applications to engineering problems.

2.4 Creep under Variable Stress

The various empirical expressions for creep of engineering materials reviewed above have been used to represent constant stress creep. If the situation involves a nonsteady stress history, the creep rate even in the secondary stage cannot be described as a function of stress alone. One approach to this problem has been to describe the creep rate as a function of strain or time in addition to stress. The strain rate equation including stress and time as variables is called a time-hardening law, while it is called a strain-hardening law if it is a function of stress and strain. Both laws have been widely used to represent creep in stress analysis of time-dependent materials. It has been observed by most investigators that the strain-hardening law usually yields a better representation and predictability of creep than the time-hardening law for both metals and plastics [2.4], [2.40].

2.5 Creep of Plastics

The creep curves for many plastics (polymers) are similar to those for some metals. However, they usually do not exhibit a pronounced secondary stage. In the following, some of the empirical equations proposed to represent the creep curves of plastics will be reviewed.

Leaderman [2.41] proposed the following equation to represent the creep of bakelite under constant torque:

$$\varepsilon = \varepsilon^0 + A \log t + Bt, \tag{2.13}$$

where ε^0, A and B are functions of stress, temperature and material.

Creep behavior in which the strain rate approaches zero at large times may be represented by [2.42]

$$\varepsilon = \varepsilon^0 + A \log t. \tag{2.14}$$

A number of investigators have also represented creep by using the equations describing creep of mechanical models consisting of linear springs and linear viscous elements (dash pots). These models are employed in linear viscoelasticity and are described in Chapter 5. A four-element model yields the following for creep.

$$\varepsilon = \varepsilon^0 + A(1-e^{-ct}) + Bt, \tag{2.15}$$

where ε^0, A, B are functions of stress and c is independent of stress.

It was observed by Findley, Khosla and Peterson [2.43–2.45] that the following empirical power function of time described creep of many different rigid plastics with good accuracy over a wide span of time:

$$\varepsilon = \varepsilon^0 + \varepsilon^+ t^n, \tag{2.16}$$

where ε^0, ε^+ are functions of stress, n is independent of stress, and ε^0, ε^+, n are functions of material. Lai and Findley [2.46] found n to be nearly independent of temperature, and generally less than one.

In [2.45] Findley and Peterson compared the predictions of (2.13), (2.14), (2.15) and (2.16) with results of long time creep of plastic laminates. The data for the first 2000 hr. were represented as well as possible by (2.13)–(2.16). Then the tests were continued for about 10 years. The prediction from (2.16) based on the data for 2000 hr. was very satisfactory. Similar results were found [2.47] for 90,000 hr. (about 10 yr.) of creep of polyethylene and poly (vinyl chloride).

It was observed in [2.45] that (2.16) gave a much more satisfactory prediction than expressions (2.13, 2.14, 2.15). This is due in part to the fact that creep of plastics, concrete and some metals under moderate stresses starts out at a very rapid rate immediately after loading and progresses at a continuously decreasing rate. These are characteristics of (2.16), but (2.13), (2.15) describe a constant rate of creep after a transition period since the contribution to the rate of creep of the second term in these equations approaches zero after a period of time. Equation (2.14) predicts zero creep rate after a period of time, and (2.15) describes a finite rate of creep at the beginning of creep.

Some of the above equations have been employed in the nonlinear range of stress by making the stress-dependent coefficients into nonlinear functions of stress. For example (2.15) may be written as

$$\varepsilon = (\sigma/E) + k\sigma^p(1-e^{-ct}) + L\sigma^q t, \tag{2.17}$$

where E, k, p, L and q are material constants at constant temperature.

Findley, Adams and Worley [2.48] employed hyperbolic sine functions in (2.16)

$$\varepsilon = \varepsilon^0 \sinh(\sigma/\sigma^0) + \varepsilon^+ t^n \sinh(\sigma/\sigma^+), \tag{2.18}$$

where ε^0, σ^0, ε^+, n and σ^+ are material constants at constant temperature. The hyperbolic sine appearing in the second term of (2.18) results from the activation energy theory of rate processes (see Chapter 11) described by Eyring [2.49], Kauzmann [2.50] and others. This factor describes the influence of stress on the thermal activation of deformational processes in the material such as rupture or interchange of atomic bonds.

The power law (or power function of stress) has been widely used by various authors to express stress-strain-time relationships for nonlinear viscoelastic materials. It has the following form which is often called the Nutting equation [2.51], and is the same as (2.6a)

$$\varepsilon^c = k\sigma^p t^n, \tag{2.19}$$

where ε^c is the creep strain, σ is stress, k, p and n are material constants. Equation (2.19) was used by Scott-Blair and his coworkers [2.22, 2.52, 2.53] for various soft materials (such as rubber). It yielded a satisfactory description of the short-time creep data for these materials.

2.6 Mathematical Representation of Creep of Materials

The stress-strain-time relations suggested for creep, reviewed above, are primarily empirical. Most were developed to fit experimental creep curves obtained under constant stress and constant temperature. The actual behavior of materials has shown that the strain at a given time depends on all of the values of the stress in the past, not just on its final value. Thus the creep phenomenon is affected by the magnitude and sequence of stresses or strains in all of the past history of the material. Based on this fact, various mathematical methods have been suggested to represent the time dependence or viscoelastic behavior of materials.

2.7 Differential Form

Viscoelastic materials exhibit a complicated time-dependent behavior, including instantaneous elasticity, delayed elasticity and viscous flow. For uniaxial stressing the stress-strain relations of these materials can be expressed in the following form, called the linear differential operator method [2.54–2.57], as described in detail in Chapter 5:

$$P\sigma = Q\varepsilon, \tag{2.20}$$

where P and Q each represent a series of linear differential operators, with respect to time, containing material constants. By properly selecting a certain number of terms of the series, the viscoelastic behavior of a specific type of linear material may be represented. Such materials, or the mechanical models of such materials may be represented as combinations of linear (Hookean) springs and linear (Newtonian) dash pots. These models have been used for more than a hundred years since they give a simple and clear physical description of the fundamentals of viscoelastic behavior. Also, they are very convenient for mathematical solutions of stress analysis problems [2.55–2.59].

The nonlinear behavior of viscoelastic materials has also been represented by means of mechanical models similar to those used for linear materials. For example, a nonlinear model may be assumed to consist of linear springs and nonlinear dash pots. The creep rate of such a non-Newtonian material may be described as follows [2.23],

$$\dot{\varepsilon} = k\sigma^p, \tag{2.21}$$

where k and p are material constants. On the other hand, Eyring [2.49] suggested a hyperbolic sine law in the following form:

$$\dot{\varepsilon} = D \sinh \frac{\sigma}{\sigma^+}, \tag{2.22}$$

where D and σ^+ are constants which characterize the nonlinear dash pot. This form was derived by considering the rate of activation of molecular movements resulting from thermal oscillation [2.49]. Nonlinear models cannot be described mathematically as conveniently and generally as linear models can. Consequently, it has been common practice to approximate nonlinear behavior by a linear model.

In stress analysis problems involving viscoelastic materials, linear differential equations (2.20) have been widely used. Some such problems have been treated by removing the time variable in the system of equations and in the boundary conditions by employing the Laplace transformation with respect to time (see Chapter 6). The viscoelastic problem thus becomes an elastic problem. After the necessary algebraic operations to effect a solution in the Laplace transform variable, the solution can be brought back to the original time variable by inverse transformation. This may be difficult to accomplish sometimes. This method can be used when the boundary conditions are specified entirely by stress or deformation. For mixed stress and deformation boundary conditions this method can be used only when the interface between stress and deformation boundaries does not change with time [2.56, 2.58]. The differential operator method has also been used to represent the propagation of waves in viscoelastic materials [2.60].

2.8 Integral Form

Besides the differential operator type of constitutive equations of linear viscoelastic behavior and associated mechanical models reviewed above, there is another means of describing behavior of viscoelastic materials, the integral operator representation. Any stress-time curve may be approximated by the sum of a series of step functions which correspond to a series of step-like increments in load. The creep compliance $J(t)$ may be defined as the creep strain resulting from unit stress. By making use of the Boltzmann superposition principle, the strain $\varepsilon(t)$ occurring during a creep test at time t may be represented as,

$$\varepsilon(t) = \int_0^t J(t-\xi)\frac{\partial\sigma(\xi)}{\partial\xi}\,d\xi, \tag{2.23}$$

where ξ is any arbitrary time between 0 and t, representing past time. This integral, called a hereditary integral, was first suggested by Volterra [2.61]. The kernel function of the integral, $J(t-\xi)$ is a memory function which describes the stress history dependence of strain. The stress relaxation of a linear viscoelastic material can be expressed [2.55, 2.62] in a similar way to that of (2.23).

Both the differential operator method, (2.20), and the integral representation, (2.23) can be easily generalized to a multiaxial state of stress [2.56, 2.58, 2.63, 2.64]. By performing a proper set of experiments the material constants and the kernel functions for a given material can be determined and used to predict behavior under other stress histories [2.65].

2.9 Development of Nonlinear Constitutive Relations

As indicated at the beginning of the chapter, if strains become greater than one or two percent (or even less) in most viscoelastic materials, the actual behavior exhibits nonlinearity. Thus it is necessary in critical situations to provide a more accurate representation than is possible by assuming linear behavior. In recent years there has been considerable effort to develop a general constitutive equation for nonlinear viscoelastic materials, see, for example [2.66–2.71]. Such constitutive equations are much more complex than the linear theory. For generality they require a large number of functions with higher order stress terms to describe creep behavior satisfactorily. In the following, developments in this field are reviewed.

Leaderman [2.41] proposed a generalization of linear equation (2.23) to the following form,

$$\varepsilon(t) = \varphi(0)\sigma(t) + \int_0^t F(t-\xi)\frac{\partial}{\partial\xi} f[\sigma(\xi)]\,d\xi, \tag{2.24}$$

where F and f are empirical functions of time and stress, respectively, and $\varphi(0)$ is the time-independent compliance. Equation (2.24) is a constitutive equation of one particular type of nonlinear viscoelastic material but not sufficiently general to describe all materials with memory.

Later Green, Rivlin and Spencer [2.66, 2.68, 2.69] proposed a more general constitutive equation for nonlinear materials by formulating stress relaxation in terms of a tensor functional* of the strain history based on invariants**. They considered that, for so-called simple materials, the current stress at a point of such materials is not only a function of the current deformation gradient but also the deformation gradients at all previous times.

The most general constitutive equation for small deformation of nonlinear materials with memory was thus expressed as follows, for a stress relaxation formulation,

$$\sigma_{ij}(t) = \underset{\xi=0}{\overset{t}{F_{ij}}}[\varepsilon_{pq}(\xi)]\,, \tag{2.25}$$

where σ_{ij} and ε_{pq} are stress and strain tensor components, respectively, F_{ij} is a continuous functional of the strain function, t is the time, ξ is any previous time between 0 and t.

A similar equation to (2.25) may be formulated for creep to express strains in terms of stress [2.72],

$$\varepsilon_{ij}(t) = \underset{\xi=0}{\overset{t}{G_{ij}}}[\sigma_{pq}(\xi)]. \tag{2.26}$$

By making use of the functional analysis, the constitutive equation (2.25) for strain relaxation was expressed by Green, Rivlin and Spencer [2.66,

* See Chapter 7 and Appendix A2 for a discussion of this term.
** See Chapter 3 for a discussion of this term.

2.68, 2.69] and by Pipkin [2.73] as a series of multiple hereditary integrals. In a similar manner by using (2.26) Onaran and Findley [2.72] described creep as a series of multiple integrals. More details will be given in Chapter 7 and Appendix A2. The theory due to Green and Rivlin has been discussed further by Noll [2.67] and Coleman and Noll [2.70]. A relaxation form of constitutive relation for an elastic fluid was proposed by Bernstein, Kearsley and Zapas [2.74] as a single integral nonlinear representation.

Another approach to characterizing nonlinear viscoelastic materials was derived from the thermodynamics of irreversible processes as described by Schapery [2.75–2.77]. In this equation the nonlinearity is contained in a "reduced time", where the reduced time is an implicit function of stress in the creep formulation.

Functional types of constitutive equations contain kernel functions which represent the time dependence (time functions) of materials. In order to use these constitutive equations for stress or displacement analysis, the kernel functions must be determined experimentally. Because of the multiple integrals this determination is rather involved, as described in Chapter 7.

Since the description of nonlinear memory-dependent behavior of materials by the multiple integral representation is a fairly recent development, studies involving the experimental evaluation of the material constants are few and some of those involve extensive simplifications of the constitutive equations. Onaran and Findley [2.72] determined the material functions for creep of unplasticized poly (vinyl chloride) in the nonlinear range. Lianis [2.78] and Ko and Blatz [2.79] described experimentally measured material functions using a simplified form of finite linear viscoelastic constitutive equation proposed by Coleman and Noll [2.70]. They performed relaxation experiments on styrene-butadiene, assumed incompressibility, neglected some of the terms to reduce the number of material functions and isolated the remaining material functions.

Nakada [2.80] considered a third order one-dimensional viscoelastic constitutive equation. He assumed that the multiple time argument functions were separable with respect to each time argument (a product form of function) and that each such time argument could be expressed as an exponential function. He also showed that it was possible to determine the material functions by performing, either step-wise or sinusoidal loadings.

Leaderman, McCrackin and Nakada [2.81] used a third order one-dimensional constitutive equation to describe the nonlinear creep of plasticized poly (vinyl chloride) with one time function and a single time argument. This was accomplished by using a suitable definition for strain and a product form for the creep function so that a complete study of the multiple time argument function was avoided.

Ward and Onat [2.82] used a one-dimensional nonlinear viscoelastic equation consisting of the first and third order integrals to describe the observed behavior of an oriented polypropylene monofilament in tensile creep.

Onaran and Findley [2.83], using two-step loadings, determined some of the kernel functions required for describing nonlinear creep under varying tensile stress and varying pure shear. Neis and Sackman [2.84] conducted tension and compression creep experiments on low density polyethylene under single step, two step and three step loading histories in order to characterize the kernel functions for uniaxial creep.

The complete determination of the kernel functions, requires very elaborate experimental programs as observed by several investigators [2.83–2.87] and greater experimental precision than is likely to be possible [2.83]. To avoid these difficulties, some other simplifying methods have been developed and compared with experiments: a modified superposition method by Findley and Lai [2.88]; a linear compressible assumption by Nolte and Findley [2.89]; limited memory forms, one of which reduced to the modified superposition principle, by Nolte and Findley [2.90]; and an additive form by Gottenberg, Bird and Agrawal [2.91].

These representations achieved a considerable simplification of the analytical and experimental work at the expense of generality. The modified superposition principle is one of the simplest methods and yields as satisfactory prediction as the others in most cases [2.83, 2.92, 2.93].

Another approach to simplifying the constitutive relation was proposed by Huang and Lee [2.94] where the multiple integral representation of the third order theory of incompressible, initially isotropic nonlinear viscoelastic materials was simplified by performing integration by parts. The integration by parts served to reduce double and triple integrals to single integrals plus remainder terms that are neglected for sufficiently short times.

The creep behavior of a material (asphalt concrete) showing both nonlinear viscous (irrecoverable) strain ε^V and nonlinear viscoelastic (recoverable) strain ε^{VE} was described by Lai and Anderson [2.95] for compressive stress σ as follows.

$$\varepsilon(t) = \varepsilon^V + \varepsilon^{VE} = \int_0^t \dot{\varepsilon}^V(\xi)\, d\xi + \int_0^t (t-\xi)^n \,[C_1 + 2C_2\sigma(\xi)]\dot{\sigma}(\xi)\, d\xi. \quad (2.27)$$

In the first term of (2.27) a strain hardening principle was employed for $\dot{\varepsilon}^V$ and in the second term the modified superposition principle was used. Creep resulting from repeated application of load and stair-step loadings were well described by (2.27).

Creep behavior of an orthotropic type of anisotropic material was described in terms of a multiple integral representation by Erman and Onaran [2.96]. This representation was applied to specimens of wood and fiber composite plastics under combined stresses and step loadings with good results.

CHAPTER 3

STATE OF STRESS AND STRAIN

3.1 State of Stress

Stress σ is defined as the ratio of the force acting on a given area in a body to the area as the size of the area tends to zero. The three-dimensional state of stress at a point in a stressed body may be described with respect to the stress on three arbitrarily chosen orthogonal planes passing through the point. These planes will be chosen normal to the coordinate axes x_1, x_2, x_3, as shown in Fig. 3.1. An infinitesimal cube whose sides are parallel to these planes will be considered in what follows.

In the most general situation there will be forces F_1, F_2, F_3, Fig. 3.1, acting at some angle to each of these planes. It is usual to resolve these forces into components parallel to the coordinate axes and consider their stress components rather than the forces, as shown in Fig. 3.1. A double subscript

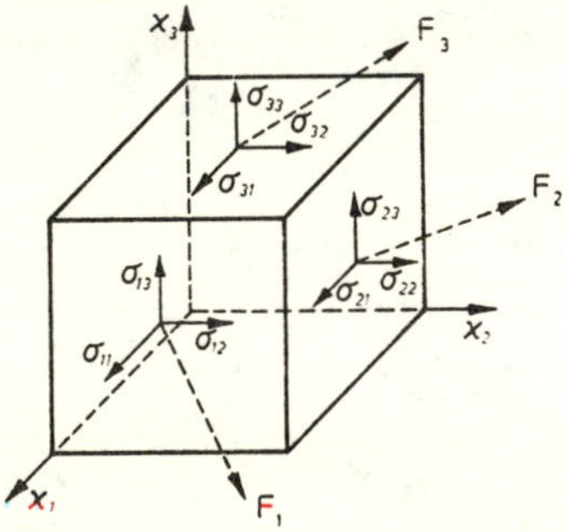

Fig. 3.1. Stress Components of a General State of Stress

notation will be employed to distinguish between the various components. The particular plane on which a stress component acts is indicated by the first subscript. The subscript 1 indicates that the stress acts on a plane normal to the x_1 direction, the subscript 2 indicates normality to x_2, etc. The second subscript indicates the direction of the vector of the stress component under consideration. Thus, considering the right-hand face of the cube in Fig. 3.1 σ_{22} is a normal stress component in the x_2 direction, σ_{21} is a shearing component in the x_1 direction, σ_{23} is a shearing component in the x_3 direction and all three components are acting on a plane which is perpendicular to the x_2 axis.

3.2 Stress Tensor

The nine components involved in describing the state of stress, as shown in Fig. 3.1, constitute a second order Cartesian tensor, called the stress tensor. The stress vector (or traction) T_i is the force per unit area acting on the surface whose normal is defined by the unit vector n_j. The traction T_i is related to the stress tensor σ_{ij} by $T_i = \sigma_{ij}n_j$. The nine components of the stress tensor σ_{ij} may be displayed in matrix form as follows:

$$\sigma_{ij} = \boldsymbol{\sigma} = \begin{vmatrix} \sigma_{11} & \sigma_{12} & \sigma_{13} \\ \sigma_{21} & \sigma_{22} & \sigma_{23} \\ \sigma_{31} & \sigma_{32} & \sigma_{33} \end{vmatrix}, \tag{3.1}$$

where the symbol σ_{ij} employs a subscript notation in which each dummy subscript i and j are understood to take values of 1, 2 and 3, as indicated in the matrix on the right-hand side of (3.1). The symbol $\boldsymbol{\sigma}$ is often used to denote the stress tensor and will be employed in subsequent chapters.

Another rule concerning the subscript notation is called the summation convention. When a letter subscript is repeated in a given term of an equation as in σ_{ii} or $\sigma_{ij}a_i$ it is to be understood that this symbol represents the sum of the three terms formed by successively substituting 1, 2 and 3 for the repeated subscript, that is, for example,

$$\sigma_{ii} = \sigma_{11}+\sigma_{22}+\sigma_{33}.$$

It will be shown from moment equilibrium in Chapter 4 that $\sigma_{12} = \sigma_{21}$, etc. It follows that the stress tensor is symmetrical about the diagonal line from σ_{11} to σ_{33}, that is $\sigma_{ij} = \sigma_{ji}$. Thus only six of the nine components of stress are independent of each other so that σ_{11}, σ_{22}, σ_{33}, σ_{12}, σ_{13}, σ_{23} completely describe the state of stress.

3.3 Unit Tensor

The Kronecker delta or unit tensor is defined as follows:

$$\delta_{ij} = \begin{cases} 1 & \text{when} \quad i = j \\ 0 & \text{when} \quad i \neq j. \end{cases}$$

It has the following matrix

$$\delta_{ij} = \mathbf{I} = \begin{vmatrix} 1 & 0 & 0 \\ 0 & 1 & 0 \\ 0 & 0 & 1 \end{vmatrix} \tag{3.2}$$

and is independent of the choice of coordinate directions.

The relationship between the components of the stress tensor and the conventional notation for stress components is as follows:

$$\sigma_{11} = \sigma_x, \quad \sigma_{22} = \sigma_y, \quad \sigma_{33} = \sigma_z,$$

$$\sigma_{12} = \sigma_{21} = \tau_{xy} = \tau_{yx}, \quad \sigma_{13} = \sigma_{31} = \tau_{xz} = \tau_{zx}, \quad \sigma_{23} = \sigma_{32} = \tau_{yz} = \tau_{zy}.$$

3.4 Principal Stresses

If the coordinate axes x_i are rotated with respect to the stress state at a given point it will be found that the components defined by σ_{ij} vary for a constant state of stress. A particular set of orthogonal planes will be found for which one of the normal stresses σ_{11}, σ_{22}, σ_{33} is a maximum and one is a minimum with respect to the rotation of coordinates. These normal stresses are the principal stresses $\sigma_{\text{I}} > \sigma_{\text{II}} > \sigma_{\text{III}}$. It happens that the shearing stresses on principal stress planes are zero.

Another set of planes will also be found on which the shearing stress components are either a maximum or a minimum with respect to rotation of the coordinate axes. The shearing stresses on these planes are the principal shear stresses $\tau_{\text{I}} > \tau_{\text{II}} > \tau_{\text{III}}$. The normal stresses on the principal shear stress planes are not zero, however. The principal shear stresses lie on planes which bisect the planes on which the associated pair of principal stresses act. For example the principal shear stress $\tau_{\text{I}} = \frac{1}{2}(\sigma_{\text{I}} - \sigma_{\text{III}})$, corresponding to σ_{13}, acts on planes which are at 45° to the planes on which principal stresses σ_{I} and σ_{III} act.

The principal stresses of the stress tensor σ_{ij} may be determined by making use of the fact that shearing stresses are zero on principal planes. Hence the total stress σ on a principal plane is a principal stress. Let $\mathbf{n}$ be a unit vector normal to the principal plane having components n_i in the directions of the coordinates x_i given by the subscripts $i, j = 1, 2, 3$ as shown in Fig. 3.2. Thus the directions of the principal stress σ and the unit vector $\mathbf{n}$ are the same and the components of the principal stress along the coordinate directions x_j are given by σn_j. The components of the stress tensor in the coordinate directions x_j are given by the vector $\sigma_{ij}n_i$. Applying the equilibrium condition to the tetrahedron in Fig. 3.2 the following results

$$\sigma_{ij}n_i - \sigma n_j = (\sigma_{ij} - \sigma\delta_{ij})n_i = 0, \tag{3.3}$$

since $\delta_{ij}n_i = n_j$.

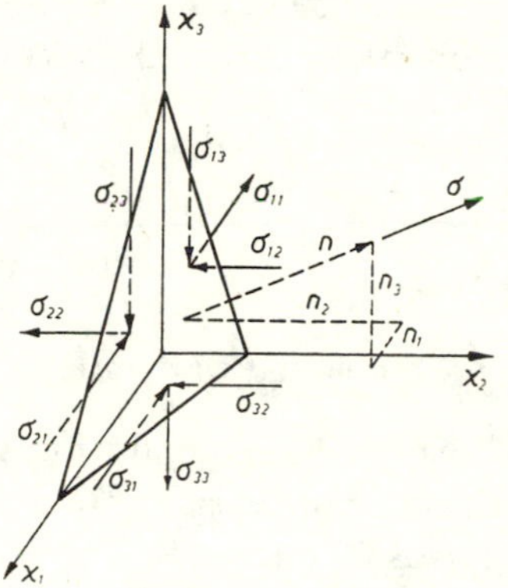

Fig. 3.2. Stress Components on a Tetrahedron Formed by a Principal Plane Normal to **n**

(3.3) represents a set of linear homogeneous equations in the components of n_i for the coordinate directions x_j as follows:

$$\begin{aligned} (\sigma_{11}-\sigma)n_1+\sigma_{21}n_2+\sigma_{31}n_3 &= 0, \qquad j=1,\\ \sigma_{12}n_1+(\sigma_{22}-\sigma)n_2+\sigma_{32}n_3 &= 0, \qquad j=2,\\ \sigma_{13}n_1+\sigma_{23}n_2+(\sigma_{33}-\sigma)n_3 &= 0, \qquad j=3. \end{aligned} \tag{3.4}$$

The non-zero solution of (3.4) is found by setting the determinant of the coefficients in (3.4) equal to zero to find the characteristic equation of the stress tensor σ_{ij},

$$\sigma^3 - I_1\sigma^2 - I_2\sigma - I_3 = 0, \tag{3.5}$$

where

$$I_1 = \sigma_{11}+\sigma_{22}+\sigma_{33} = \sigma_{ii}, \tag{3.6}$$

$$\begin{aligned} I_2 &= (\sigma_{12}^2+\sigma_{23}^2+\sigma_{31}^2)-(\sigma_{11}\sigma_{22}+\sigma_{22}\sigma_{33}+\sigma_{33}\sigma_{11})\\ &= \tfrac{1}{2}(\sigma_{ij}\sigma_{ji}-\sigma_{ii}\sigma_{jj}), \end{aligned} \tag{3.7}$$

$$\begin{aligned} I_3 &= \sigma_{11}\sigma_{22}\sigma_{33}+2\sigma_{12}\sigma_{23}\sigma_{13}-(\sigma_{11}\sigma_{23}^2+\sigma_{22}\sigma_{13}^2+\sigma_{33}\sigma_{12}^2)\\ &= \tfrac{1}{6}(\sigma_{ii}\sigma_{jj}\sigma_{kk}+2\sigma_{ij}\sigma_{jk}\sigma_{ki}-3\sigma_{ij}\sigma_{ji}\sigma_{kk}), \qquad i,j = 1, 2, 3, \end{aligned} \tag{3.8}$$

and the symmetry condition $\sigma_{ij} = \sigma_{ji}$ has been employed. I_1, I_2 and I_3 are called the invariants of the stress tensor, see Article 3.6.

The three values of σ, ($\sigma_{\mathrm{I}} > \sigma_{\mathrm{II}} > \sigma_{\mathrm{III}}$) which are roots of (3.5) are the principal stresses. Methods for determining principal planes and magnitudes for principal stresses and principal shear stresses are discussed in many books on elasticity and related topics and will not be reviewed here.

If the coordinate axes are made to coincide with the normals to principal planes (principal directions) the matrix of the stress tensor becomes

$$\sigma_{ij} = \begin{vmatrix} \sigma_{\mathrm{I}} & 0 & 0 \\ 0 & \sigma_{\mathrm{II}} & 0 \\ 0 & 0 & \sigma_{\mathrm{III}} \end{vmatrix}. \tag{3.9}$$

3.5 Mean Normal Stress Tensor and Deviatoric Stress Tensor

The stress tensor σ_{ij} may be divided into two parts, a mean normal stress tensor $\sigma_v\delta_{ij}$ and the remainder or deviation from the mean stress, called the deviatoric stress tensor s_{ij} as follows

$$\sigma_{ij} = \begin{vmatrix} \sigma_{11} & \sigma_{12} & \sigma_{13} \\ \sigma_{21} & \sigma_{22} & \sigma_{23} \\ \sigma_{31} & \sigma_{32} & \sigma_{33} \end{vmatrix} = \begin{vmatrix} \sigma_v & 0 & 0 \\ 0 & \sigma_v & 0 \\ 0 & 0 & \sigma_v \end{vmatrix} + \begin{vmatrix} s_{11} & s_{12} & s_{13} \\ s_{21} & s_{22} & s_{23} \\ s_{31} & s_{32} & s_{33} \end{vmatrix}, \tag{3.10}$$

or

$$\sigma_{ij} = \sigma_v\delta_{ij} + s_{ij}, \tag{3.11}$$

where the addition applies to the corresponding term of each matrix. That is $s_{11} = \sigma_{11} - \sigma_v$, $\sigma_{12} = s_{12}$ for example. Thus the components with mixed subscripts (shear components) are equal in the stress tensor σ_{ij} and the deviatoric tensor s_{ij}.

Choosing σ_v to be the average normal stress of the stress tensor σ_{ij}

$$\sigma_v = \tfrac{1}{3}\sigma_{kk} = \tfrac{1}{3}(\sigma_{11} + \sigma_{22} + \sigma_{33}), \tag{3.12}$$

where σ_{kk} makes use of the summation convention. The average normal stress σ_v may be shown to be independent of the choice of the coordinate axes.

Solving equations (3.11) and (3.12) for the deviatoric stress s_{ij} yields

$$s_{ij} = \sigma_{ij} - \tfrac{1}{3}\sigma_{kk}\,\delta_{ij}, \tag{3.13}$$

and

$$s_{ij} = \begin{vmatrix} s_{11} & s_{12} & s_{13} \\ s_{21} & s_{22} & s_{23} \\ s_{31} & s_{32} & s_{33} \end{vmatrix} = \begin{vmatrix} \sigma_{11} - \sigma_v & \sigma_{12} & \sigma_{13} \\ \sigma_{21} & \sigma_{22} - \sigma_v & \sigma_{23} \\ \sigma_{31} & \sigma_{32} & \sigma_{33} - \sigma_v \end{vmatrix}, \tag{3.14}$$

respectively.

From (3.12) and (3.13) it can be shown that the sum of the normal deviatoric stress components s_{ii} is zero if as assumed σ_v is the average of the normal components of σ_{ij}. That is

$$s_{11}+s_{22}+s_{33} = 0. \tag{3.15}$$

It may also be shown from (3.10) and (3.14) that σ_{ij} and s_{ij} have the same principal directions.

When the coordinate axes coincide with the principal axes, the mean normal stress tensor $\sigma_v\delta_{ij}$ and the deviatoric stress tensor s_{ij} become

$$\sigma_v\delta_{ij} = \tfrac{1}{3}(\sigma_{\mathrm{I}}+\sigma_{\mathrm{II}}+\sigma_{\mathrm{III}})\delta_{ij} \tag{3.16}$$

and

$$s_{ij} = \begin{vmatrix} s_{\mathrm{I}} & 0 & 0 \\ 0 & s_{\mathrm{II}} & 0 \\ 0 & 0 & s_{\mathrm{III}} \end{vmatrix} = \begin{vmatrix} \sigma_{\mathrm{I}}-\sigma_v & 0 & 0 \\ 0 & \sigma_{\mathrm{II}}-\sigma_v & 0 \\ 0 & 0 & \sigma_{\mathrm{III}}-\sigma_v \end{vmatrix}. \tag{3.17}$$

Writing (3.12) in terms of principal directions as in (3.16) and substituting in (3.17) yields

$$s_{ij} = \frac{1}{3}\begin{vmatrix} (2\sigma_{\mathrm{I}}-\sigma_{\mathrm{II}}-\sigma_{\mathrm{III}}) & 0 & 0 \\ 0 & (2\sigma_{\mathrm{II}}-\sigma_{\mathrm{I}}-\sigma_{\mathrm{III}}) & 0 \\ 0 & 0 & (2\sigma_{\mathrm{III}}-\sigma_{\mathrm{I}}-\sigma_{\mathrm{II}}) \end{vmatrix}. \tag{3.18}$$

Division of the state of stress into a volume stress component and a deviatoric component is often convenient because of differences in mechanical behavior of materials under volumetric stressing in which there are no shearing stresses versus states of stress involving shearing stress. For example, note the following differences: brittle cleavage fracture under tri-axial tension versus ductile fracture under shear stressing; and different rates of wave propagation between so-called dilatation and shear waves.

The relationships governing the determination of the principal stresses of the stress deviator $s_{\mathrm{I}} > s_{\mathrm{II}} > s_{\mathrm{III}}$ may be found in a manner exactly analogous to that employed for the stress tensor. This results in the following expression similar to (3.5),

$$s^3-J_1s^2-J_2s-J_3 = 0, \tag{3.19}$$

where $J_1 = s_{11}+s_{22}+s_{33} = 0$ in accordance with (3.15), (3.20)

$$J_2 = -(s_{11}s_{22}+s_{22}s_{33}+s_{33}s_{11})+s_{12}^2+s_{23}^2+s_{31}^2, = \tfrac{1}{2}s_{ij}s_{ji}, \tag{3.21}$$

$$J_3 = s_{11}s_{22}s_{33}+2s_{12}s_{31}s_{32}-(s_{11}s_{23}^2+s_{22}s_{31}^2+s_{33}s_{12}^2)$$
$$= \tfrac{1}{3}s_{ij}s_{jk}s_{ki}. \tag{3.22}$$

Table 3.1

Invariants and Traces for Different States of Stress

Column	1	2	3	4	5	6
Stress state, $\boldsymbol{\sigma} = \sigma_{ij} =$	$\begin{vmatrix} \sigma & 0 & 0 \\ 0 & 0 & 0 \\ 0 & 0 & 0 \end{vmatrix}$	$\begin{vmatrix} 0 & \tau & 0 \\ \tau & 0 & 0 \\ 0 & 0 & 0 \end{vmatrix}$	$\begin{vmatrix} \sigma & \tau & 0 \\ \tau & 0 & 0 \\ 0 & 0 & 0 \end{vmatrix}$	$\begin{vmatrix} \sigma_1 & \tau & 0 \\ \tau & \sigma_2 & 0 \\ 0 & 0 & 0 \end{vmatrix}$	$\begin{vmatrix} \sigma_I & 0 & 0 \\ 0 & \sigma_{II} & 0 \\ 0 & 0 & 0 \end{vmatrix}$	$\begin{vmatrix} \sigma_I & 0 & 0 \\ 0 & \sigma_{II} & 0 \\ 0 & 0 & \sigma_{III} \end{vmatrix}$
$\boldsymbol{\sigma\sigma} =$	$\begin{vmatrix} \sigma^2 & 0 & 0 \\ 0 & 0 & 0 \\ 0 & 0 & 0 \end{vmatrix}$	$\begin{vmatrix} \tau^2 & 0 & 0 \\ 0 & \tau^2 & 0 \\ 0 & 0 & 0 \end{vmatrix}$	$\begin{vmatrix} (\sigma^2+\tau^2) & \sigma\tau & 0 \\ \sigma\tau & \tau^2 & 0 \\ 0 & 0 & 0 \end{vmatrix}$	$\begin{vmatrix} (\sigma_1^2+\tau^2) & (\sigma_1\tau+\sigma_2\tau) & 0 \\ (\sigma_1\tau+\sigma_2\tau) & (\sigma_2^2+\tau^2) & 0 \\ 0 & 0 & 0 \end{vmatrix}$	$\begin{vmatrix} \sigma_I^2 & 0 & 0 \\ 0 & \sigma_{II}^2 & 0 \\ 0 & 0 & 0 \end{vmatrix}$	$\begin{vmatrix} \sigma_I^2 & 0 & 0 \\ 0 & \sigma_{II}^2 & 0 \\ 0 & 0 & \sigma^2_{II} \end{vmatrix}$
$\boldsymbol{\sigma\sigma\sigma} =$	$\begin{vmatrix} \sigma^3 & 0 & 0 \\ 0 & 0 & 0 \\ 0 & 0 & 0 \end{vmatrix}$	$\begin{vmatrix} 0 & \tau^3 & 0 \\ \tau^3 & 0 & 0 \\ 0 & 0 & 0 \end{vmatrix}$	$\begin{vmatrix} (\sigma^3+2\sigma\tau^2) & (\sigma^2\tau+\tau^3) & 0 \\ (\sigma^2\tau+\tau^3) & \sigma\tau^2 & 0 \\ 0 & 0 & 0 \end{vmatrix}$	$\begin{vmatrix} (\sigma_1^3+2\sigma_1\tau^2 +\sigma_2\tau^2) & (\sigma_1^2\tau+\sigma_1\sigma_2\tau +\sigma_2^2\tau+\tau^3) & 0 \\ (\sigma_1^2\tau+\tau^3 +\sigma_1\sigma_2\tau+\sigma_2^2\tau) & (\sigma_1\tau^2 +2\sigma_2\tau^2+\sigma_2^3) & 0 \\ 0 & 0 & 0 \end{vmatrix}$	$\begin{vmatrix} \sigma_I^3 & 0 & 0 \\ 0 & \sigma_{II}^3 & 0 \\ 0 & 0 & 0 \end{vmatrix}$	$\begin{vmatrix} \sigma_I^3 & 0 & 0 \\ 0 & \sigma_{II}^3 & 0 \\ 0 & 0 & \sigma_{III}^3 \end{vmatrix}$

Table 3.1 (cont'd)

Column	1	2	3	4	5	6
$\mathrm{tr}(\sigma_{ij}) = \bar{\boldsymbol{\sigma}} =$	σ	0	σ	$\sigma_1+\sigma_2$	$\sigma_I+\sigma_{II}$	$\sigma_I+\sigma_{II}+\sigma_{III}$
$\overline{\boldsymbol{\sigma\sigma}} =$	σ^2	$2\tau^2$	$\sigma^2+2\tau^2$	$\sigma_1^2+\sigma_2^2+2\tau^2$	$\sigma_I^2+\sigma_{II}^2$	$\sigma_I^2+\sigma_{II}^2+\sigma_{III}^2$
$\overline{\boldsymbol{\sigma\sigma\sigma}} =$	σ^3	0	$\sigma^3+3\sigma\tau^2$	$\sigma_1^3+\sigma_2^3+3(\sigma_1+\sigma_2)\tau^2$	$\sigma_I^3+\sigma_{II}^3$	$\sigma_I^3+\sigma_{II}^3+\sigma_{III}^3$
$\sigma_v = \frac{1}{3}\sigma_{kk} =$	$\sigma/3$	0	$\sigma/3$	$(\sigma_1+\sigma_2)/3$	$(\sigma_I+\sigma_{II})/3$	$(\sigma_I+\sigma_{II}+\sigma_{III})/3$
$I_1 =$	σ	0	σ	$\sigma_1+\sigma_2$	$\sigma_I+\sigma_{II}$	$\sigma_I+\sigma_{II}+\sigma_{III}$
$I_2 =$	0	τ^2	τ^2	$\tau^2-\sigma_1\sigma_2$	$-\sigma_I\sigma_{II}$	$-(\sigma_I\sigma_{II}+\sigma_{II}\sigma_{III}+\sigma_I\sigma_{III})$
$I_3 =$	0	0	0	0	0	$\sigma_I\sigma_{II}\sigma_{III}$
$J_1 =$	0	0	0	0	0	0
$J_2 =$	$\sigma^2/3$	τ^2	$(\sigma^2/3)+\tau^2$	$[(\sigma_1^2-\sigma_1\sigma_2+\sigma_2^2)/3]+\tau^2$	$(\sigma_I^2-\sigma_I\sigma_{II}+\sigma_{II}^2)/3$	$(\sigma_I^2+\sigma_{II}^2+\sigma_{III}^2-\sigma_I\sigma_{II}$ $-\sigma_{II}\sigma_{III}-\sigma_I\sigma_{III})/3$
$J_3 =$	$2\sigma^3/27$	0	$(2\sigma^3/27)+(\sigma\tau^2/3)$	$2(\sigma_1+\sigma_2)^3/27$ $+(\sigma_1+\sigma_2)(\tau^2-\sigma_1\sigma_2)/3$	$\frac{2}{27}(\sigma_I+\sigma_{II})^3$ $-\frac{1}{3}(\sigma_I+\sigma_{II})\sigma_I\sigma_{II}$	$\frac{2}{27}(\sigma_I+\sigma_{II}+\sigma_{III})^3$ $-\frac{1}{3}(\sigma_I+\sigma_{II}+\sigma_{III})(\sigma_I\sigma_{II}$ $+\sigma_I\sigma_{III}+\sigma_{II}\sigma_{III})$ $+\sigma_I\sigma_{II}\sigma_{III}$

J_1, J_2 and J_3 are called the invariants of the stress deviator, see Article 3.6. Substituting the components of s_{ij} in terms of σ_{ij} from (3.14) into (3.21) and (3.22), making use of (3.16) and noting that $s_{12} = \sigma_{12}$ etc., it may be shown that

$$J_2 = I_2 + 3\sigma_v^2 = I_2 + \tfrac{1}{3}\sigma_{kk}^2, \tag{3.23}$$

$$J_3 = I_3 + I_2\sigma_v + 2\sigma_v^3. \tag{3.24}$$

3.6 Invariants of Stress

A stress invariant is a function of the components of the state of stress the magnitude of which function is independent of (or invariant with) the choice of orientation of the coordinate axes at the point in question. It is possible to formulate a large number of functions which have this property. Generally these may be expressed in terms of three primary functions I_1, I_2, I_3 as given in (3.6, 3.7, 3.8), called the invariants of the stress tensor. Also a stress-deviator invariant is a function whose magnitude is independent of the choice of coordinates. It may be expressed in terms of the invariants of the stress deviator $J_1 = 0$, J_2, J_3 as in (3.20, 3.21, 3.22).

That the coefficients in the cubic equations of (3.5) and (3.19) are invariant with rotation of coordinate axes may be demonstrated from the fact that the solutions of these equations are principal stresses. As such they are independent of the coordinate system. Hence it follows that the coefficients in these equations must also be invariants and are scalars. Values of the invariants for several special states of stress are given in Table 3.1.

3.7 Traces of Tensors and Products of Tensors

The trace of a tensor is a scalar equal to the sum of the diagonal terms of the matrix and is denoted by the symbol $\mathrm{tr}\sigma_{ij}$ or by $\bar{\sigma}_{ij}$ for a stress tensor for example. Thus the trace of the stress tensor σ_{ij} (3.1) is

$$\mathrm{tr}\sigma_{ij} = \bar{\sigma}_{ij} = \bar{\boldsymbol{\sigma}} = \sigma_{ii} = \sigma_{11}+\sigma_{22}+\sigma_{33}, \tag{3.25}$$

where $\bar{\sigma}_{ij}$ and $\bar{\boldsymbol{\sigma}}$ are alternate notations.

To find the trace of a product of a tensor by itself first consider the tensor product:

$$\sigma_{ip}\sigma_{pj} = \boldsymbol{\sigma\sigma} = \begin{vmatrix} \sigma_{11} & \sigma_{12} & \sigma_{13} \\ \sigma_{21} & \sigma_{22} & \sigma_{23} \\ \sigma_{31} & \sigma_{32} & \sigma_{33} \end{vmatrix} \begin{vmatrix} \sigma_{11} & \sigma_{12} & \sigma_{13} \\ \sigma_{21} & \sigma_{22} & \sigma_{23} \\ \sigma_{31} & \sigma_{32} & \sigma_{33} \end{vmatrix}. \tag{3.26}$$

The product of two tensors is a tensor composed of all products containing one component of each of the two tensors. Thus a given term in the matrix of the product will be the sum of the products of the components of the first tensor for which the first subscript is the same with the components of the second tensor for which the second subscript is the same. For example the term in the product in row 1, column 1 is formed of the successive products of components in row 1 of the first tensor (for which the first subscript is 1) with the components in column 1 of the second tensor (for which the second subscript is 1). Similarly the component of the product in row 2, column 1 is the sum of products of terms in row 2 of the first tensor times the terms in column 1 of the second tensor. Thus

$$\sigma_{ip}\sigma_{pj} = \begin{vmatrix} (\sigma_{11}\sigma_{11}+\sigma_{12}\sigma_{21}+\sigma_{13}\sigma_{31}) & (\sigma_{11}\sigma_{12}+\sigma_{12}\sigma_{22}+\sigma_{13}\sigma_{32}) & (\sigma_{11}\sigma_{13}+\sigma_{12}\sigma_{23}+\sigma_{13}\sigma_{33}) \\ (\sigma_{21}\sigma_{11}+\sigma_{22}\sigma_{21}+\sigma_{23}\sigma_{31}) & (\sigma_{21}\sigma_{12}+\sigma_{22}\sigma_{22}+\sigma_{23}\sigma_{32}) & (\sigma_{21}\sigma_{13}+\sigma_{22}\sigma_{23}+\sigma_{23}\sigma_{33}) \\ (\sigma_{31}\sigma_{11}+\sigma_{32}\sigma_{21}+\sigma_{33}\sigma_{31}) & (\sigma_{31}\sigma_{12}+\sigma_{32}\sigma_{22}+\sigma_{33}\sigma_{32}) & (\sigma_{31}\sigma_{13}+\sigma_{32}\sigma_{23}+\sigma_{33}\sigma_{33}) \end{vmatrix}. \tag{3.27}$$

The trace of the tensor of the product (3.27) is a scalar equal to the sum of the three diagonal terms:

$$\operatorname{tr}(\sigma_{ip}\sigma_{pj}) = (\sigma_{11}\sigma_{11}+\sigma_{12}\sigma_{21}+\sigma_{13}\sigma_{31})+(\sigma_{21}\sigma_{12}+\sigma_{22}\sigma_{22}+\sigma_{23}\sigma_{32}) + (\sigma_{31}\sigma_{13}+\sigma_{32}\sigma_{23}+\sigma_{33}\sigma_{33}). \tag{3.28}$$

Examination of (3.28) discloses that it can be represented in subscript notation as follows

$$\operatorname{tr}(\sigma_{ip}\sigma_{pj}) = \sigma_{ij}\sigma_{ji}. \tag{3.29}$$

In a similar manner the product of (3.27) and (3.1) may be obtained. Let

$$\sigma_{ip}\sigma_{pq}\sigma_{qj} = \Omega_{ij}. \tag{3.30}$$

Then the terms of the matrix of the triple product Ω_{ij} are as follows after making use of the symmetry of the stress tensor and collecting terms

$$\begin{aligned}
\Omega_{11} &= \sigma_{11}^3+2\sigma_{11}(\sigma_{12}^2+\sigma_{13}^2)+\sigma_{22}\sigma_{12}^2+\sigma_{33}\sigma_{13}^2+2\sigma_{12}\sigma_{13}\sigma_{23},\\
\Omega_{22} &= \sigma_{22}^3+2\sigma_{22}(\sigma_{12}^2+\sigma_{23}^2)+\sigma_{11}\sigma_{12}^2+\sigma_{33}\sigma_{23}^2+2\sigma_{12}\sigma_{13}\sigma_{23},\\
\Omega_{33} &= \sigma_{33}^3+2\sigma_{33}(\sigma_{13}^2+\sigma_{23}^2)+\sigma_{11}\sigma_{13}^2+\sigma_{22}\sigma_{23}^2+2\sigma_{12}\sigma_{13}\sigma_{23},\\
\Omega_{12} = \Omega_{21} &= \sigma_{12}^3+\sigma_{12}(\sigma_{11}^2+\sigma_{22}^2+\sigma_{13}^2+\sigma_{23}^2)\\
&\quad +\sigma_{13}\sigma_{23}(\sigma_{11}+\sigma_{22}+\sigma_{33})+\sigma_{11}\sigma_{22}\sigma_{12},\\
\Omega_{23} = \Omega_{32} &= \sigma_{23}^3+\sigma_{23}(\sigma_{22}^2+\sigma_{33}^2+\sigma_{12}^2+\sigma_{13}^2)\\
&\quad +\sigma_{12}\sigma_{13}(\sigma_{11}+\sigma_{22}+\sigma_{33})+\sigma_{22}\sigma_{33}\sigma_{23},\\
\Omega_{13} = \Omega_{31} &= \sigma_{13}^3+\sigma_{13}(\sigma_{11}^2+\sigma_{33}^2+\sigma_{12}^2+\sigma_{23}^2)\\
&\quad +\sigma_{12}\sigma_{23}(\sigma_{11}+\sigma_{22}+\sigma_{33})+\sigma_{11}\sigma_{33}\sigma_{13}.
\end{aligned} \tag{3.30a}$$

The components of (3.30) may also be obtained directly as follows. To obtain Ω_{11} for example let $i = j = 1$ in (3.30), i.e., $\Omega = \sigma_{1p}\sigma_{pq}\sigma_{q1}$. Then using the summation convention let p and q take values 1, 2, 3 successively and add all resulting terms. This yields the first of equations (3.30a).

The trace of (3.30) is the sum of the first three terms of (3.30a),

$$\begin{aligned}\mathrm{tr}(\sigma_{ip}\sigma_{pq}\sigma_{qj}) &= \Omega_{ii} = \Omega_{11}+\Omega_{22}+\Omega_{33}\\ &= \sigma_{11}^3+\sigma_{22}^3+\sigma_{33}^3+3\sigma_{12}^2(\sigma_{11}+\sigma_{22})+3\sigma_{13}^2(\sigma_{11}+\sigma_{33})\\ &\quad +3\sigma_{23}^2(\sigma_{22}+\sigma_{33})+6\sigma_{12}\sigma_{13}\sigma_{23},\end{aligned} \tag{3.31}$$

or

$$\mathrm{tr}(\sigma_{ip}\sigma_{pq}\sigma_{qj}) = \sigma_{ip}\sigma_{pq}\sigma_{qi}. \tag{3.31a}$$

The values of the traces given by (3.25, 3.28, 3.31) are shown in Table 3.1 for several special states of stress of frequent interest.

3.8 Invariants in Terms of Traces

A comparison of (3.25) and (3.6) shows that the trace of a tensor $\mathrm{tr}\sigma_{ij}$ is the same as the first invariant I_1,

$$I_1 = \sigma_{ii} = \mathrm{tr}\sigma_{ij} = \bar{\boldsymbol{\sigma}}, \tag{3.32}$$

where a bar over a tensor, such as $\bar{\boldsymbol{\sigma}}$ indicates the trace of the tensor.

Making use of (3.25, 3.28, 3.31) in the expressions for I_2 (3.7) and I_3 (3.8), these invariants may also be expressed in terms of traces of the stress tensor or traces of products as follows:

$$I_2 = \tfrac{1}{2}(\sigma_{ij}\sigma_{ji}-\sigma_{ii}\sigma_{jj}) = \tfrac{1}{2}[\mathrm{tr}(\sigma_{ip}\sigma_{pj})-\mathrm{tr}(\sigma_{ij})\mathrm{tr}(\sigma_{ij})] = \tfrac{1}{2}(\overline{\boldsymbol{\sigma\sigma}}-\bar{\boldsymbol{\sigma}}\,\bar{\boldsymbol{\sigma}}), \tag{3.33}$$

$$\begin{aligned}I_3 &= \tfrac{1}{6}(\sigma_{ii}\sigma_{jj}\sigma_{kk}+2\sigma_{ip}\sigma_{pq}\sigma_{qi}-3\sigma_{ij}\sigma_{ji}\sigma_{kk})\\ &= \tfrac{1}{6}[\mathrm{tr}\sigma_{ij}\mathrm{tr}\sigma_{ij}\mathrm{tr}\sigma_{ij}+2\mathrm{tr}(\sigma_{ip}\sigma_{pq}\sigma_{qj})-3\mathrm{tr}(\sigma_{ip}\sigma_{pj})\mathrm{tr}\sigma_{ij}]\\ &= \tfrac{1}{6}(\bar{\boldsymbol{\sigma}}\bar{\boldsymbol{\sigma}}\bar{\boldsymbol{\sigma}}+2\overline{\boldsymbol{\sigma\sigma\sigma}}-3\overline{\boldsymbol{\sigma\sigma}}\bar{\boldsymbol{\sigma}}),\end{aligned} \tag{3.34}$$

where a bar continuously over two or more tensors such as $\overline{\boldsymbol{\sigma\sigma}}$ indicates the trace of the products of the tensors covered by the continuous bar. Where a bar covers a single tensor the trace of that tensor is indicated. Values of the invariants given by (3.32, 3.33, 3.34) are shown in Table 3.1 for several special states of stress.

For *biaxial states of stress* I_3 is zero as shown in Table 3.1. As a consequence of this fact, the following expression relating traces of third order products of biaxial stress tensors is found by setting $I_3 = 0$ in (3.34)

$$\overline{\boldsymbol{\sigma\sigma\sigma}} = \tfrac{1}{2}(3\overline{\boldsymbol{\sigma\sigma}}\bar{\boldsymbol{\sigma}}-\bar{\boldsymbol{\sigma}}\bar{\boldsymbol{\sigma}}\bar{\boldsymbol{\sigma}}). \tag{3.35}$$

Similarly, the invariants of the stress deviators (3.21, 3.22) may be expressed in terms of traces as follows:

$$J_2 = \tfrac{1}{2}s_{ij}s_{ji} = \tfrac{1}{2}\overline{\mathbf{ss}}, \tag{3.36}$$

$$J_3 = \tfrac{1}{3}s_{ij}s_{jk}s_{ki} = \tfrac{1}{3}\overline{\mathbf{sss}}. \tag{3.37}$$

Values of the invariants J_2, J_3 are also shown in Table 3.1 for several states of stress.

3.9 Hamilton-Cayley Equation

It may be shown [3.1] that the matrix of the product with itself of a stress tensor expressed with respect to principal axes is a tensor expressed in terms of the products of the corresponding principal stresses and has the same principal axes as the stress tensor. Also each principal stress σ_{I}, σ_{II}, σ_{III} must satisfy the characteristic equation (3.5). In view of these conditions the three equations found by substituting successively the three principal stresses into (3.5) may be combined into one tensor equation as follows:

$$\begin{vmatrix} \sigma_{\mathrm{I}} & 0 & 0 \\ 0 & \sigma_{\mathrm{II}} & 0 \\ 0 & 0 & \sigma_{\mathrm{III}} \end{vmatrix}^3 - I_1 \begin{vmatrix} \sigma_{\mathrm{I}} & 0 & 0 \\ 0 & \sigma_{\mathrm{II}} & 0 \\ 0 & 0 & \sigma_{\mathrm{III}} \end{vmatrix}^2 - I_2 \begin{vmatrix} \sigma_{\mathrm{I}} & 0 & 0 \\ 0 & \sigma_{\mathrm{II}} & 0 \\ 0 & 0 & \sigma_{\mathrm{III}} \end{vmatrix} - I_3 \begin{vmatrix} 1 & 0 & 0 \\ 0 & 1 & 0 \\ 0 & 0 & 1 \end{vmatrix} = 0.$$

The above equation involves the stress tensor in terms of principal stresses. It may be shown [3.2] that this equation may be written for the general stress state σ_{ij} in a similar manner as follows,

$$\sigma_{ij}^3 - I_1\sigma_{ij}^2 - I_2\sigma_{ij} - I_3\delta_{ij} = 0. \tag{3.38}$$

Inserting (3.32, 3.33, 3.34) and substituting the notation $\boldsymbol{\sigma}$ for σ_{ij}, and $\mathbf{I}$ for δ_{ij} yields

$$\boldsymbol{\sigma}^3 = \bar{\boldsymbol{\sigma}}\boldsymbol{\sigma}\boldsymbol{\sigma} + \tfrac{1}{2}\overline{\boldsymbol{\sigma}\boldsymbol{\sigma}}\boldsymbol{\sigma} - \tfrac{1}{2}\bar{\boldsymbol{\sigma}}\bar{\boldsymbol{\sigma}}\boldsymbol{\sigma} + \tfrac{1}{3}\mathbf{I}\overline{\boldsymbol{\sigma}\boldsymbol{\sigma}\boldsymbol{\sigma}} - \tfrac{1}{2}\mathbf{I}\overline{\boldsymbol{\sigma}\boldsymbol{\sigma}}\bar{\boldsymbol{\sigma}} + \tfrac{1}{6}\mathbf{I}\bar{\boldsymbol{\sigma}}\bar{\boldsymbol{\sigma}}\bar{\boldsymbol{\sigma}}. \tag{3.39}$$

This is the Hamilton-Cayley equation for a 3×3 matrix. It provides a linear relation between $\boldsymbol{\sigma}^3$ versus $\boldsymbol{\sigma}^2$, $\boldsymbol{\sigma}$, and $\mathbf{I}$ thus making it possible to reduce the order of stress terms in a polynomial in stress containing terms of the third order. It may also be shown [3.1] that all higher orders may be expressed as linear combinations of $\boldsymbol{\sigma}^2$, $\boldsymbol{\sigma}$, and $\mathbf{I}$.

Biaxial states of stress σ_{ij} have the following 3×3 matrix:

$$\sigma_{ij} = \begin{vmatrix} \sigma_{11} & \sigma_{12} & 0 \\ \sigma_{21} & \sigma_{22} & 0 \\ 0 & 0 & 0 \end{vmatrix}, \qquad i, j = 1, 2, 3. \tag{3.40}$$

Thus I_3 in (3.8) is zero for biaxial stress and (3.38) becomes

$$\sigma_{ij}^3 = I_1\sigma_{ij}^2 + I_2\sigma_{ij}. \tag{3.41}$$

Inserting (3.32) and (3.33) in (3.41) with $\boldsymbol{\sigma}$ for σ_{ij} and $\mathbf{I}$ for δ_{ij} yields

$$\boldsymbol{\sigma\sigma\sigma} = \boldsymbol{\sigma\sigma}\bar{\boldsymbol{\sigma}} + \tfrac{1}{2}\boldsymbol{\sigma}\overline{\boldsymbol{\sigma\sigma}} - \tfrac{1}{2}\boldsymbol{\sigma}\bar{\boldsymbol{\sigma}}\bar{\boldsymbol{\sigma}}. \tag{3.42}$$

Taking the trace of (3.42) gives

$$\overline{\boldsymbol{\sigma\sigma\sigma}} = \tfrac{3}{2}\bar{\boldsymbol{\sigma}}\overline{\boldsymbol{\sigma\sigma}} - \tfrac{1}{2}\bar{\boldsymbol{\sigma}}\bar{\boldsymbol{\sigma}}\bar{\boldsymbol{\sigma}}, \tag{3.43}$$

which is the same as (3.35) as obtained by setting $I_3 = 0$ in (3.34).

Considering biaxial states of stress as a 2×2 matrix

$$\sigma_{ij} = \begin{vmatrix} \sigma_{11} & \sigma_{12} \\ \sigma_{21} & \sigma_{22} \end{vmatrix}, \qquad i, j = 1, 2. \tag{3.44}$$

As in Section 3.4 the characteristic equation corresponding to (3.44) is found from

$$\begin{vmatrix} \sigma_{11}-\sigma & \sigma_{12} \\ \sigma_{21} & \sigma_{22}-\sigma \end{vmatrix} = 0, \tag{3.45}$$

by setting the determinant of (3.45) equal to zero,

$$(\sigma_{11}\sigma_{22} - \sigma_{12}^2) - (\sigma_{11} + \sigma_{22})\sigma + \sigma^2 = 0. \tag{3.46}$$

(3.46) may be written

$$-I_2 - I_1\sigma + \sigma^2 = 0, \tag{3.47}$$

where

$$I_1 = \sigma_{ii}, \quad I_2 = \tfrac{1}{2}(\sigma_{ij}\sigma_{ji} - \sigma_{ii}\sigma_{jj}), \qquad i, j = 1, 2. \tag{3.48}$$

I_1 and I_2 are as given by (3.6) and (3.7) except that i and j have only two values 1 and 2.

A Hamilton-Cayley equation for a 2×2 matrix (3.44) may be derived from (3.47) in a similar manner to the derivation of (3.38) with the result

$$\sigma_{ij}^2 - I_1\sigma_{ij} - I_2\delta_{ij} = 0, \qquad i, j = 1, 2. \tag{3.49}$$

Substituting (3.32) and (3.33) with the notation $\boldsymbol{\sigma}$ for σ_{ij} and $\mathbf{I}$ for δ_{ij} and the restriction that $i, j = 1, 2$ yields the biaxial Hamilton-Cayley equation

$$\boldsymbol{\sigma\sigma} = \bar{\boldsymbol{\sigma}}\boldsymbol{\sigma} + \tfrac{1}{2}\mathbf{I}(\overline{\boldsymbol{\sigma\sigma}} - \bar{\boldsymbol{\sigma}}\bar{\boldsymbol{\sigma}}), \tag{3.50}$$

where in the corresponding subscript notation i, j take values 1 and 2 only. Thus $\boldsymbol{\sigma}$ is given by (3.44) and

$$\mathbf{I} = \begin{vmatrix} 1 & 0 \\ 0 & 1 \end{vmatrix}. \tag{3.51}$$

Taking the trace of (3.50)

$$\overline{\boldsymbol{\sigma}\boldsymbol{\sigma}} = \bar{\boldsymbol{\sigma}}\bar{\boldsymbol{\sigma}}+\overline{\boldsymbol{\sigma}\boldsymbol{\sigma}}-\bar{\boldsymbol{\sigma}}\bar{\boldsymbol{\sigma}} = \overline{\boldsymbol{\sigma}\boldsymbol{\sigma}},$$

since from (3.51) the trace of $\mathbf{I}$ is 2,

$$\mathrm{tr}\mathbf{I} = 2, \qquad i, j = 1, 2. \tag{3.52}$$

Multiplying (3.50) by $\boldsymbol{\sigma}$ and noting that

$$\mathbf{I}\boldsymbol{\sigma} = \boldsymbol{\sigma} \tag{3.53}$$

(this may be demonstrated by expanding the product of the matrices of $\mathbf{I}$ and $\boldsymbol{\sigma}$) yields

$$\boldsymbol{\sigma}\boldsymbol{\sigma}\boldsymbol{\sigma} = \bar{\boldsymbol{\sigma}}\boldsymbol{\sigma}\boldsymbol{\sigma}+\tfrac{1}{2}\overline{\boldsymbol{\sigma}\boldsymbol{\sigma}}\boldsymbol{\sigma}-\tfrac{1}{2}\bar{\boldsymbol{\sigma}}\bar{\boldsymbol{\sigma}}\boldsymbol{\sigma}, \tag{3.54}$$

which is the same as (3.42). Note, however, that (3.42) may be used for a biaxial 3×3 matrix (3.40) whereas (3.50) is restricted to a 2×2 matrix (3.44). Also note that whereas multiplication of (3.50) by $\boldsymbol{\sigma}$ to obtain (3.42) is allowable the inverse is not allowable since the tensor (3.40) does not have an inverse.

3.10 State of Strain

Strain is a measure of the intensity of deformation at a point. Linear strain* is the change in length of a line segment per unit of length as the length of the line segment becomes infinitesimal. Engineering shear strain is the change in angle at the point of intersection of two lines originally at right angles. To describe the state of strain at a point mathematically consider the change in configuration of a deformed body. The deformation of the body may be described by the functional relation between the undeformed and the deformed states.

Consider the position of a point P (Fig. 3.3) and a neighboring point P′ both in the undeformed state of a body. The positions of P and P′ are described by fixed rectangular Cartesian coordinates X_i and $X_i+\Delta X_i$ respectively, where $i = 1, 2, 3$, see Fig. 3.3.

After deformation takes place, at time t the material points P and P′ deform into p and p′ where the position of p and p′ are described in the same coordinate system by x_i and $x_i+\Delta x_i$, ($i = 1, 2, 3$). The deformation of the body is assumed to be continuous and can be described by the functional

* Linear strain or unit elongation will be referred to as "normal strain" in parallel with normal stress in subsequent chapters.

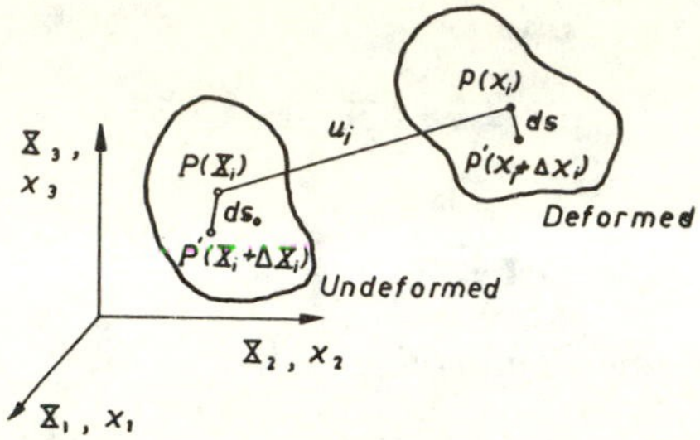

Fig. 3.3. Deformation

dependence of x_i on X_i and t:

$$x_i = x_i[X_q, t]. \tag{3.55}$$

It is further assumed that equation (3.55) has a single-valued inverse

$$X_i = X_i[x_q, t], \tag{3.56}$$

so that the deformation from P to p is one-to-one.

The squares of the lengths ds_0 of the segment PP′ and ds of pp′ in the undeformed and in the deformed states, respectively, are

$$(ds_0)^2 = dX_i dX_i = \frac{\partial X_i}{\partial x_k}\frac{\partial X_i}{\partial x_l}\,dx_k\,dx_l \tag{3.57}$$

$$(ds)^2 = dx_i dx_i = \frac{\partial x_i}{\partial X_k}\frac{\partial x_i}{\partial X_l}\,dX_k dX_l. \tag{3.58}$$

The difference between the square of the length elements can be written as

$$(ds)^2-(ds_0)^2 = \left(\frac{\partial x_i}{\partial X_k}\frac{\partial x_i}{\partial X_l}-\delta_{kl}\right)dX_k dX_l = 2E_{kl}\,dX_k dX_l, \tag{3.59}$$

or

$$(ds)^2-(ds_0)^2 = \left(\delta_{kl}-\frac{\partial X_i}{\partial x_k}\frac{\partial X_i}{\partial x_l}\right)dx_k dx_l = 2e_{kl}\,dx_k dx_l, \tag{3.60}$$

where E_{kl}, e_{kl} are defined as

$$E_{kl} = \tfrac{1}{2}\left(\frac{\partial x_i}{\partial X_k}\frac{\partial x_i}{\partial X_l}-\delta_{kl}\right), \tag{3.61}$$

$$e_{kl} = \tfrac{1}{2}\left(\delta_{kl}-\frac{\partial X_i}{\partial x_k}\frac{\partial X_i}{\partial x_l}\right). \tag{3.62}$$

The strain tensors E_{kl} and e_{kl} thus defined are called Green's* strain tensor

* e_{kl} is sometimes called Almansi's strain tensor and in hydrodynamics (3.61) and (3.62) are called Lagrange and Euler tensors, respectively.

and Cauchy's* strain tensor respectively. In the former, the strain is measured with respect to the undeformed state (gage length in uniaxial strain); in the latter it is measured with respect to the deformed state (sometimes called the true strain for uniaxial strain). These two tensors are symmetric with respect to the coordinates.

3.11 Strain-Displacement Relation

The displacement vector u_i (see Fig. 3.3) is defined by

$$u_i(X, t) = x_i(X, t) - X_i, \tag{3.63}$$

or

$$u_i(x, t) = x_i - X_i(x, t), \tag{3.64}$$

where X and x, representing X_1, X_2, X_3 and x_1, x_2, x_3 respectively, are the coordinates of the point in question. All three coordinates of the point appear in the expression for each value of i in u_i.

Inserting (3.63) for x_i in (3.61) and (3.64) for X_i in (3.62) yields after simplification

$$E_{ij} = \tfrac{1}{2}\left[\frac{\partial u_i}{\partial X_j} + \frac{\partial u_j}{\partial X_i} + \frac{\partial u_\alpha}{\partial X_i}\frac{\partial u_\alpha}{\partial X_j}\right], \tag{3.65}$$

$$e_{ij} = \tfrac{1}{2}\left[\frac{\partial u_i}{\partial x_j} + \frac{\partial u_j}{\partial x_i} - \frac{\partial u_\alpha}{\partial x_i}\frac{\partial u_\alpha}{\partial x_j}\right]. \tag{3.66}$$

If the components of displacement u_i are such that their first derivatives are so small that the products of the partial derivatives of u_i in the third term of (3.66) are negligible compared to the first two terms, then Cauchy's strain tensor reduces to Cauchy's infinitesimal strain tensor,

$$\varepsilon_{ij} = \tfrac{1}{2}\left[\frac{\partial u_i}{\partial x_j} + \frac{\partial u_j}{\partial x_i}\right]. \tag{3.67}$$

In ordinary notation (x, y, z for x_1, x_2, x_3; u, v, w, for u_1, u_2, u_3) the components of (3.67) become as follows:

$$\begin{aligned}
\varepsilon_{xx} &= \frac{\partial u}{\partial x}, & \varepsilon_{xy} &= \tfrac{1}{2}\left(\frac{\partial u}{\partial y} + \frac{\partial v}{\partial x}\right) = \varepsilon_{yx}, \\
\varepsilon_{yy} &= \frac{\partial v}{\partial y}, & \varepsilon_{xz} &= \tfrac{1}{2}\left(\frac{\partial u}{\partial z} + \frac{\partial w}{\partial x}\right) = \varepsilon_{zx}, \\
\varepsilon_{zz} &= \frac{\partial w}{\partial z}, & \varepsilon_{yz} &= \tfrac{1}{2}\left(\frac{\partial v}{\partial z} + \frac{\partial w}{\partial y}\right) = \varepsilon_{zy}.
\end{aligned} \tag{3.68}$$

In the case of infinitesimal displacement, the distinction between Green's and Cauchy's strain tensors disappears, since then it is immaterial whether the derivatives of displacements are with respect to deformed or undeformed coordinates. For a geometric derivation and interpretation of (3.68) see [3.3].

It should be noted that the terms in parentheses in (3.68) define the change in angle between two mutually perpendicular reference lines, which is the engineering shear strain γ. Thus it follows that the tensor shear strain is one half the engineering shear strain,

$$\varepsilon_{xy} = \tfrac{1}{2}\gamma_{xy}, \quad \varepsilon_{xz} = \tfrac{1}{2}\gamma_{xz}, \quad \varepsilon_{yz} = \tfrac{1}{2}\gamma_{yz}. \tag{3.69}$$

3.12 Strain Tensor

The nine components of the strain tensor ε_{ij} comprise a second order Cartesian tensor whose matrix is as follows

$$\varepsilon_{ij} = \boldsymbol{\epsilon} = \begin{vmatrix} \varepsilon_{11} & \varepsilon_{12} & \varepsilon_{13} \\ \varepsilon_{21} & \varepsilon_{22} & \varepsilon_{23} \\ \varepsilon_{31} & \varepsilon_{32} & \varepsilon_{33} \end{vmatrix}, \tag{3.70}$$

where the subscript notation has the same significance as in the stress tensor except that its relationship to engineering strain is not as direct as for stress in view of (3.69). The strain tensor is also symmetrical about the diagonal line from ε_{11} to ε_{33}.

In view of the similarity of the strain tensor (3.70) and the stress tensor (3.1) it follows that all of the stress relationships derived above from the stress tensor (3.1) have a counterpart for strain. It is only necessary to substitute ε for σ in the above equations as in the following example.

Volumetric Strain, Deviatoric Strain: In a manner similar to that employed in (3.10) the strain tensor ε_{ij} may be divided into a volumetric strain ε_v and a deviatoric strain d_{ij} as follows:

$$\varepsilon_{ij} = \begin{vmatrix} \varepsilon_{11} & \varepsilon_{12} & \varepsilon_{13} \\ \varepsilon_{21} & \varepsilon_{22} & \varepsilon_{23} \\ \varepsilon_{31} & \varepsilon_{32} & \varepsilon_{33} \end{vmatrix} = \begin{vmatrix} \varepsilon_v & 0 & 0 \\ 0 & \varepsilon_v & 0 \\ 0 & 0 & \varepsilon_v \end{vmatrix} + \begin{vmatrix} d_{11} & \varepsilon_{12} & \varepsilon_{13} \\ \varepsilon_{21} & d_{22} & \varepsilon_{23} \\ \varepsilon_{31} & \varepsilon_{32} & d_{33} \end{vmatrix}, \tag{3.71}$$

where $d_{11} = \varepsilon_{11} - \varepsilon_v$, $d_{22} = \varepsilon_{22} - \varepsilon_v$, $d_{33} = \varepsilon_{33} - \varepsilon_v$.

As was true for the stress tensor, the components in (3.71) with mixed subscripts are the same in the first and last matrix of (3.71). The second matrix of (3.71) is the volumetric strain ε_v and the third matrix is the deviatoric strain.

Taking ε_v to be the average linear strain of the strain tensor,

$$\varepsilon_v = \tfrac{1}{3}(\varepsilon_{11}+\varepsilon_{22}+\varepsilon_{33}) = \tfrac{1}{3}\varepsilon_{kk}, \tag{3.72}$$

makes the volumetric strain d_v of the deviatoric tensor zero,

$$d_v = \tfrac{1}{3}(d_{11}+d_{22}+d_{33}) = 0.$$

Thus the volumetric strain ε_v describes a pure dilatation and the strain deviator describes pure distortion at constant volume. For small strains the change in volume per unit volume (dilatation) equals ε_{kk}.

Using the subscript notation the deviatoric strain is,

$$d_{ij} = \varepsilon_{ij}-\tfrac{1}{3}\delta_{ij}\varepsilon_{kk}. \tag{3.73}$$

CHAPTER 4

MECHANICS OF STRESS AND DEFORMATION ANALYSES

4.1 Introduction

The mechanical behavior of any continuous material is governed by certain physical laws. Some of them are common for all continuous materials, while others are intrinsic properties of each group or each individual material. These laws include conservation of mass, balance of momentum, balance of moment of momentum, conservation of energy, constitutive relations and principles governing thermodynamics.

Design of load bearing members of a structure or machine requires a knowledge of the stresses, strains and displacements resulting from the external forces or imposed displacements acting on it. In some situations only the magnitude and distribution of the stresses are of interest while in others, in which stiffness and deflections are important or where deformation affects the stress distribution, the strains and displacements must also be determined.

In any solid body subjected to external forces and/or displacements the stresses and strains at every point including points on the boundary must simultaneously satisfy three basic equations relating to: (1) equilibrium; (2) kinematics; and (3) material constitution (constitutive equations). The first two equations are independent of the type of material or whether the material behaves in a linear manner or not. The third equation depends on the material and represents its mechanical behavior.

This chapter includes a brief review of the equations needed in the analysis of the stress and deformation of elastic solid bodies having prescribed boundaries with prescribed stresses and/or displacements at the boundaries. In order to demonstrate the application of these equations the analysis of stress in a linear, isotropic elastic material is discussed in a very general way. For application to specific types of load-carrying members the reader is referred to the many texts on elasticity and strength of materials.

4.2 Law of Motion

Newton's law of motion states that the rate of change of linear momentum is equal to the resultant of the applied forces and that the rate of change of the moment of momentum with respect to the origin of the coordinates is equal to the resultant moment of applied forces about the same origin. This law of motion holds for particles as well as for bodies of all kinds. There are various ways of transforming this law into differential equations for use in the mechanics of deformable bodies. A direct approach is used in the following.

4.3 Equations of Equilibrium

Consider the static equilibrium state of an infinitesimal parallelepiped of sides dx_1, dx_2, dx_3 with surfaces parallel to the coordinate planes. The stresses acting on the various surfaces and the body forces are shown in Fig. 4.1, where σ_{11}, σ_{22}, σ_{33} are normal stresses, σ_{12}, σ_{13}, σ_{23}, etc., are shearing stresses and F_1, F_2, F_3 are the components of the body force per unit volume in the direction of the coordinate axes. Examples of body forces are gravity,

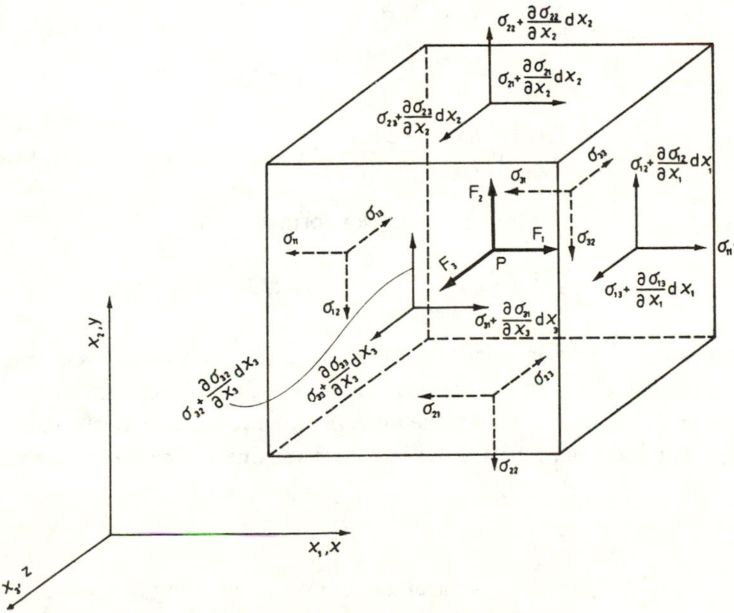

Fig. 4.1. Equilibrium Stress Components

centrifugal force and magnetic force. The body force may be different at different positions x_i and may change with time $F(x_i, t)$.

The static equilibrium of the body requires that the resultant of all forces is zero. The equilibrium equation along x_1 or x direction is then obtained by summing all the forces acting on the element in the x_1 or x direction. The forces are obtained by multiplying the stress by the area on which the stress acts. The sum of these forces is:

$$\left(\sigma_{11}+\frac{\partial \sigma_{11}}{\partial x_1}\,dx_1\right) dx_2\,dx_3-\sigma_{11}\,dx_2\,dx_3$$
$$+\left(\sigma_{21}+\frac{\partial \sigma_{21}}{\partial x_2}\,dx_2\right) dx_1\,dx_3-\sigma_{21}\,dx_1\,dx_3$$
$$+\left(\sigma_{31}+\frac{\partial \sigma_{31}}{\partial x_3}\,dx_3\right) dx_1\,dx_2-\sigma_{31}\,dx_1\,dx_2+F_1\,dx_1\,dx_2\,dx_3 = 0. \quad (4.1)$$

The other two equilibrium equations along x_2 and x_3 (or y and z) coordinates can be obtained in the same way. Dividing the equations by $dx_1\,dx_2\,dx_3$ yields the following three equilibrium equations oriented along the coordinate axes

$$\frac{\partial \sigma_{11}}{\partial x_1}+\frac{\partial \sigma_{21}}{\partial x_2}+\frac{\partial \sigma_{31}}{\partial x_3}+F_1 = 0,$$
$$\frac{\partial \sigma_{12}}{\partial x_1}+\frac{\partial \sigma_{22}}{\partial x_2}+\frac{\partial \sigma_{32}}{\partial x_3}+F_2 = 0,$$
$$\frac{\partial \sigma_{13}}{\partial x_1}+\frac{\partial \sigma_{23}}{\partial x_2}+\frac{\partial \sigma_{33}}{\partial x_3}+F_3 = 0. \quad (4.2)$$

Equations (4.2) can be written in the tensor form as follows

$$\frac{\partial \sigma_{ij}}{\partial x_i}+F_j=0 \qquad (i, j = 1, 2, 3). \quad (4.3)$$

The equation of motion under dynamic conditions can be derived in the same way by including the inertia term, rate of change of momentum. Thus the right-hand side of (4.2) equals the mass per unit volume times the second derivation of the displacement u_j with respect to time t. The tensor form is

$$\frac{\partial \sigma_{ij}}{\partial x_i}+F_j = \varrho\frac{\partial^2 u_j}{\partial t^2}, \quad (4.4)$$

where $u_j, j = 1, 2, 3$, are the displacement components along the coordinate axes.

4.4 Equilibrium of Moments

The equilibrium of an element requires also that the resultant rate of change of moment of momentum be zero. Considering moments about a line through the center P of the parallelepiped and parallel to the x_1 axis of all the force components as shown in Fig. 4.1 the sum is

$$-\left(\sigma_{32}+\frac{\partial\sigma_{32}}{\partial x_3}\,\mathrm{d}x_3\right)\mathrm{d}x_1\,\mathrm{d}x_2\frac{\mathrm{d}x_3}{2}-\sigma_{32}\,\mathrm{d}x_1\,\mathrm{d}x_2\frac{\mathrm{d}x_3}{2}$$
$$+\left(\sigma_{23}+\frac{\partial\sigma_{23}}{\partial x_2}\,\mathrm{d}x_2\right)\mathrm{d}x_1\,\mathrm{d}x_3\frac{\mathrm{d}x_2}{2}+\sigma_{23}\,\mathrm{d}x_1\,\mathrm{d}x_3\frac{\mathrm{d}x_2}{2}=0.$$

Dividing by $\mathrm{d}x_1\,\mathrm{d}x_2\,\mathrm{d}x_3$ and taking the limit as $\mathrm{d}x_2 \to 0, \mathrm{d}x_3 \to 0$, the following result is obtained:

$$\sigma_{32} = \sigma_{23}. \tag{4.5}$$

The moment of forces about lines through P parallel to x_2, x_3 yields two similar relations

$$\sigma_{12} = \sigma_{21}, \quad \sigma_{13} = \sigma_{31},$$

so in tensor form

$$\sigma_{ij} = \sigma_{ji}. \tag{4.6}$$

Equation (4.6) shows that the stress tensor is symmetric. In deriving (4.6), it has been assumed that the material is nonpolar, namely, that there are no external moments proportional to a volume (such as ferrous material under electromagnetic field) and that there are no couple stresses. (See references [4.1, 4.2], for example, for discussion of couple stresses.)

4.5 Kinematics

The kinematic relationships include the strain-displacement relations described in Chapter 3, see (3.67) and (3.68), and the functional relationships among strains required to insure that all strains acting at a given point are mutually compatible as described in the next section.

4.6 Compatibility Equations

The six strain components (3.68) of the strain tensor are functions of only three displacement components u, v, w, (u_j). It follows that the six strains cannot all be independent. If the six strains were specified independently, continuity of the material would be violated such that holes would have to result or two or more portions of material would have to occupy the same space. Neither is permissible in a continuous material. Thus in determining the displacements from (3.68), not all six strain components can be arbitrarily assigned, as there are six equations which have only three unknowns. Therefore, the six strain components must satisfy certain conditions in order to assure the integrability of (3.68) and that the displacement components are continuous and single valued. The compatibility relations between strain components are obtained (in a sense) by eliminating the displacement components in (3.68). For example, consider the three components of strain from (3.68) lying in the x-y plane

$$\varepsilon_{xx} = \frac{\partial u}{\partial x}, \quad \varepsilon_{yy} = \frac{\partial v}{\partial y}, \quad \varepsilon_{xy} = \tfrac{1}{2}\left(\frac{\partial u}{\partial y} + \frac{\partial v}{\partial x}\right).$$

Taking the second derivative of ε_{xx} with respect to y, the second derivative of ε_{yy} with respect to x, and the second derivative of ε_{xy} once with respect to x and once with respect to y yields

$$\frac{\partial^2 \varepsilon_{xx}}{\partial y^2} = \frac{\partial^3 u}{\partial x\,\partial y^2}, \quad \frac{\partial^2 \varepsilon_{yy}}{\partial x^2} = \frac{\partial^3 v}{\partial y\,\partial x^2}, \quad \frac{\partial^2 \varepsilon_{xy}}{\partial x\,\partial y} = \tfrac{1}{2}\left(\frac{\partial^3 u}{\partial x\,\partial y^2} + \frac{\partial^3 v}{\partial x^2\,\partial y}\right). \tag{4.7}$$

Substituting the first two expressions of (4.7) in the third eliminates the displacement terms and yields

$$2\frac{\partial^2 \varepsilon_{xy}}{\partial x\,\partial y} = \frac{\partial^2 \varepsilon_{xx}}{\partial y^2} + \frac{\partial^2 \varepsilon_{yy}}{\partial x^2}. \tag{4.8}$$

This is one of a set of six independent compatibility relations which must be satisfied to insure compatibility of strains. For a plane strain problem (4.8) is a complete statement of compatibility. For plane stress of a thin plate (4.8) is sufficient. For the general case the following set of six equations is required, see for example [4.3].

$$\frac{\partial^2 \varepsilon_{xx}}{\partial y\,\partial z} = \frac{\partial}{\partial x}\left(-\frac{\partial \varepsilon_{yz}}{\partial x} + \frac{\partial \varepsilon_{zx}}{\partial y} + \frac{\partial \varepsilon_{xy}}{\partial z}\right)$$

$$\frac{\partial^2 \varepsilon_{yy}}{\partial z\,\partial x} = \frac{\partial}{\partial y}\left(-\frac{\partial \varepsilon_{zx}}{\partial y} + \frac{\partial \varepsilon_{xy}}{\partial z} + \frac{\partial \varepsilon_{yz}}{\partial x}\right)$$

$$\frac{\partial^2 \varepsilon_{zz}}{\partial x\,\partial y} = \frac{\partial}{\partial z}\left(-\frac{\partial \varepsilon_{xy}}{\partial z} + \frac{\partial \varepsilon_{yz}}{\partial x} + \frac{\partial \varepsilon_{zx}}{\partial y}\right)$$

$$2\frac{\partial^2 \varepsilon_{xy}}{\partial x\,\partial y} = \frac{\partial^2 \varepsilon_{xx}}{\partial y^2} + \frac{\partial^2 \varepsilon_{yy}}{\partial x^2}$$

$$2\frac{\partial^2 \varepsilon_{yz}}{\partial y\,\partial z} = \frac{\partial^2 \varepsilon_{yy}}{\partial z^2} + \frac{\partial^2 \varepsilon_{zz}}{\partial y^2}$$

$$2\frac{\partial^2 \varepsilon_{zx}}{\partial z\,\partial x} = \frac{\partial^2 \varepsilon_{zz}}{\partial x^2} + \frac{\partial^2 \varepsilon_{xx}}{\partial z^2}. \tag{4.9}$$

These six equations are known as the equations of compatibility and were obtained by Saint-Venant in 1860.

4.7 Constitutive Equations

The entire content of the previous sections is applicable to all continuous materials. However, material bodies of the same mass and the same geometric shape respond to the same external forces or displacements (excitation) in different ways. This difference in response is attributed to the difference of the internal constitution of the materials. Equations characterizing the individual material and its reaction to external excitations are called "constitutive equations". Since constitutive equations describe the intrinsic characteristics of the materials, these equations must satisfy certain physical requirements if they are to faithfully represent material behavior. Thus in formulation of constitutive equations it is desirable to satisfy as many of these requirements as possible. For a discussion of these physical requirements see, for example, Eringen [4.4] and Rivlin [4.5].

Real materials behave in such complex ways that when the entire range of possible temperature and deformation is considered it is presently impossible to write down a single equation which will describe accurately the behavior of a real material over the entire range of the variables. Instead, an idealization approach is adopted where separate constitutive equations are formulated to describe various kinds of idealized material response. Each of these equations is a mathematical formulation designed to approximately describe the observed response of a real material over a certain restricted range of the variables involved.

To illustrate the constitutive equations discussed in this section, the stress-strain relation of an idealized Hookean elastic solid is discussed in the next section.

4.8 Linear Elastic Solid

A linear (or Hookean) elastic solid is a solid that obeys Hooke's law, that the stress tensor is linearly proportional to the strain tensor, i.e.,

$$\sigma_{ij} = C_{ijkl}\varepsilon_{kl}, \tag{4.10}$$

where σ_{ij} and ε_{kl} are the stress and strain tensors, respectively, as discussed in Chapter 3, and C_{ijkl} is a 4th order tensor describing the elastic moduli. C_{ijkl} is independent of stress and strain and therefore is a constant when all other variables such as temperature are constant. C_{ijkl}, in general, has $3^4 = 81$ elements; but in view of symmetry of the stress tensor and strain tensors ($\sigma_{ij} = \sigma_{ji}$ and $\varepsilon_{kl} = \varepsilon_{lk}$) C_{ijkl} has at most 36 independent constants.

For most elastic solids the number of independent elastic constants is much less than 36. This reduction is due to the existence of material symmetry. The reader is referred to the books by Love [4.6], Sokolnikoff [4.3] and Zener [4.7] for a more complete discussion of anisotropic elastic constants.

When the material exhibits complete isotropy such that the elastic properties are identical in all directions, it can be shown that there remain only two independent constants in C_{ijkl}. The stress-strain relation thus has the following form

$$\sigma_{ij} = \lambda\varepsilon_{kk}\delta_{ij} + 2G\varepsilon_{ij}. \tag{4.11}$$

The constants λ and G are called Lamé constants, where

$$G = \frac{E}{2(1+\nu)} \tag{4.12}$$

is the modulus of elasticity in shear, E is the elastic modulus, ν is Poisson's ratio, and

$$\lambda = \frac{E\nu}{(1+\nu)(1-2\nu)} = \frac{2G\nu}{1-2\nu}. \tag{4.13}$$

In terms of the elastic modulus E and Poisson's ratio ν (4.11) becomes

$$\sigma_{ij} = \frac{E\nu}{(1+\nu)(1-2\nu)}\,\varepsilon_{kk}\,\delta_{ij} + \frac{E}{(1+\nu)}\,\varepsilon_{ij}. \tag{4.14a}$$

Inverting (4.14a) to solve for the strain tensor ε_{ij} yields

$$\varepsilon_{ij} = \frac{1+\nu}{E}\sigma_{ij} - \frac{\nu}{E}\sigma_{kk}\delta_{ij}. \tag{4.14b}$$

Table 4.1

Relations Among Elastic Moduli for an Isotropic Linear Elastic Solid

	(G, E)	(G, K)	(E, K)	(G, ν)	(E, ν)	(K, ν)	(λ, G)
E	–	$\frac{9GK}{3K+G}$	–	$2G(1+\nu)$	–	$3K(1-2\nu)$	$\frac{G(3\lambda+2G)}{G+\lambda}$
G	–	–	$\frac{3EK}{9K-E}$	–	$\frac{E}{2(1+\nu)}$	$\frac{3K(1-2\nu)}{2(1+\nu)}$	–
K	$\frac{EG}{9G-3E}$	–	–	$\frac{2G(1+\nu)}{3(1-2\nu)}$	$\frac{E}{3(1-2\nu)}$	–	$\frac{2G}{3}+\lambda$
ν	$\frac{E}{2G}-1$	$\frac{3K-2G}{2[3K+G]}$	$\frac{1}{2}\left[1-\frac{E}{3K}\right]$	–	–	–	$\frac{\lambda}{2(G+\lambda)}$
λ	$\frac{(E-2G)G}{3G-E}$	$K-\frac{2G}{3}$	$\frac{3K(3K-E)}{9K-E}$	$\frac{2G\nu}{1-2\nu}$	$\frac{E\nu}{(1+\nu)(1-2\nu)}$	$\frac{3K\nu}{1+\nu}$	–

The bulk modulus K is another elastic constant. It represents the ratio of the isotropic normal stress $\sigma_v = -p$ (where p is pressure) to the change in volume per unit volume (the dilatation) ε_{kk}. Equations (4.11) or (4.14a) and (4.14b) can be expressed in terms of K and G as follows

$$\sigma_{ij} = \left(K - \frac{2G}{3}\right) \varepsilon_{kk}\delta_{ij} + 2G\varepsilon_{ij}, \tag{4.15a}$$

$$\varepsilon_{ij} = \left(\frac{1}{9K} - \frac{1}{6G}\right) \sigma_{kk}\, \delta_{ij} + \frac{1}{2G}\, \sigma_{ij}. \tag{4.15b}$$

Table 4.1 shows some of the relationships among the elastic constants for isotropic linearly elastic materials.

4.9 **Boundary Conditions**

The following three types of boundary conditions are observed:

(a) *Specified traction.* The tractions T_j (forces per unit of surface area) may be given on the entire surface Γ_a of the body. Often the tractions will be specified as being zero over much of the body.

(b) *Specified displacements.* The surface displacements U_j may be given on the entire surface Γ_a of the body. This includes the possibility that the displacements may be specified to be zero over part or all of the surface.

(c) *Mixed tractions and displacements.* The tractions T_j may be given on some portions of the surface Γ_T and the displacements U_j are prescribed on the rest of the surface Γ_u of the body. As in (a) and (b) the forces and/or displacements may be specified to be zero on some portions of the surface.

Thus the boundary conditions may be written

$$T_j = \sigma_{ij} n_i \quad \text{on} \quad \Gamma_T, \tag{4.16}$$

$$U_j = u_j \quad \text{on} \quad \Gamma_u, \tag{4.17}$$

where n_i is a unit vector denoting the outward normal to the boundary surface.

4.10 **The Stress Analysis Problem in a Linear Isotropic Elastic Solid**

The stress and deformation analysis of a linear elastic body is composed of the solution of the system of equations described above and reviewed as follows:

(a) Equations of equilibrium, (4.2),

(b) Strain displacement relations, (3.68),

(c) Compatibility relations, (4.9),
(d) Stress-strain relations, (4.11 or 4.14),
(e) Boundary conditions, (4.16, 4.17).

According to the types of boundary conditions discussed in Art. 4.9, elasticity problems can be classified as first boundary value problems, second boundary value problems and mixed boundary value problems, respectively, for the traction boundary condition, displacement boundary condition and mixed traction and displacement boundary conditions.

In all three situations, the quantities sought are generally the stresses (6 components), strains (6 components) and displacements (3 components) at every point in the body, with the result that there are fifteen unknowns in the general formulation.

In the solution of first boundary value problems, the system of basic equations is reduced to only six independent differential equations containing only six stress components. These six equations are known as the Beltrami-Michel compatibility equations. In the solution of second boundary value problems, the system of basic equations is reduced to only three independent differential equations containing only three displacement components. These equations are known as Navier equations.

It is not generally possible to employ a straightforward method to obtain a solution to the set of equations for a particular situation. Thus, different special techniques are often required for each situation. These are considered in texts on theory of elasticity.

Detailed derivations of equations discussed in this chapter may be found in [4.3, 4.6, 4.8, 4.9] for example.

CHAPTER 5

LINEAR VISCOELASTIC CONSTITUTIVE EQUATIONS

5.1 Introduction

Viscoelasticity is concerned with materials which exhibit strain rate effects in response to applied stresses. These effects are manifested by the phenomena of creep under constant stress and stress relaxation under constant strain. These time-dependent phenomena may have a considerable effect on the stress distribution developed in a member, such as a thick tube made of viscoelastic material subjected to prescribed loads or prescribed surface displacements. The stress and/or strain at a specific point in the material may vary significantly with time even though the applied forces are constant. In order to be able to predict the change in stress and strain distribution with time, a viscoelastic stress analysis method is needed.

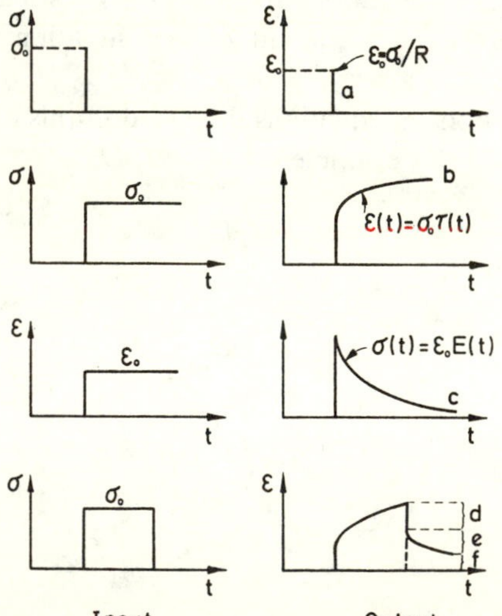

Fig. 5.1. Phenomena Common to Many Viscoelastic Materials: *a*. Instantaneous elasticity, *b*. Creep under constant stress, *c*. Stress relaxation under constant strain, *d*. Instantaneous recovery, *e*. Delayed recovery, *f*. Permanent set.

Most materials exhibit linear, or nearly linear, behavior under small stress levels while the same material may have a nonlinear behavior at high stress levels. The linear theory of viscoelasticity yields a mathematically tractable representation for stress-strain-time relations which permits reasonably simple solutions for many stress analysis problems. Therefore, there has been considerable activity in this area in recent years to develop new mathematical representations of linear viscoelastic behavior and new methods for linear viscoelastic analysis; see for example [5.1, 5.2, 5.3, 5.4]. By employing those mathematical methods a great number of stress analysis and wave propagation problems in linear viscoelasticity have been solved which are in reasonable agreement with experiments.

This book is primarily devoted to nonlinear viscoelasticity, but to maintain completeness the basic concepts of linear viscoelasticity, including constitutive equations and techniques of stress analysis, will be discussed in this and the next chapter.

There are some phenomena which are common to many viscoelastic materials. These are the following, as illustrated in Fig. 5.1:

a. Instantaneous elasticity,
b. Creep under constant stress,
c. Stress relaxation under constant strain,
d. Instantaneous recovery,
e. Delayed recovery,
f. Permanent set.

In general there are two alternative forms used to represent the stress-strain-time relations of viscoelastic materials. They are called the differential operator method and the integral representation. The differential operator method has been widely used for analysis since the mathematical processes [5.1, 5.2] required are reasonably simple. On the other hand, the integral representation is able to describe the time dependence more generally, but it sometimes leads to difficult mathematics in stress analysis. The differential operator form and the integral representation are discussed in Sections 5.9 and 5.14, respectively, in this chapter.

5.2 Viscoelastic Models

In the following discussion of mechanical models, stress σ (force per unit area) and strain ε (deformation per unit length) instead of force and deformation of the model will be used to compare the stress-strain-time relation of viscoelastic materials with the viscoelastic models considered.

5.3 The Basic Elements: Spring and Dashpot

All linear viscoelastic models are made up of linear springs and linear viscous dashpots. Inertia effects are neglected in such models. In the linear spring shown in Fig. 5.2a

$$\sigma = R\varepsilon, \tag{5.1}$$

where R in (5.1) can be interpreted as a linear spring constant or a Young's modulus. The spring element exhibits instantaneous elasticity and instantaneous recovery as shown in Fig. 5.2b.

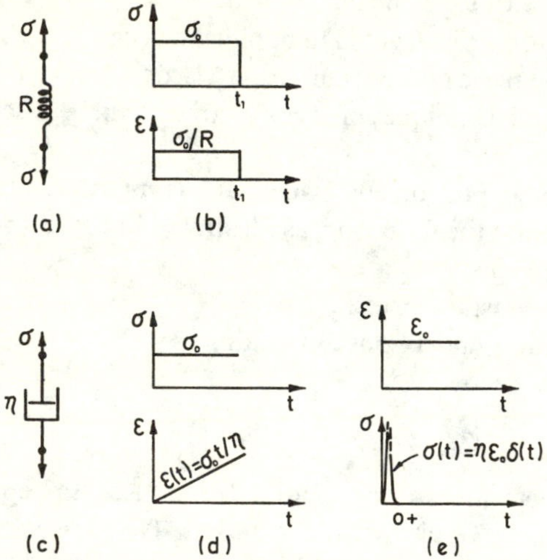

Fig. 5.2. Behavior of a Linear Spring and Linear Dashpot.

A linear viscous dashpot element is shown in Fig. 5.2c where

$$\sigma = \eta \frac{d\varepsilon}{dt} = \eta\dot{\varepsilon}, \tag{5.2}$$

and the constant η is called the coefficient of viscosity. Equation (5.2) states that the strain rate $\dot{\varepsilon}$ is proportional to the stress or, in other words, the dashpot will be deformed continuously at a constant rate when it is subjected to a step of constant stress as shown in Fig. 5.2d. On the other hand, when a step of constant strain is imposed on the dashpot the stress will have an infinite value at the instant when the constant strain is imposed and the stress will then rapidly diminish with time to zero at $t = 0+$ and will remain zero, as shown in Fig. 5.2e. This behavior for a step change in strain is indicated ing ε_1 and ε_2 from these equations. Inserting (5.5) and the time derivative of (5.4) into (5.6) ε_1 and ε_2 can be eliminated and the following stress-strain rate relation for the Maxwell model is obtained:

$$\dot{\varepsilon} = \frac{\dot{\sigma}}{R} + \frac{\sigma}{\eta}. \tag{5.7}$$

The strain-time relations under various stress conditions and stress-time relations under given strain input can be obtained by solving differential equation (5.7).

For example, applying a constant stress $\sigma = \sigma_0$ at $t = 0$ (5.7) becomes a first order differential equation of ε. The following strain-time relation can be obtained after applying integration together with the initial condition $\sigma = \sigma_0$ at $t = t_0$

$$\varepsilon(t) = \frac{\sigma_0}{R} + \frac{\sigma_0}{\eta}\, t. \tag{5.8}$$

This result is shown in Fig. 5.3b. In fact, for such a simple model, the result can be obtained readily from physical arguments by directly adding the strain response shown in Fig. 5.2b and Fig. 5.2d. If the stress is removed from the Maxwell model at time t_1, the elastic strain σ_0/R in the spring returns to zero at the instant the stress is removed, while $(\sigma_0/\eta)\, t_1$ represents a permanent strain which does not disappear.

If the Maxwell model is subjected to a constant strain ε_0 at time $t = 0$, for which the initial value of stress is σ_0, the stress response can be obtained by integrating (5.7) for these initial conditions with the following result

$$\sigma(t) = \sigma_0 e^{-Rt/\eta} = R\varepsilon_0 e^{-Rt/\eta}, \tag{5.9}$$

where ε_0 is the initial strain at $t = 0+$, and $0+$ refers to the time just after application of the strain. Equation (5.9) describes the stress relaxation phenomenon for a Maxwell model under constant strain. This phenomenon is shown in Fig. 5.3c. The rate of stress change is given by the derivative of (5.9)

$$\dot{\sigma} = -(\sigma_0 R/\eta) e^{-Rt/\eta}. \tag{5.10}$$

Thus the initial rate of change in stress at $t = 0+$ is $\dot{\sigma} = -\sigma_0 R/\eta$. If the stress were to decrease continuously at this initial rate, the relaxation equation would have the following form:

$$\sigma = -(\sigma_0 Rt/\eta) + \sigma_0. \tag{5.11}$$

According to (5.11), the stress would then reach zero at time $t_R = \eta/R$, which is called the relaxation time of the Maxwell model. The relaxation time mathematically by the Dirac delta function* $\delta(t)$, where $\delta(t) = 0$ for $t \neq 0$ $\delta(t) = \infty$ for $t = 0$. Thus the stress resulting from applying a step change in strain ε_0 to (5.2) is indicated as follows

$$\sigma(t) = \eta\varepsilon_0\,\delta(t). \tag{5.3}$$

An infinite stress is impossible in reality. It is therefore impossible to impose instantaneously any finite deformation on the dashpot.

5.4 Maxwell Model

The Maxwell model is a two-element model consisting of a linear spring element and a linear viscous dashpot element connected in series as shown in Fig. 5.3a. The stress-strain relations of spring and dashpot follow from (5.1) and (5.2), respectively:

$$\sigma = R\varepsilon_2, \tag{5.4}$$

$$\sigma = \eta\dot{\varepsilon}_1. \tag{5.5}$$

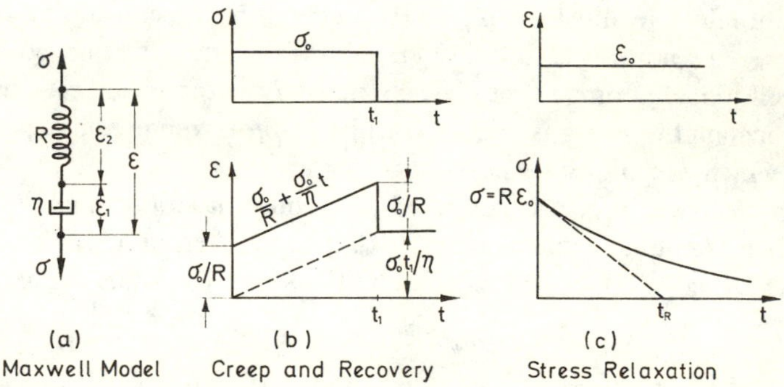

Fig. 5.3. Behavior of a Maxwell Model.

Since both elements are connected in series, the total strain is

$$\varepsilon = \varepsilon_1 + \varepsilon_2,$$

or the strain rate is

$$\dot{\varepsilon} = \dot{\varepsilon}_1 + \dot{\varepsilon}_2. \tag{5.6}$$

Equations (5.4), (5.5) and (5.6) contain four unknowns σ, ε, ε_1, ε_2. Among these ε_1 and ε_2 are internal variables and σ and ε are the external variables. Therefore the stress-strain relation of the model can be obtained by eliminat-

* See Appendix A3.

characterizes one of the viscoelastic properties of the material. Actually most of the relaxation of stress occurred before time t_R since the variable factor e^{-t/t_R} in (5.9) converges toward zero very rapidly for $t < t_R$. For example at $t = t_R$, $\sigma(t) = \sigma_0/e = 0.37\sigma_0$. Thus only 37% of the initial stress remains at $t = t_R$.

5.5 Kelvin Model

The Kelvin model is shown in Fig. 5.4a where a spring element and dashpot element are connected in parallel. The spring and dashpot have the following

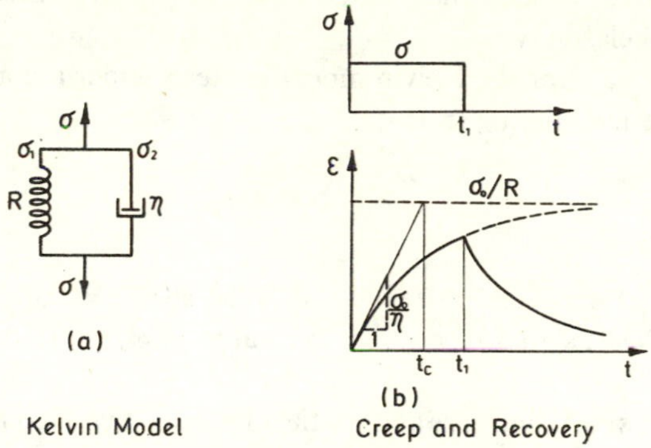

Fig. 5.4. Behavior of a Kelvin Model.

stress-strain relations according to (5.1, 5.2)

$$\sigma_1 = R\varepsilon, \tag{5.12}$$

$$\sigma_2 = \eta\dot{\varepsilon}. \tag{5.13}$$

Since both elements are connected in parallel, the total stress is

$$\sigma = \sigma_1 + \sigma_2. \tag{5.14}$$

Again, (5.12), (5.13) and (5.14) contain four unknowns. Eliminating σ_1 and σ_2 among these equations yields the following equation between stress σ and strain ε:

$$\dot{\varepsilon} + \frac{R}{\eta}\varepsilon = \frac{\sigma}{\eta}. \tag{5.15}$$

The solution of (5.15) may be shown to have the following form for creep under a constant stress σ_0 applied at $t = 0$,

$$\varepsilon = \frac{\sigma_0}{R}(1 - e^{-Rt/\eta}). \tag{5.16}$$

As shown in Fig. 5.4b, the strain described by (5.16) increases with a decreasing rate and approaches asymptotically the value of σ_0/R when t tends to infinity. The response of this model to an abruptly applied stress is that the stress is at first carried entirely by the viscous element, η. Under the stress the viscous element then elongates, thus transferring a greater and greater portion of the load to the elastic element R. Thus, finally the entire stress is carried by the elastic element. The behavior just described is appropriately called delayed elasticity.

The strain rate $\dot{\varepsilon}$ for the Kelvin model in creep under a constant stress σ_0 is found by differentiating (5.16):

$$\dot{\varepsilon} = \frac{\sigma_0}{\eta} e^{-Rt/\eta}. \tag{5.17}$$

Thus, the initial strain rate at $t = 0+$ is finite, $\dot{\varepsilon}(0+) = \sigma_0/\eta$, and the strain rate approaches asymptotically to the value $\dot{\varepsilon}(\infty) = 0$ when t tends to infinity.

If the strain were to increase at its initial rate σ_0/η, it would cross the asymptotic value σ_0/R at time $t_c = \eta/R$, called the retardation time. Actually, most of the total strain σ_0/R occurs within the retardation time period since $e^{-Rt/\eta}$ converges toward the asymptotic value rapidly for $t < t_c$. At $t = t_c$, $\varepsilon = (\sigma_0/R)(1 - 1/e) = 0.63\sigma_0/R$. Thus only 37% of the asymptotic strain remains to be accomplished after $t = t_c$.

If the stress is removed at time t_1 the strain following stress removal can be determined by the superposition principle. The strain ε_a in the Kelvin model resulting from stress σ_0 applied at $t = 0$ is from (5.16)

$$\varepsilon_a = \frac{\sigma_0}{R}(1 - e^{-Rt/\eta}). \tag{5.18}$$

The strain ε_b resulting from applying a stress $(-\sigma_0)$ independently at time $t = t_1$ is

$$\varepsilon_b = -\frac{\sigma_0}{R}(1 - e^{-R(t-t_1)/\eta}). \tag{5.19}$$

If the stress σ_0 is applied at $t = 0$ and removed at $t = t_1$ ($-\sigma_0$ added) the superposition principle yields the strain ε for $t > t_1$ during recovery

$$\varepsilon = \varepsilon_a + \varepsilon_b = \frac{\sigma_0}{R} e^{-Rt/\eta}[e^{Rt_1/\eta} - 1] \qquad t > t_1. \tag{5.20}$$

When t tends to ∞ the recovery tends toward zero as shown by (5.20) and illustrated in Fig. 5.4. Some real materials show full recovery while others only partial recovery.

The Kelvin model does not show a time-dependent relaxation. Owing to the presence of the viscous element an abrupt change in strain ε_0 can be accomplished only by an infinite stress. Having achieved the change in strain either by infinite stress (if that were possible) or by slow application of strain the stress carried by the viscous element drops to zero but a constant stress remains in the spring. These results are obtained from (5.15) by using the Heaviside* H(t) and Dirac $\delta(t)$ functions to describe the step change in strain, $\varepsilon(t) = \varepsilon_0 H(t)$, $\dot{\varepsilon}(t) = \varepsilon_0 \delta(t)$. Thus,

$$\varepsilon_0 \delta(t) + \frac{R}{\eta} \varepsilon_0 H(t) = \frac{\sigma}{\eta}, \tag{5.21}$$

where the first term describes the infinite stress pulse on application of the strain and the second the change in stress in the spring.

Neither the Maxwell nor Kelvin model described above accurately represents the behavior of most viscoelastic materials. For example, the Kelvin model does not exhibit time-independent strain on loading or unloading, nor does it describe a permanent strain after unloading. The Maxwell model shows no time-dependent recovery and does not show the decreasing strain rate under constant stress which is a characteristic of primary creep. Both models show a finite initial strain rate whereas the apparent initial strain rate for many materials is very rapid.

5.6 **Burgers or Four-element Model**

The Burgers model is shown in Fig. 5.5a where a Maxwell and a Kelvin model are connected in series. The constitutive equation for a Burgers model can be derived by considering the strain response under constant stress of each of the elements coupled in series as shown in Fig. 5.5a. The total strain at time t will be the sum of the strain in the three elements, where

* See Appendix A3.

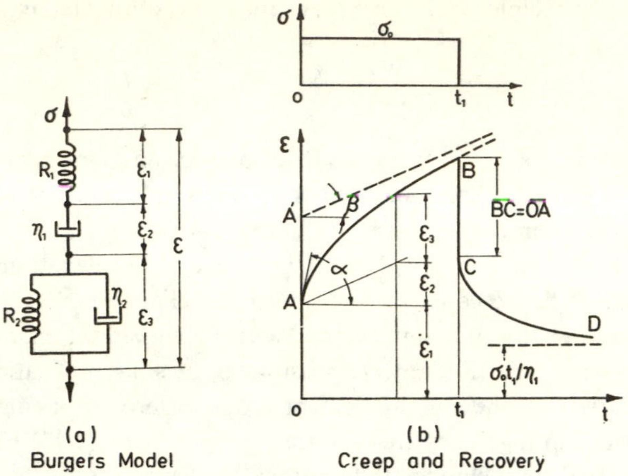

Fig. 5.5. Behavior of a Burgers Model.

the spring and dashpot in the Maxwell model are considered as two elements:

$$\varepsilon = \varepsilon_1 + \varepsilon_2 + \varepsilon_3, \tag{5.22}$$

where ε_1 is the strain of the spring

$$\varepsilon_1 = \frac{\sigma}{R_1}, \tag{5.23}$$

ε_2 is the strain in the dashpot

$$\dot{\varepsilon}_2 = \frac{\sigma}{\eta_1}, \tag{5.24}$$

and ε_3 is the strain in the Kelvin unit which can be derived from (5.15),

$$\dot{\varepsilon}_3 + \frac{R_2}{\eta_2}\varepsilon_3 = \frac{\sigma}{\eta_2}. \tag{5.25}$$

Equations (5.22–5.25) contain five unknowns ε, σ, ε_1, ε_2, ε_3, where ε and σ are external variables and ε_1, ε_2, ε_3 are internal variables. In principle, ε_1, ε_2 and ε_3 can be eliminated from these four equations to yield a constitutive equation between σ and ε for the Burgers model with the following result

$$\sigma + \left(\frac{\eta_1}{R_1} + \frac{\eta_1}{R_2} + \frac{\eta_2}{R_2}\right)\dot{\sigma} + \frac{\eta_1\eta_2}{R_1 R_2}\ddot{\sigma} = \eta_1\dot{\varepsilon} + \frac{\eta_1\eta_2}{R_2}\ddot{\varepsilon}. \tag{5.26}$$

There are many ways of obtaining (5.26) from (5.22–5.25). The following method using Laplace transforms and their inverse (See Appendix A4 for review of Laplace transforms) has the advantages of simplicity and consistency. Applying the Laplace transformation to (5.22–5.25) reduces all of these to algebraic equations, even though (5.24, 5.25) are differential equations. Assuming $\varepsilon_2 = \varepsilon_3 = 0$ at $t = 0-$ (that is, before straining) yields, after transformation,

$$\hat{\varepsilon} = \hat{\varepsilon}_1 + \hat{\varepsilon}_2 + \hat{\varepsilon}_3, \tag{5.27}$$

$$\hat{\varepsilon}_1 = \frac{\hat{\sigma}}{R_1}, \tag{5.28}$$

$$s\hat{\varepsilon}_2 = \frac{\hat{\sigma}}{\eta_1}, \tag{5.29}$$

$$\left(s + \frac{R_2}{\eta_2}\right)\hat{\varepsilon}_3 = \frac{\hat{\sigma}}{\eta_2}, \tag{5.30}$$

where a caret (^) over a symbol indicates that the quantity has been transformed and is now a function of the complex variable s instead of time t.

Inserting (5.28–5.30) into (5.27) yields

$$\hat{\varepsilon} = \frac{\hat{\sigma}}{R_1} + \frac{\hat{\sigma}}{\eta_1 s} + \frac{\hat{\sigma}}{\eta_2(s + R_2/\eta_2)}. \tag{5.31}$$

After multiplying both sides of (5.31) by $\dfrac{\eta_1\eta_2}{R_2} s\left(s + \dfrac{R_2}{\eta_2}\right)$ and rearranging terms, the following equation can be obtained

$$\hat{\sigma} + \left(\frac{\eta_1}{R_1} + \frac{\eta_1}{R_2} + \frac{\eta_2}{R_2}\right)s\hat{\sigma} + \frac{\eta_1\eta_2}{R_1R_2}\, s^2\hat{\sigma} = \eta_1 s\hat{\varepsilon} + \frac{\eta_1\eta_2}{R_2}\, s^2\hat{\varepsilon}. \tag{5.32}$$

The inverse Laplace transformation of (5.32) back to the time variable t yields (5.26).

The creep behavior of the Burgers model under constant stress σ_0 can be obtained from (5.26) by solving this second order differential equation with two initial conditions

$$\varepsilon = \varepsilon_1 = \frac{\sigma_0}{R_1}, \quad \varepsilon_2 = \varepsilon_3 = 0, \qquad t = 0 \tag{5.33}$$

$$\dot{\varepsilon} = \frac{\sigma_0}{\eta_1} + \frac{\sigma_0}{\eta_2}, \qquad t = 0. \tag{5.34}$$

The Laplace transformation method of solving a differential equation as shown in Appendix A4 is recommended here. Thus the creep behavior may be found to be as follows and as illustrated in Fig. 5.5b.

$$\varepsilon(t) = \frac{\sigma_0}{R_1} + \frac{\sigma_0}{\eta_1} t + \frac{\sigma_0}{R_2} (1 - e^{-R_2 t/\eta_2}). \tag{5.35}$$

Comparison of (5.35) with (5.8) and (5.16) indicates that the creep behavior of the Burgers model is the sum of the creep behavior of the Maxwell and Kelvin models. The first two terms on the right-hand side of (5.35) represent instantaneous elastic strain and viscous flow, and the last term represents delayed elasticity of the Kelvin model.

Differentiating (5.35) yields the creep rate $\dot{\varepsilon}$ as follows

$$\dot{\varepsilon} = \frac{\sigma_0}{\eta_1} + \frac{\sigma_0}{\eta_2} e^{-R_2 t/\eta_2}. \tag{5.36}$$

Thus the creep rate starts at $t = 0+$ with a finite value

$$\dot{\varepsilon}(0+) = \left(\frac{1}{\eta_1} + \frac{1}{\eta_2}\right) \sigma_0 = \tan \alpha, \tag{5.37}$$

see Fig. 5.5b, and approaches asymptotically to the value

$$\dot{\varepsilon}(\infty) = \sigma_0/\eta_1 = \tan \beta, \tag{5.38}$$

see Fig. 5.5b. It may also be observed from Fig. 5.5b that $\overline{OA} = \sigma_0/R_1$ and $\overline{AA'} = \sigma_0/R_2$. Thus in theory the material constants R_1, R_2, η_1, η_2 may be determined from a creep experiment by measuring α, β, $\overline{OA}$ and $\overline{AA'}$ as in Fig. 5.5b.

However, few if any viscoelastic materials have creep characteristics which are described closely enough by (5.35) to permit direct determination of α, β, $\overline{OA}$ and $\overline{AA'}$. As shown by the creep curves for polycarbonate in torsion [5.5] in Fig. 5.6, the initial creep rate is so rapid that a value of α other than 90° would be hard to justify. For the same reason point A and, hence $\overline{OA}$ and $\overline{AA'}$, is almost impossible to determine directly. Also, the creep of many plastics, for example, have a continously decreasing creep rate. Thus there is no asymptote to the creep curve. An apparent asymptote in Fig. 5.6, for example, is merely a tangent. Hence β cannot be determined directly for such materials. Of course, a Burgers model may nevertheless be employed to approximate the real behavior of such materials over a given time span.

If the stress σ_0 is removed at time t_1, the recovery behavior of the Burgers model can be obtained from (5.35) and the superposition principle by considering that at $t = t_1$ a constant stress $\sigma = -\sigma_0$ is added. According to the

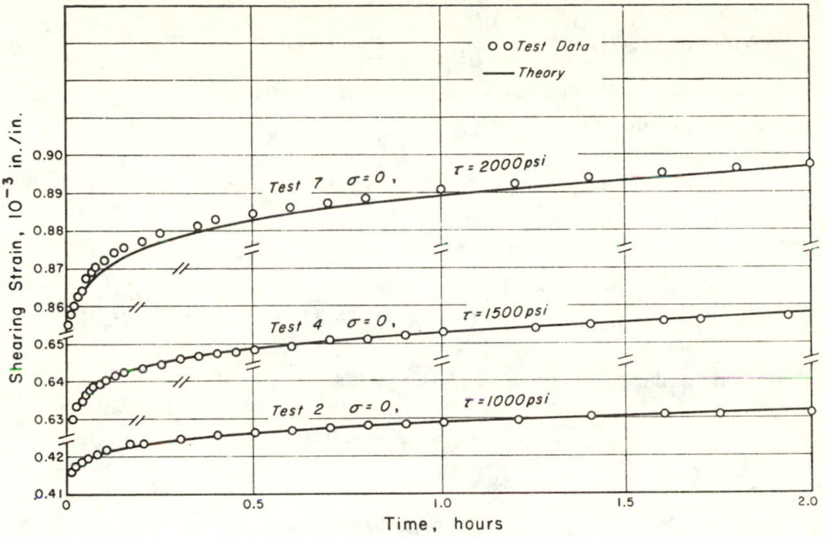

Fig. 5.6. Creep Curves for Torsion Strain in Pure Torsion Creep Tests of Polycarbonate at 75°F, 50 per cent R. H. Theory shown is Equation 2.16. From [5.5], courtesy Polymer Engineering and Science.

superposition principle the recovery strain $\varepsilon(t)$, $t > t_1$ is the sum of these two independent actions:

$$\varepsilon(t) = \frac{\sigma_0}{R_1} + \frac{\sigma_0}{\eta_1} t + \frac{\sigma_0}{R_2}(1 - e^{-R_2 t/\eta_2})$$

$$- \left[\frac{\sigma_0}{R_1} + \frac{\sigma_0}{\eta_1}(t - t_1) + \frac{\sigma_0}{R_2}(1 - e^{-R_2(t-t_1)/\eta_2}) \right] \qquad t > t_1$$

or

$$\varepsilon(t) = \frac{\sigma_0}{\eta_1} t_1 + \frac{\sigma_0}{R_2}(e^{R_2 t_1/\eta_2} - 1)e^{-R_2 t/\eta_2}, \qquad t > t_1. \tag{5.39}$$

Recovery is also illustrated in Fig. 5.5b. The recovery has an instantaneous elastic recovery followed by creep recovery at a decreasing rate, as shown in (5.39). The second term of (5.39) decreases toward zero for large times, while the first term represents a permanent strain due to the viscous flow of η_1. Thus the recovery approaches asymptotically to $\varepsilon(\infty) = (\sigma_0/\eta_1)\, t_1$ as t approaches infinity.

The stress relaxation behavior of the Burgers model may be obtained from (5.26). For relaxation resulting from a step of strain to ε_0 at $t = 0+$,

$\varepsilon = \varepsilon_0 H(t)$, $\dot{\varepsilon} = \varepsilon_0 \delta(t)$, $\ddot{\varepsilon} = \varepsilon_0 \dfrac{d\delta(t)}{dt}$, where $H(t)$ and $\delta(t)$ are the Heaviside and Dirac delta functions, respectively (see Appendix A3).

Thus (5.26) becomes

$$\sigma + p_1\dot{\sigma} + p_2\ddot{\sigma} = q_1\varepsilon_0\,\delta(t) + q_2\varepsilon_0\frac{d\delta(t)}{dt}, \tag{5.40}$$

where

$$p_1 = \frac{\eta_1}{R_1} + \frac{\eta_1}{R_2} + \frac{\eta_2}{R_2}, \quad p_2 = \frac{\eta_1\eta_2}{R_1R_2}, \quad q_1 = \eta_1, \quad q_2 = \frac{\eta_1\eta_2}{R_2}.$$

Taking the Laplace transform of (5.40) yields

$$\hat{\sigma} + p_1 s\hat{\sigma} + p_2 s^2\hat{\sigma} = q_1\varepsilon_0 + q_2\varepsilon_0 s.$$

Solving for $\hat{\sigma}$

$$\hat{\sigma} = \frac{\varepsilon_0(q_1 + q_2 s)}{1 + p_1 s + p_2 s^2}. \tag{5.41}$$

Expanding (5.41) by partial fractions and performing the inverse Laplace transformation yields the stress relaxation,

$$\sigma(t) = \frac{\varepsilon_0}{A}\left[(q_1 - q_2 r_1)e^{-r_1 t} - (q_1 - q_2 r_2)e^{-r_2 t}\right], \tag{5.42}$$

where

$$r_1 = (p_1 - A)/2p_2, \quad r_2 = (p_1 + A)/2p_2, \quad A = \sqrt{p_1^2 - 4p_2}.$$

There are several possible combinations of three-element and four-element models. Their viscoelastic behavior can be studied by similar methods to those discussed above. Several of these models are described by the same type of constitutive equation. Thus they are mechanically equivalent and quantitatively the same with proper choice of constants. They are classified into four groups as shown in Fig. 5.7. Model (a) in Group I is called "the standard solid" and is widely used especially to describe shear wave propagation in viscoelastic materials [5.6, 5.7]. The constitutive equation of the standard solid can be obtained from (5.32) by multiplying the equation by $R_1R_2/\eta_1(R_1+R_2)s$ then considering $\eta_1 \to \infty$, which yields after inverse transformation

$$\sigma + \frac{\eta_2}{R_1 + R_2}\dot{\sigma} = \frac{R_1R_2}{R_1 + R_2}\varepsilon + \frac{R_1\eta_2}{R_1 + R_2}\dot{\varepsilon}. \tag{5.43}$$

Group I exhibits a solid-like character with retarded elasticity (instantaneous elastic deformation and delayed elastic deformation). Group II has a liquid-like character, viscous flow, plus delayed elasticity. Group III exhibits

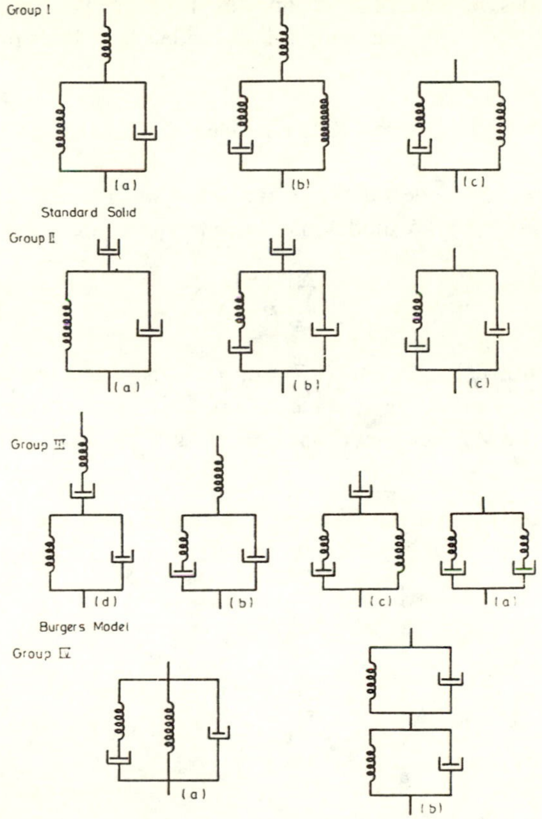

Fig. 5.7. Various Combinations of Three-element and Four-element Models.

an instantaneous elastic response followed by viscous flow and delayed elasticity. Model (d) in Group III consists of a Kelvin and a Maxwell model in series as in Fig. 5.5. Group IV shows delayed elasticity with two retardation times. The same form of constitutive equation is found for each model in Group I, Group II, Group III and Group IV.

The models discussed cannot describe the behavior of many viscoelastic materials over a wide range of the variables especially for both large and small values of time. In addition to that, the constants of some materials, such as concrete and some plastics may change with age due to significant changes in composition and/or structure in the course of a creep test. Except for Group IV the models described above have only one relaxation or retardation time, whereas real materials often behave as though they have several relaxation times, because of their complex structure. To deal with this situation several types of complex mechanical models have been proposed.

5.7 Generalized Maxwell and Kelvin Models

Several Maxwell models in series are shown in Fig. 5.8a. The constitutive equation of this complex model may have the following form

$$\dot{\varepsilon} = \dot{\sigma} \sum_{i=1}^{N} \frac{1}{R_i} + \sigma \sum_{i=1}^{N} \frac{1}{\eta_i}. \tag{5.44}$$

This equation is equivalent to (5.7) and describes the same mechanical behavior. Consequently a chain of elements consisting of springs and dashpots is identical to a Maxwell model as seen in Fig. 5.8b.

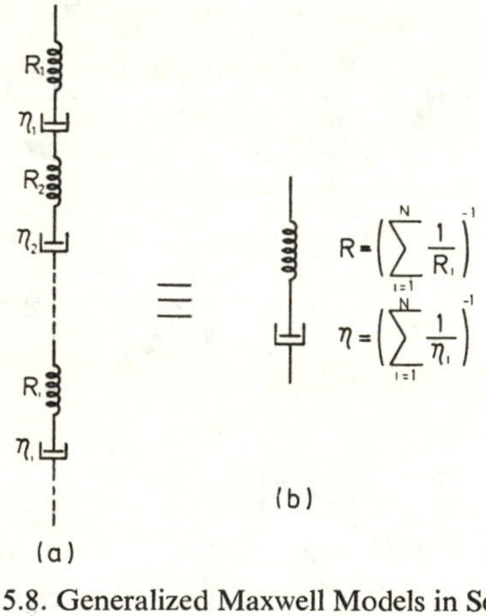

Fig. 5.8. Generalized Maxwell Models in Series.

If several Kelvin models are connected in parallel, they do not exhibit any different behavior than an equivalent Kelvin model as shown in Fig. 5.9. The constitutive equation of this model has the form:

$$\sigma = \varepsilon \sum_{i=1}^{N} R_i + \dot{\varepsilon} \sum_{i=1}^{N} \eta_i. \tag{5.45}$$

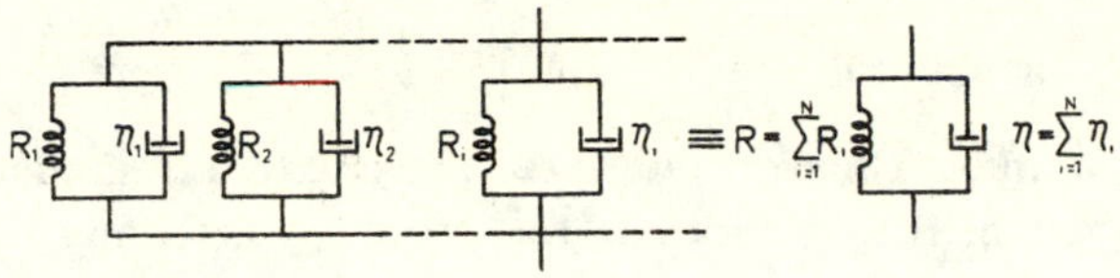

Fig. 5.9. Generalized Kelvin Models in Parallel.

On the other hand several Maxwell models connected in parallel represent instantaneous elasticity, delayed elasticity with various retardation times, stress relaxation with various relaxation times and also viscous flow. To predict the stress associated with a prescribed strain variation, the generalized Maxwell model, as shown in Fig. 5.10, is rather convenient since the

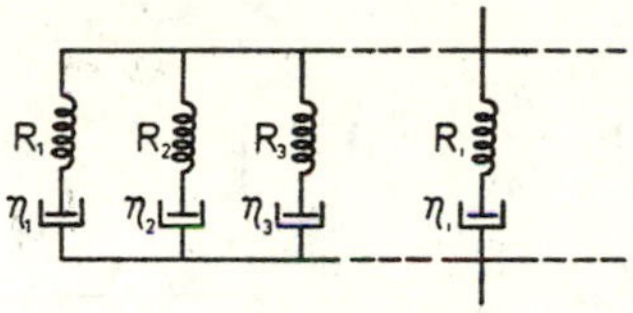

Fig. 5.10. Generalized Maxwell Models in Parallel.

same prescribed strain is applied to each individual element, and also the resulting stress is the sum of the individual contributions. From (5.7) the contribution of the first element is

$$\mathrm{D}\varepsilon = \frac{\mathrm{D}\sigma_1}{R_1} + \frac{\sigma_1}{\eta_1},$$

where D is the differential operator with respect to time $\mathrm{D} = \frac{\mathrm{d}}{\mathrm{d}t}$, from which

$$\sigma_1 = \frac{\mathrm{D}}{\dfrac{\mathrm{D}}{R_1} + \dfrac{1}{\eta_1}}\,\varepsilon.$$

The *i*-th element yields

$$\sigma_i = \frac{\mathrm{D}}{\dfrac{\mathrm{D}}{R_i} + \dfrac{1}{\eta_i}}\,\varepsilon. \tag{5.46}$$

The sum of both sides of (5.46) yields

$$\sigma = \sum_{i=1}^{a} \sigma_i = \left(\sum_{i=1}^{a} \frac{\mathrm{D}}{\dfrac{\mathrm{D}}{R_i} + \dfrac{1}{\eta_i}} \right) \varepsilon. \tag{5.47}$$

By multiplying both sides of (5.47) by $\prod_{i=1}^{a}\left(\frac{\mathrm{D}}{R_i}+\frac{1}{\eta_i}\right)$, where $\prod_{i=1}^{a}$ denotes the product of a-terms, the time operator D can be removed from the denominator of the equation. The open form of this process is as follows

$$\left[\left(\frac{\mathrm{D}}{R_1}+\frac{1}{\eta_1}\right)\left(\frac{\mathrm{D}}{R_2}+\frac{1}{\eta_2}\right)\cdots\right]\sigma$$
$$=\left[\left(\frac{\mathrm{D}}{R_1}+\frac{1}{\eta_1}\right)\left(\frac{\mathrm{D}}{R_2}+\frac{1}{\eta_2}\right)\cdots\right]\left[\frac{\mathrm{D}}{\frac{\mathrm{D}}{R_1}+\frac{1}{\eta_1}}+\frac{\mathrm{D}}{\frac{\mathrm{D}}{R_2}+\frac{1}{\eta_2}}+\cdots\right]\varepsilon. \quad (5.48)$$

The least common denominator of the last term in square brackets on the right is identical with the first term in square brackets on the right; hence these terms cancel and (5.48) becomes

$$\left[\left(\frac{\mathrm{D}}{R_1}+\frac{1}{\eta_1}\right)\left(\frac{\mathrm{D}}{R_2}+\frac{1}{\eta_2}\right)\left(\frac{\mathrm{D}}{R_3}+\frac{1}{\eta_3}\right)\cdots\right]\sigma$$
$$=\left[\mathrm{D}\left(\frac{\mathrm{D}}{R_2}+\frac{1}{\eta_2}\right)\left(\frac{\mathrm{D}}{R_3}+\frac{1}{\eta_3}\right)\cdots+\mathrm{D}\left(\frac{\mathrm{D}}{R_1}+\frac{1}{\eta_1}\right)\left(\frac{\mathrm{D}}{R_3}+\frac{1}{\eta_3}\right)\cdots+\cdots\right]\varepsilon. \quad (5.49)$$

Another generalized form of the basic models may be obtained by connecting various Kelvin models in series as in Fig. 5.11. The strain contribu-

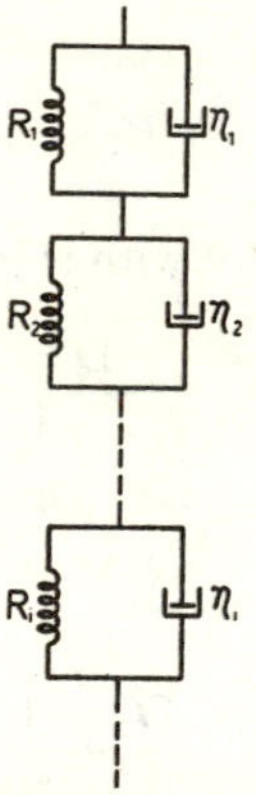

Fig. 5.11. Generalized Kelvin Models in Series.

tion of the first unit can be obtained from (5.15) as

$$\varepsilon_1=\frac{1}{\mathrm{D}\eta_1+R_1}\,\sigma.$$

The sum of the strain contribution of a-elements is

$$\varepsilon = \sum_{i=1}^{a} \varepsilon_i = \left(\sum_{i=1}^{a} \frac{1}{D\eta_i + R_i} \right) \sigma. \tag{5.50}$$

The time operator D can be removed from the denominator of (5.50) by multiplying both sides by $\prod_{i=1}^{a} (D\eta_i + R_i)$. In a similar manner to that employed in (5.48) the open form of this process results in the following:

$$\begin{aligned} &[(D\eta_1 + R_1)(D\eta_2 + R_2)(D\eta_3 + R_3)\ldots]\varepsilon \\ &\quad = [(D\eta_2 + R_2)(D\eta_3 + R_3)\ldots + (D\eta_1 + R_1)(D\eta_3 + R_3)\ldots + \ldots]\sigma. \end{aligned} \tag{5.51}$$

The generalized Kelvin model (Fig. 5.11) is more convenient than the generalized Maxwell model (Fig. 5.10) for viscoelastic analysis in cases where the stress history is prescribed, whereas the generalized Maxwell model is the more convenient in cases where the strain history is prescribed. Because of the range of different relaxation times which can be brought into play, both of these models permit a close description of real behavior over a wider time span than is possible with simpler models.

The spring constants R_i and dashpot constants η_i of the generalized models should vary over a large range in order to obtain the best representation for real materials, since real materials usually have a rather large spectrum of

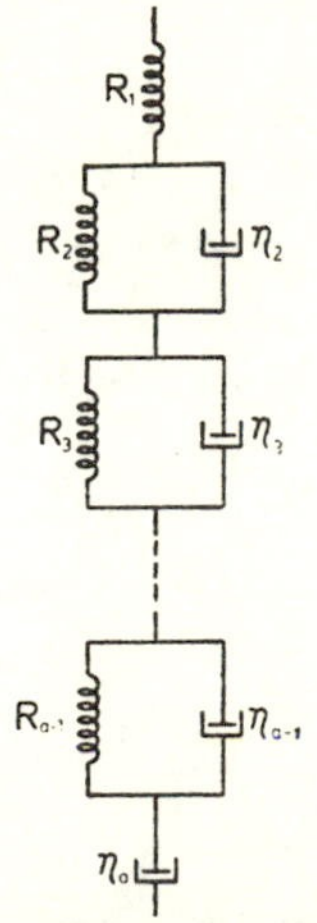

Fig. 5.12. Generalized Model for Elastic Response, Viscous Flow and Delayed Elasticity with Multiple Retardation Times.

relaxation times (or retardation times) [5.8, 5.9]. By considering limiting values of the constants R or η, one model may be converted to another type. For example, a Maxwell model with infinite spring constant or a Kelvin model with zero spring constant becomes a dashpot. Similarly, a Maxwell model with infinite viscosity or a Kelvin model with zero viscosity is converted to a spring. Consequently the generalized models may be arranged such that they either represent a solid character or a liquid character even though they are represented by the same type of constitutive equation as (5.50). Figure 5.12 shows a model in which the spring R_1 exhibits elastic response, η_a is viscous flow and the Kelvin elements in series represent delayed elasticity with multiple retardation times.

The creep strain of a generalized Kelvin model (Fig. 5.11) under constant stress σ_0 can be obtained directly, instead of solving (5.51), by considering that the total strain is the sum of the creep strain of each individual Kelvin model which is given in (5.16). Thus, the creep strain of the generalized Kelvin model has the following form

$$\varepsilon(t) = \sigma_0 \sum_{i=1}^{a} \varphi_i (1 - e^{-t/t_c^i}), \tag{5.52}$$

where $\varphi_i = \dfrac{1}{R_i}$ is the reciprocal modulus (or compliance), and $t_c^i = \eta_i/R_i$ is the retardation time.

If the number of Kelvin elements in the generalized Kelvin model increases indefinitely, (5.52) becomes

$$\varepsilon = \sigma_0 \int_0^\infty \varphi(t_c)(1 - e^{-t/t_c})\, dt_c \tag{5.53}$$

and $\varphi(t_c)$ is called the "retardation spectrum"–a distribution function of retardation times.

Similarly, the stress relaxation of a generalized Maxwell model, comprising several Maxwell models in parallel (Fig. 5.10), under constant strain ε_0 can be expressed as

$$\sigma(t) = \varepsilon_0 \sum_{i=1}^{a} R_i e^{-t/t_R^i}, \tag{5.54}$$

where $t_R^i = \eta_i/R_i$ is the relaxation time. Again (5.54) can be obtained directly from (5.9). For a continuous distribution of relaxation times (5.54) takes the following form

$$\sigma(t) = \varepsilon_0 \int_0^\infty R(t_R) e^{-t/t_R}\, dt_R . \tag{5.55}$$

$R(t_R)$ is called the "relaxation spectrum"—a distribution function of relaxation times.

5.8 Retardation Spectrum for t^n

As indicated in Chapter 2 a power function of time as given by (2.16) describes the creep behavior of many materials quite well. As this form will be used extensively in subsequent chapters, it is of interest to show that this function represents a particular retardation spectrum. Considering one Kelvin element of the generalized Kelvin model to degenerate to a spring R_1 (5.53) becomes

$$\varepsilon = \sigma_0 \left[R_1 + \int_0^\infty \varphi(t_c)(1-e^{-t/t_c})\, dt_c \right]. \tag{5.56}$$

For linear behavior the creep equation (2.16) may be expressed as follows

$$\varepsilon = \varepsilon^0 + \varepsilon^+ t^n = \sigma_0 [R_1 + m t^n], \tag{5.57}$$

where R_1, m, n are constants independent of stress. Comparing (5.56) and (5.57) yields the relation

$$m t^n = \int_0^\infty \varphi(t_c)(1-e^{-t/t_c})\, dt_c , \tag{5.58}$$

which must be satisfied by the desired retardation spectrum $\varphi(t_c)$. Introducing the following change of variables in (5.58)

$$\frac{1}{t_c} = x, \quad dt_c = -\frac{dx}{x^2}, \tag{5.59}$$

making the appropriate change in limits and interchanging limits to remove a negative sign yields

$$m t^n = \int_0^\infty \left[\frac{1}{x^2} \varphi\left(\frac{1}{x}\right) \right] [1-e^{-tx}]\, dx. \tag{5.60}$$

Letting $F(x) = \varphi\left(\frac{1}{x}\right)$ and taking the time derivative of (5.60) yields

$$nmt^{n-1} = \int_0^\infty \frac{F(x)}{x} e^{-xt}\, dx. \tag{5.61}$$

The definition of the gamma function of $1-n$, $\Gamma(1-n)$, may be written

$$\Gamma(1-n) = \int_0^\infty (tx)^{-n} e^{-tx}\, d(tx)$$

or

$$\frac{\Gamma(1-n)}{t^{1-n}} = \int_0^\infty x^{-n} e^{-tx}\, dx. \tag{5.62}$$

Eliminating t^{1-n} between (5.61) and (5.62) leaves

$$\int_0^\infty \frac{F(x)}{x} e^{-tx}\, dx = \int_0^\infty \frac{nm}{\Gamma(1-n)} x^{-n} e^{-tx}\, dx. \tag{5.63}$$

The integrands of both sides of (5.63) are equal. Thus, it follows that

$$F(x) = \frac{nm}{\Gamma(1-n)} x^{1-n}. \tag{5.64a}$$

Since $F(x) = \varphi\left(\frac{1}{x}\right)$ and $x = \frac{1}{t_c}$ the desired retardation spectrum corresponding to t^n is as follows from (5.64a)

$$\varphi(t_c) = \frac{nmt_c^{n-1}}{\Gamma(1-n)}. \tag{5.64b}$$

For most materials $0 < n < 1$. Thus $\varphi(t_c)$ approaches infinity at small values of t_c, decreases as t_c increases and approaches zero as t_c becomes very large. The corresponding curve of log $\varphi(t_c)$ vs. log t_c is a straight line of negative slope. Hence, the time function $\varepsilon^+ t^n$ for creep in the equation

$$\varepsilon = \varepsilon^0 + \varepsilon^+ t^n$$

may be considered to describe a material having a continuous spectrum of retardation times proportional to t_c^{n-1}.

5.9 Differential Form of Constitutive Equations for Simple Stress States

The constitutive equation of a rate sensitive linear material for a simple stress state, such as uniaxial stress or pure shear, may be expressed as a linear function of stress σ, strain ε and their time derivatives as follows

$$f(\sigma, \dot{\sigma}, \ddot{\sigma}, \ldots; \quad \varepsilon, \dot{\varepsilon}, \ddot{\varepsilon}, \ldots) = 0, \tag{5.65}$$

where $\sigma = \sigma(t)$ describes the variation of stress with time, $\varepsilon = \varepsilon(t)$ describes the variation of strain with time. The dots represent the derivatives with respect to time. Equation (5.65) is commonly written in a more compact form as follows

$$\mathrm{P}\sigma = \mathrm{Q}\varepsilon, \tag{5.66}$$

where P and Q are linear differential operators with respect to time, as shown in the following

$$\mathrm{P} = \sum_{r=0}^{a} p_r \frac{\partial^r}{\partial t^r}, \quad \mathrm{Q} = \sum_{r=0}^{b} q_r \frac{\partial^r}{\partial t^r}. \tag{5.67}$$

A differential operator form of constitutive equation is obtained by combining (5.66) and (5.67) as follows

$$\begin{aligned} \mathrm{P}\sigma &= p_0\sigma + p_1\dot{\sigma} + p_2\ddot{\sigma} + \ldots + p_a \frac{\partial^a \sigma}{\partial t^a} \\ &= q_0\varepsilon + q_1\dot{\varepsilon} + q_2\ddot{\varepsilon} + \ldots + q_b \frac{\partial^b \varepsilon}{\partial t^b} = \mathrm{Q}\varepsilon, \end{aligned} \tag{5.68}$$

where dots denote the derivative of the variable with respect to time and $p_0, p_1, p_2, \ldots$ and $q_0, q_1, q_2, \ldots$ are material constants. In (5.68), there is no loss of generality by assuming $p_0 = 1$. Taking the Laplace transform of (5.68) for zero initial conditions yields

$$\begin{aligned} \hat{\mathrm{P}}(s)\hat{\sigma}(s) &= (p_0 + p_1 s + p_2 s^2 + \ldots + p_a s^a)\hat{\sigma}(s) \\ &= \hat{\mathrm{Q}}(s)\hat{\varepsilon}(s) = (q_0 + q_1 s + q_2 s^2 + \ldots + q_b s^b)\hat{\varepsilon}(s), \end{aligned} \tag{5.69}$$

where s is the transform variable.

From (5.69)

$$\frac{\hat{\mathrm{Q}}(s)}{\hat{\mathrm{P}}(s)} = \frac{\hat{\sigma}}{\hat{\varepsilon}}. \tag{5.70}$$

For the linear case p_r and q_r in (5.68) are independent of stress and strain, but may depend on time. Certain combinations of the terms of (5.68) may be used to describe particular idealized viscoelastic behavior. It is convenient in connection with certain mechanical models composed of linear elastic springs and linear viscous dashpots, as discussed in Sections 5.4, 5.5 and 5.6, to employ the corresponding differential operator forms of linear viscoelastic constitutive equations as follows.

A comparison of (5.68) and the results of Sections 5.4–5.6 shows that (5.68) is capable of describing the behavior of various viscoelastic models by properly choosing the coefficients. By choosing

$$p_0 = 1, \quad p_1 = \frac{\eta}{R}, \quad q_1 = \eta, \tag{5.71}$$

and the other coefficients equal to zero, (5.68) represents a constitutive relation described by the Maxwell model. On the other hand, by choosing

$$p_0 = 1, \quad q_0 = R, \quad q_1 = \eta, \tag{5.72}$$

and the other coefficients equal to zero, (5.68) represents a constitutive relation described by the Kelvin model, see also Table 5.1.

However, by retaining only three coefficients, either p_0, p_1, q_1 or p_0, q_0, q_1, the resulting constitutive equations, as discussed in Sections 5.4, 5.5 for Maxwell or Kelvin models, respectively, are not capable of accurately representing the behavior of real materials. For example, the Kelvin unit does not exhibit instantaneous elastic strain on loading or a permanent strain following unloading, while the Maxwell unit is not able to represent the decreasing rate behavior which is characteristic of the primary stage of creep of many materials.

By retaining more terms, (5.68) is capable of representing the behavior of real materials somewhat better. For example, retaining p_0, p_1, q_0, q_1, permits the equation to describe the material behavior with instantaneous deformation and time-dependent creep under constant load like those mechanical models shown in Group I of Fig. 5.7, see (5.43). Again, retaining p_0, p_1, p_2, q_0, q_1, q_2 only permits (5.68) to describe the material behavior depicted by the Burgers model in Group III, see Fig. 5.7, (5.26), and Table 5.1.

Comparison of (5.49), (5.51) with (5.68) shows that those equations are equivalent, which implies that (5.68) is capable of describing materials with a large number of relaxation or retardation times or a continuous spectrum of relaxation or retardation times.

However, if a large number of terms of (5.68) are required in order to describe the material behavior accurately, as many real materials do, see [5.8, 5.9], it is not practical and may be impossible to determine the coeffi-

Table 5.1
Properties of Viscoelastic Models

Column	I	II	III	IV
Model	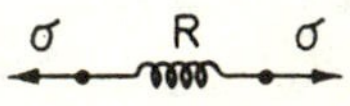 Hook	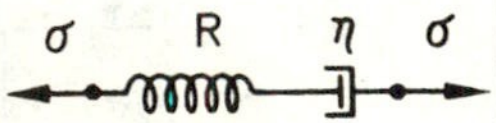 Maxwell	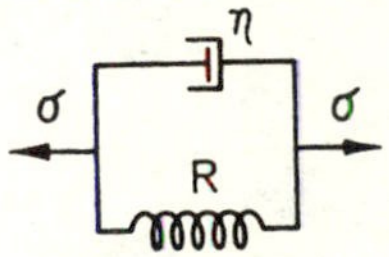 Kelvin	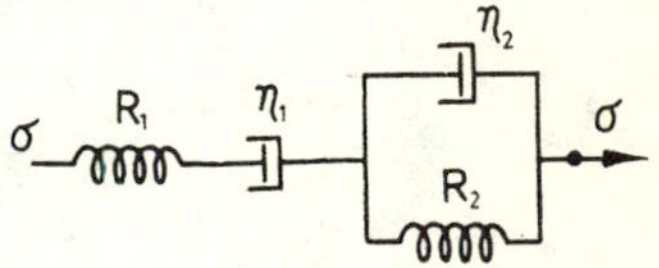 Burgers
General Operator Equation	$P\sigma = p_0\sigma + p_1\dot{\sigma} + p_2\ddot{\sigma} + \ldots = Q\varepsilon = q_0\varepsilon + q_1\dot{\varepsilon} + q_2\ddot{\varepsilon} + \ldots$ (5.68)†			
Transformed	$\hat{P}(s)\hat{\sigma}(s) = (p_0 + p_1 s + p_2 s^2 + \ldots)\hat{\sigma}(s) = \hat{Q}(s)\hat{\varepsilon}(s) = (q_0 + q_1 s + q_2 s^2 + \ldots)\hat{\varepsilon}(s)$ (5.69)			
Differential Equation	$\sigma = R\varepsilon$ (5.1)	$\sigma + \frac{\eta}{R}\dot{\sigma} = \eta\dot{\varepsilon}$ (5.7)	$\sigma = R\varepsilon + \eta\dot{\varepsilon}$ (5.15)	$\sigma + p_1\dot{\sigma} + p_2\ddot{\sigma} = q_1\dot{\varepsilon} + q_2\ddot{\varepsilon}$ (5.26)
Operators	$P = 1, \quad Q = R$	$P = 1 + \frac{\eta}{R}\frac{\partial}{\partial t}$, $Q = \eta\frac{\partial}{\partial t}$	$P = 1, Q = R + \eta\frac{\partial}{\partial t}$	$P = 1 + p_1\frac{\partial}{\partial t} + p_2\frac{\partial^2}{\partial t^2}, \quad Q = q_1\frac{\partial}{\partial t} + q_2\frac{\partial^2}{\partial t^2}$
	$p_0 = 1, \quad q_0 = R$	$p_0 = 1, \quad p_1 = \eta/R,$ $q_1 = \eta$ (5.71)	$p_0 = 1, \quad q_0 = R,$ $q_1 = \eta$ (5.72)	$p_0 = 1, \quad p_1 = \frac{\eta_1}{R_1} + \frac{\eta_1}{R_2} + \frac{\eta_2}{R_2},$ $p_2 = \frac{\eta_1\eta_2}{R_1R_2},$ $q_1 = \eta_1, \quad q_2 = \eta_1\eta_2/R_2$ (5.26)

Table 5.1 (cont'd)

Column	I	II	III	IV
Deviatoric Operators	$P_1 = 1, \quad Q_1 = 2G$	$P_1 = 1+\frac{\eta'}{R'}\frac{\partial}{\partial t}$, $Q_1 = \eta'\frac{\partial}{\partial t}$ (5.93)	$P_1 = 1$, $Q_1 = R'+\eta'\frac{\partial}{\partial t}$	$P_1 = 1+p_1'\frac{\partial}{\partial t}+p_2'\frac{\partial^2}{\partial t^2}$, $Q_1 = q_1'\frac{\partial}{\partial t}+q_2'\frac{\partial}{\partial t^2}$
Transformed	$\hat{P}_1(s) = 1$, $\hat{Q}_1(s) = 2G$	$\hat{P}_1(s) = 1+\frac{\eta'}{R'}s$ $\hat{Q}_1(s) = \eta' s$	$\hat{P}_1(s) = 1$ $\hat{Q}_1(s) = R'+\eta' s$	$\hat{P}_1(s) = 1+p_1's+p_2's^2$ $\hat{Q}_1(s) = q_1's+q_2's^2$
Dilatational Operators	$P_2 = 1$, $Q_2 = 3K$ (5.94)	$P_2 = 1+\frac{\eta''}{R''}\frac{\partial}{\partial t}$, $Q_2 = \eta''\frac{\partial}{\partial t}$	$P_2 = 1$, $Q_2 = R''+\eta''\frac{\partial}{\partial t}$	$P_2 = 1+p_1''\frac{\partial}{\partial t}+p_2''\frac{\partial^2}{\partial t^2}$, $Q_2 = q_1''\frac{\partial}{\partial t}+q_2''\frac{\partial^2}{\partial t^2}$
Transformed	$\hat{P}_2(s) = 1$ $\hat{Q}_2(s) = 3K$	$\hat{P}_2(s) = 1+\frac{\eta''}{3K}s$ $\hat{Q}_2(s) = \eta'' s$	$\hat{P}_2(s) = 1$ $\hat{Q}_2(s) = R''+\eta'' s$	$\hat{P}_2(s) = 1+p_1''s+p_2''s^2$ $\hat{Q}_2(s) = q_1''s+q_2''s^2$
Creep Compliance J (t)	$1/R$	$\frac{1}{R}+\frac{t}{\eta}$	$\frac{1}{R}(1-e^{-Rt/\eta})$	$\frac{1}{R_1}+\frac{1}{R_2}(1-e^{-R_2t/\eta_2})+\frac{t}{\eta_1}$
Transformed $\hat{J}(s)$	$1/Rs$	$\frac{1}{Rs}+\frac{1}{\eta s^2}$	$\frac{1/\eta}{s(s+R/\eta)} = \frac{1}{s(R+\eta s)}$	$\frac{1}{R_1 s}+\frac{1/\eta_2}{s(s+R_2/\eta_2)}+\frac{1/\eta_1}{s^2}$

Table 5.1 (cont'd)

Column	I	II	III	IV
Model	 Hook	 Maxwell	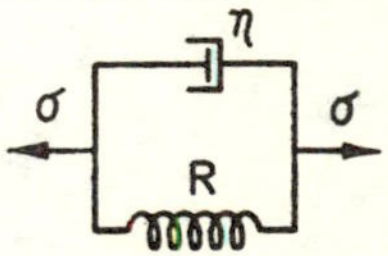 Kelvin	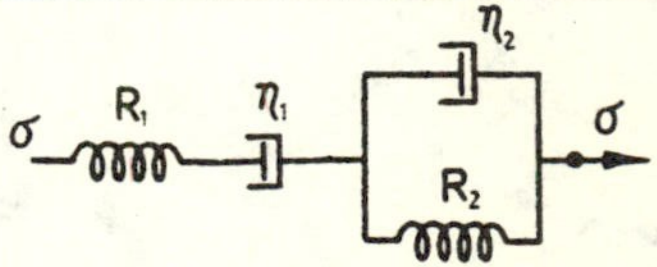 Burgers
Relaxation Modulus $E(t)$	R	$Re^{-Rt/\eta}$ (5.9)	$R+\eta\delta(t)$	$[(q_1-q_2r_1)e^{-r_1t}-(q_1-q_2r_2)e^{-r_2t}]/A$ $r_1,\quad r_2=(p_1\mp A)/2p_2\ A=\sqrt{p_1^2-4p_2}$
Transformed $\hat{E}(s)=\dfrac{1}{s^2\hat{J}(s)}$	R/s	$\dfrac{R}{s+(R/\eta)}$	$(R/s)+\eta$	$\dfrac{q_1+q_2s}{1+p_1s+p_2s^2}$
Transformed Modulus $\dfrac{\hat{\sigma}}{\hat{\varepsilon}}=\dfrac{\hat{Q}(s)}{\hat{P}(s)}=$ $=s\hat{E}(s)=\dfrac{1}{s\hat{J}(s)}$	R	$\dfrac{Rs}{s+(R/\eta)}$	$R+\eta s$	$\dfrac{q_1s+q_2s^2}{p_0+p_1s+p_2s^2}$
Complex Compliance (5.153) $J^*=J_1-iJ_2$ Real Part $J_1(\omega)$	$1/R$	$1/R$	$\dfrac{R}{R^2+\eta^2\omega^2}$	$\dfrac{1}{R_1}+\dfrac{R_2}{R_2^2+\eta_2^2\omega^2}$
Imaginary Part $J_2(\omega)$	0	$1/\eta\omega$	$\dfrac{\eta\omega}{R^2+\eta^2\omega^2}$	$\dfrac{1}{\eta_1\omega}+\dfrac{\eta_2\omega}{R_2^2+\eta_2^2\omega^2}$

Table 5.1 (cont'd)

Column	I	II	III	IV
Complex Modulus (5.147) $E^* = E_1 + iE_2 = 1/J^*$ (5.158)		(5.171)	(5.175)	(5.181)
Real Part $E_1(\omega)$	R	$\frac{\eta^2\omega^2/R}{1+\eta^2\omega^2/R^2}$	R	$\frac{p_1q_1\omega^2 - q_2\omega^2(1-p_2\omega^2)}{p_1^2\omega^2+(1-p_2\omega^2)^2}$
Imaginary Part $E_2(\omega)$	0	$\frac{\eta\omega}{1+\eta^2\omega^2/R^2}$	$\eta\omega$	$\frac{p_1q_2\omega^3 + q_1\omega(1-p_2\omega^2)}{p_1^2\omega^2+(1-p_2\omega^2)^2}$

†Numbers in parentheses indicate equation numbers in text.

cients of each term. Instead, the following method is more practical and has been used widely.

From (5.69) with $p_0 = 1$ the following may be obtained

$$\hat{\sigma}(s) = \left[\frac{q_0+q_1s+q_2s^2+\ldots+q_bs^b}{1+p_1s+p_2s^2+\ldots+p_as^a}\right]\hat{\varepsilon}(s), \tag{5.73}$$

or

$$\hat{\varepsilon}(s) = \left[\frac{1+p_1s+p_2s^2+\ldots+p_as^a}{q_0+q_1s+q_2s^2+\ldots+q_bs^b}\right]\hat{\sigma}(s), \tag{5.74}$$

where s is the transform variable.

Expanding (5.74) as a sum of partial fractions (see any calculus book) yields

$$\hat{\varepsilon}(s) = \left[\sum_{i=1}^{c}\frac{\alpha_i}{\beta_i+s}\right]\hat{\sigma}(s) = \sum_{i=1}^{c}A_i(s)\hat{\sigma}(s), \tag{5.75}$$

where α_i, β_i are functions of p_i, q_i and $A_i(s) = \alpha_i/(\beta_i+s)$. Then taking the inverse Laplace transform of each product $A_i(s)\hat{\sigma}(s)$ using the convolution form of the inverse transform (see Appendix A4) yields

$$\varepsilon(t) = \int_0^t \sum_{i=1}^{c} A_i(t-\xi)\sigma(\xi)\,\mathrm{d}\xi, \tag{5.76}$$

where the inverse transform $A_i(t-\xi)$ of $A_i(s) = \alpha_i/(\beta_i+s)$ is

$$A_i(t-\xi) = \alpha_i \exp\left[-\beta_i(t-\xi)\right], \tag{5.77}$$

and where t is time, ξ is time prior to t.

Usually α_i and β_i in (5.77) are easier to determine from experimental measurements of material behavior than p_i, q_i in (5.68). Furthermore, if c is a very large number then $\sum_{i=1}^{c} A_i(t-\xi)$ can be approximated by

$$\sum_{i=1}^{c} A_i(t-\xi) \simeq B(t-\xi) = \int_0^\infty \alpha(\beta)\exp\left[-\beta(t-\xi)\right]\mathrm{d}\xi. \tag{5.78}$$

Thus $B(t-\xi)$ in (5.78) is the Laplace transform of $\alpha(\beta)$ if β is taken to be the Laplace transform variable; and function $\alpha(\beta)$ is the inverse Laplace transform of function $B(t-\xi)$. In principle $B(t-\xi)$ can be obtained from experiments. A detailed discussion of this subject has been given by Williams [5.10].

5.10 Differential Form of Constitutive Equations for Multiaxial Stress States

In the first part of this chapter simple stress and strain systems, such as uniaxial and pure shearing, were considered. The time operator method discussed there can be generalized to describe the constitutive relations of linear rate dependent materials under multiaxial stress-multiaxial strain states. In describing the linear viscoelastic behavior under multiaxial stress states it is convenient to separate the shear (or deviatoric) from the dilatational effects, since the response of viscoelastic materials to shear stress and to dilatational stress are different. Thus, the constitutive equations of linear viscoelastic materials under multiaxial stress states, analogous to (5.66) and (5.68), have the following form

$$\mathrm{P}_1 s_{ij}(t) = \mathrm{Q}_1 d_{ij}(t), \tag{5.79}$$

or

$$\left[p_0'+p_1'\frac{\partial}{\partial t}+p_2'\frac{\partial^2}{\partial t^2}+\ldots+p_a'\frac{\partial^a}{\partial t^a}\right]s_{ij}(t) = \left[q_0'+q_1'\frac{\partial}{\partial t}+q_2'\frac{\partial^2}{\partial t^2}+\ldots q_b'\frac{\partial^b}{\partial t^b}\right]d_{ij}(t), \tag{5.80}$$

and

$$\mathrm{P}_2 \sigma_{ii}(t) = \mathrm{Q}_2 \varepsilon_{ii}(t), \tag{5.81}$$

or

$$\left[p_0''+p_1''\frac{\partial}{\partial t}+p_2''\frac{\partial^2}{\partial t^2}+\ldots+p_a''\frac{\partial^a}{\partial t^a}\right]\sigma_{ii}(t) = \left[q_0''+q_1''\frac{\partial}{\partial t}+q_2''\frac{\partial^2}{\partial t^2}+\ldots+q_b''\frac{\partial^b}{\partial t^b}\right]\varepsilon_{ii}(t), \tag{5.82}$$

where P_1, Q_1, P_2 and Q_2 are time operators as in (5.66) and (5.68). s_{ij} and d_{ij} are the deviatoric stress and strain tensors and σ_{ii} and ε_{ii} are the dilatational (or mean normal) stress and strain, respectively, as discussed in Sections 3.5 and 3.12. p_r', q_r', p_r'' and q_r'' are material constants determined from experimental measurements. They may be expressed in terms of R and η for mechanical models as in Table 5.1.

By taking the Laplace transform of (5.80), (5.82), they can be expressed in terms of the transform variable s as follows

$$[p_0'+p_1's+p_2's^2+\ldots+p_a's^a]\hat{s}_{ij}(s) = [q_0'+q_1's+q_2's^2+\ldots+q_b's^b]\,\hat{d}_{ij}(s), \tag{5.83}$$

$$[p_0''+p_1''s+p_2''s^2+\ldots+p_a''s^a]\hat{\sigma}_{ii}(s) = [q_0''+q_1''s+q_2''s^2+\ldots+q_b''s^b]\hat{\varepsilon}_{ii}(s). \tag{5.84}$$

Since the elastic solid is a limiting case of viscoelastic material there is an analogy between the constitutive equations of elastic and viscoelastic materials. The stress-strain relation of an isotropic elastic material can be expressed by two equations by separating deviatoric s_{ij}, d_{ij} from dilatational σ_{ii}, ε_{ii} stress and strain as follows

$$s_{ij} = 2G\, d_{ij}, \tag{5.85}$$

$$\sigma_{ii} = 3K\varepsilon_{ii}, \tag{5.86}$$

where G is the shear modulus and K the bulk modulus or compressibility. By considering the similarities (analogy) between (5.79, 5.81) and (5.85, 5.86) and also employing the superposition principle for both linear elastic and linear viscoelastic solids, relationships among the viscoelastic coefficients and the time operators can be established as in the following. The equivalence between (5.79) and (5.85) yields:

$$G = \tfrac{1}{2}\frac{\mathrm{Q}_1}{\mathrm{P}_1}. \tag{5.87}$$

Similarly, from (5.81) and (5.86)

$$K = \tfrac{1}{3}\frac{\mathrm{Q}_2}{\mathrm{P}_2}. \tag{5.88}$$

The modulus of elasticity E and Poisson's ratio ν of an elastic solid are related to the elastic constants K, G as follows, see for example [5.11] and Table 4.1:

$$E = \frac{9KG}{3K+G}, \tag{5.89}$$

$$\nu = \frac{3K-2G}{6K+2G}. \tag{5.90}$$

Inserting (5.87) and (5.88) into (5.89) the equivalent to a Young's modulus E_v for a viscoelastic material can be obtained in terms of time operators as follows

$$E_v(t) = \frac{3\dfrac{\mathrm{Q}_1}{\mathrm{P}_1}\dfrac{\mathrm{Q}_2}{\mathrm{P}_2}}{\dfrac{\mathrm{Q}_1}{\mathrm{P}_1}+\dfrac{2\mathrm{Q}_2}{\mathrm{P}_2}} = \frac{3\mathrm{Q}_1\mathrm{Q}_2}{\mathrm{P}_2\mathrm{Q}_1+2\mathrm{P}_1\mathrm{Q}_2}, \tag{5.91}$$

where E_v has the same meaning as for an elastic solid, that is $E_v(t) = \sigma(t)/\varepsilon(t)$, except that it is time dependent and is usually called the relaxation modulus.

Similarly by introducing (5.87) and (5.88) into (5.90) the equivalent of Poisson's ratio ν_v for viscoelastic materials may be expressed in terms of time operators (Poisson's ratio is the ratio of lateral strain to axial strain in uniaxial stressing)

$$\nu_v(t) = \frac{\dfrac{\mathrm{Q}_2}{\mathrm{P}_2} - \dfrac{\mathrm{Q}_1}{\mathrm{P}_1}}{\dfrac{2\mathrm{Q}_2}{\mathrm{P}_2} + \dfrac{\mathrm{Q}_1}{\mathrm{P}_1}} = \frac{\mathrm{P}_1\mathrm{Q}_2 - \mathrm{P}_2\mathrm{Q}_1}{\mathrm{P}_2\mathrm{Q}_1 + 2\mathrm{P}_1\mathrm{Q}_2}. \tag{5.92}$$

Again ν_v is time dependent in general. Hence, Poisson's ratio ν_v is not a constant for viscoelastic material.

By employing (5.87–5.92) the relationship among the various methods of representing viscoelastic behavior involving tension, shear and volumetric effects can be determined.

It has been observed for some viscoelastic materials that hydrostatic pressure, in the range of values less than the tensile strength, produces a linear elastic volume change with a time-dependent volume change much smaller than the time-dependent distortion resulting from shearing stresses applied to the same material [5.12, 5.13]. Thus for such materials the dilatational stress tensor may be considered to cause only elastic strain while the stress deviator causes elastic plus time-dependent strain. When the dilatational strains are negligible compared to the deviatoric strains this simplification makes stress analysis much easier.

As examples for constitutive equations of a few mechanical models consider the following. For a material whose deviatoric behavior is of Maxwell type as given by (5.7) $p_0' = 1$, $p_1' = \eta'/R'$ and $q_1' = \eta'$, where R' and η' are the stiffness of the spring and viscosity of the dashpot in a Maxwell material in shear. Hence

$$\mathrm{P}_1 = 1 + \frac{\eta'}{R'}\frac{\partial}{\partial t}, \quad \mathrm{Q}_1 = \eta'\frac{\partial}{\partial t}. \tag{5.93}$$

If the material is elastic under hydrostatic pressure $p_0'' = 1$, $q_0'' = 3K$ in accordance with (5.86). Thus for dilatational behavior

$$\mathrm{P}_2 = 1, \quad \mathrm{Q}_2 = 3K. \tag{5.94}$$

Values of P_1, Q_1, P_2, Q_2 for some other mechanical models are shown in Table 5.1.

5.11 Integral Representation of Viscoelastic Constitutive Equations

In the previous section viscoelastic stress-strain relations were described by differential operator forms. The stress-strain relations will now be described by another means, the hereditary integrals. The advantage of using integral representation over differential operator forms is in the flexibility of representation of the actually measured viscoelastic material properties. The integral representation also can be extended readily to describe the behavior of aging material; and it is easier to incorporate temperature effects.

5.12 Creep Compliance

In a creep test a step of constant stress $\sigma = \sigma_0 H(t)$* is applied and the time-dependent strain $\varepsilon(t)$ is measured. For linear material, the strain can be represented by

$$\varepsilon(t) = \sigma_0 J(t), \tag{5.95}$$

or

$$J(t) = \varepsilon(t)/\sigma_0. \tag{5.96}$$

The function $J(t)$ is called the *creep compliance*, that is, the creep strain per unit of applied stress. Creep compliance is a material property. Thus each material has its own value of creep compliance. For example, $J(t) = \left(\frac{1}{R} + \frac{1}{\eta} t\right)$ for a Maxwell material, see (5.8) and Table 5.1. The creep compliance for several models is shown in Table 5.1.

5.13 Relaxation Modulus

In a relaxation test a step of constant strain $\varepsilon = \varepsilon_0 H(t)$ is applied and the stress $\sigma(t)$ is measured. If the material behavior is linear the stress can be represented by

$$\sigma(t) = \varepsilon_0 E(t), \tag{5.97}$$

or

$$E(t) = \sigma(t)/\varepsilon_0. \tag{5.98}$$

The function $E(t)$ thus obtained is called the *relaxation modulus*. It is the stress per unit of applied strain and is different for each material (a material

* $H(t)$ is the Heaviside unit funticon, see Appendix A3.

property). For example

$$E(t) = Re^{-Rt/\eta}$$

for a Maxwell material, see (5.9).

5.14 Boltzmann's Superposition Principle and Integral Representation

If a constant stress σ_1 is applied at $t = \xi_1$ then $\sigma(t) = \sigma_1 H(t-\xi_1)$* and the corresponding creep strain will be

$$\varepsilon(t) = \sigma_1 J(t-\xi_1)H(t-\xi_1). \tag{5.99}$$

If $\sigma_1 = \sigma_0$ this output will be precisely the same as given in (5.95) but displaced in time by the amount of ξ_1, see Fig. 5.13.

If stress σ_0 is applied at time $t = 0$ to a linearly viscoelastic material and then at time $t = \xi_1$, σ_1 is applied as shown in Fig. 5.13 the strain output at

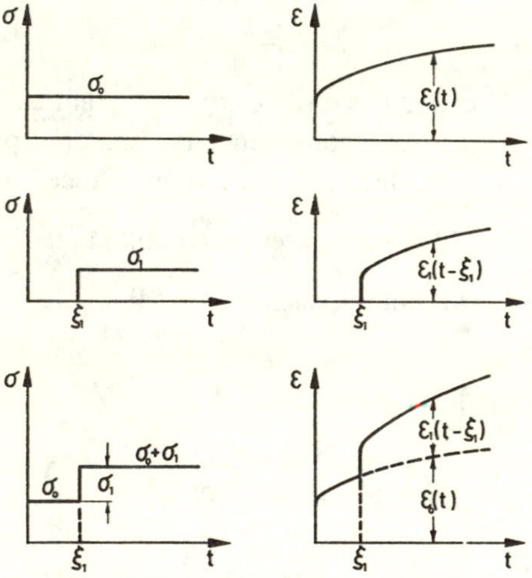

Fig. 5.13. Boltzmann Superposition Principle.

any time subsequent to ξ_1 is given by the sum of the strains at that time due to the two stresses computed as though each were acting separately, i.e., (5.95) and (5.99). This is the Boltzmann superposition principle.

* $H(t)$ is the Heaviside unit function, see Appendix A3.

If the stress input $\sigma(t)$ is arbitrary (variable with time) instead of a constant, this arbitrary stress input can be approximated by the sum of a series of constant stress inputs as shown in Fig. 5.14 and described by

$$\sigma(t) = \sum_{i=1}^{r} \Delta\sigma_i H(t-\xi_i). \tag{5.100}$$

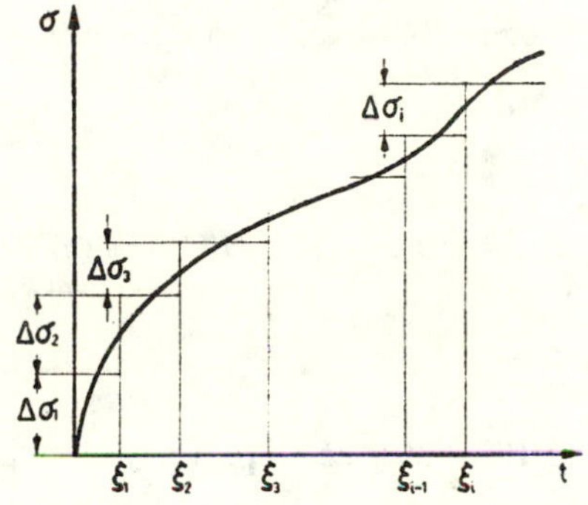

Fig. 5.14. Variable Stress Input Approximated by the Sum of a Series of Constant Stress Inputs.

The Boltzmann superposition principle states that the sum of the strain outputs resulting from each component of stress input is the same as the strain output resulting from the combined stress input. Therefore the strain output under variable stress $\sigma(t)$ equals

$$\varepsilon(t) = \sum_{i=1}^{r} \varepsilon_i(t-\xi_i) = \sum_{i=1}^{r} \Delta\sigma_i J(t-\xi_i)H(t-\xi_i). \tag{5.101}$$

If the number of the steps tends to infinity, the total strain can be expressed by an integral representation as

$$\varepsilon(t) = \int_0^t J(t-\xi)H(t-\xi)\,\mathrm{d}[\sigma(\xi)]. \tag{5.102}$$

This equation is a Stieltjes integral. If the stress history is differentiable and since the dummy variable ξ is always less than or equal to t, the function $H(t-\xi)$ is therefore always unity in the range of integration. Hence (5.102) reduces to the following form

$$\varepsilon(t) = \int_0^t J(t-\xi)\frac{\partial\sigma(\xi)}{\partial\xi}\,\mathrm{d}\xi, \tag{5.103}$$

where $d[\sigma(\xi)]$ has been replaced by $\frac{\partial\sigma(\xi)}{\partial\xi}\,d\xi$ in order that time ξ may be the independent variable.

This equation, an integral representation of creep, can be used to describe (and to predict) the creep strains under any given stress history provided the creep compliance $J(t)$ is known.

An alternate form for (5.103) may be obtained by employing integration by parts, taking $u = J(t-\xi)$ and $dv = \frac{\partial\sigma(\xi)}{\partial\xi}\,d\xi$, with the result

$$\varepsilon(t) = \sigma(t)J(0) - \int_0^t J'(t-\xi)\sigma(\xi)\,d\xi, \qquad (5.103a)$$

where $J'(t-\xi) = \frac{\partial J(t-\xi)}{\partial\xi}$.

Either (5.103) or (5.103a) may be employed depending on which is easier for the problem at hand.

If the creep compliance $J(t)$ is separated into a time-independent (elastic) compliance J_0 and a time-dependent creep function $\varphi(t)$ (5.103) becomes

$$\varepsilon(t) = J_0\sigma(t) + \int_0^t \varphi(t-\xi)\frac{\partial\sigma(\xi)}{\partial\xi}\,d\xi. \qquad (5.104)$$

In principle $J(t)$ can be determined from a constant stress creep test. Equation (5.103) with $J(t)$ determined can be used to predict the stress $\sigma(t)$ from a prescribed strain history $\varepsilon(t)$. However, the problem becomes involved, since it is necessary to solve an integral equation instead of integrating directly.

Exactly the same arguments apply when step changes or arbitrary changes in strain are applied and the resulting change in stress as a function of time is determined. Thus the Boltzmann superposition principle can be restated by substituting stress for strain and strain for stress. Hence (5.101–5.103) can be rewritten by substituting σ for ε and ε for σ and using a different symbol $E(t)$, the relaxation modulus, for $J(t)$ the creep compliance. Thus the following equation is obtained for stress relaxation under arbitrarily prescribed strain ε

$$\sigma(t) = \int_0^t E(t-\xi)\frac{\partial\varepsilon(\xi)}{\partial\xi}\,d\xi. \qquad (5.105)$$

Equations (5.110 to 5.112) define the relationship between the creep compliance $J(t)$ and the relaxation modulus $E(t)$ for linear materials.

By using the initial value theorem and the final value theorem of the Laplace transformation as discussed in Appendix A4, it can be proven that

$$E(0)J(0) = 1, \tag{5.113}$$

and

$$E(\infty)J(\infty) = 1, \tag{5.114}$$

where $E(0)$ and $J(0)$ stand for the initial (time-independent or glassy) relaxation modulus and creep compliance, respectively, and $E(\infty)$ and $J(\infty)$ stand for the long-time (or rubbery) relaxation modulus and creep compliance, respectively. For times between these extremes (5.111, 5.112) apply. In (5.111 or 5.112) if the creep compliance is given, then the relaxation modulus can be obtained by solving the integral equation, and vice versa, see also Table 5.1. It is to be noted that only for certain simple forms of creep compliance (or relaxation modulus) is an analytical integration possible. For example, if the creep compliance can be represented by a power function of time with negligible elastic compliance

$$J(t) = At^n, \tag{5.115}$$

the Laplace transform of J becomes

$$\hat{J}(s) = \frac{A\Gamma(n+1)}{s^{1+n}}, \tag{5.116}$$

where $\Gamma(n+1)$ is the gamma function. Inserting (5.116) into (5.110) and solving for $E(s)$ yields

$$\hat{E}(s) = \frac{1}{A\Gamma(n+1)} \frac{1}{s^{1-n}}. \tag{5.117}$$

The inverse Laplace transform of (5.117) yields the relaxation modulus

$$E(t) = \frac{1}{A\Gamma(1+n)\Gamma(1-n)} t^{-n}, \tag{5.118}$$

which of course predicts an infinite stiffness (modulus) at $t = 0$. Equations (5.115–5.118) have been used by Lai and Fitzgerald [5.14] in the prediction of the relaxation modulus from the creep compliance of asphalt concrete.

It is rather rare that an analytical form of $E(t)$ can be obtained from (5.111 or 5.112) for the prediction of relaxation from creep or creep from relaxation, except for some idealized material such as the one shown above.

Separating the relaxation modulus $E(t)$ into a time-independent (elast
modulus E_0 and a time-dependent stress relaxation function $\psi(t)$ (5.10
becomes

$$\sigma(t) = E_0\varepsilon(t) - \int_0^t \psi(t-\xi)\,\frac{\partial\varepsilon(\xi)}{\partial\xi}\,d\xi. \qquad (5.106)$$

Under a given strain history, it is easier to use (5.105) to predict the stress relaxation behavior than (5.103), since it involves only the direct integration of (5.105), but it requires determination of $E(t)$ from relaxation experiments not $J(t)$ from creep experiments.

5.15 Relation Between Creep Compliance and Relaxation Modulus

Since creep and stress relaxation phenomena are two aspects of the same viscoelastic behavior of materials, they should be related. In other words, one should be predictable if the other is known. Applying the Laplace transform to (5.103) and (5.105) yields algebraic equations in the transform variable s:

$$\hat{\varepsilon}(s) = s\hat{J}(s)\hat{\sigma}(s), \qquad (5.107)$$

$$\hat{\sigma}(s) = s\hat{E}(s)\hat{\varepsilon}(s). \qquad (5.108)$$

From (5.107, 5.108)

$$\frac{\hat{\sigma}(s)}{\hat{\varepsilon}(s)} = s\hat{E}(s) = \frac{1}{s\hat{J}(s)}, \qquad (5.109)$$

or

$$\hat{J}(s)\hat{E}(s) = \frac{1}{s^2}. \qquad (5.110)$$

Applying the inverse Laplace transform to (5.110) yields*

$$\int_0^t J(t-\xi)E(\xi)\,d\xi = t, \qquad (5.111)$$

or

$$\int_0^t E(t-\xi)J(\xi)\,d\xi = t. \qquad (5.112)$$

---

* See Appendix A4, Table A4.1, items 9 and 29.

This is due either to the difficulty of expressing the creep function $J(t)$ in analytical form or due to the difficulty of solving the integral equation even when the analytical form of $J(t)$ can be obtained. However, a numerical solution of the integral equations (5.111) or (5.112) for $J(t)$ given in analytical or tabular form is always possible.

Hopkins and Hamming [5.15] used a numerical method to solve for the creep function $J(t)$ from (5.112) using a given tabulated form of $E(t)$ of polyisobutylene at 25°C. Secor and Monismith [5.16] utilized the same method to calculate the relaxation modulus from a given creep compliance of asphalt concrete.

5.16 Generalization of the Integral Representation to Three Dimensions

So far the integral representation of only simple deformations, such as shear, tension, has been considered. The behavior of viscoelastic material as described for simple cases can be generalized on the basis of the similarity between linear viscoelasticity and elasticity as shown for the differential operator form in Section 5.9. In the most general case the integral representation may take the following form

$$\varepsilon_{ij}(t) = \int C_{ijkl}(t-\xi)\dot{\sigma}_{kl}(\xi)\, d\xi, \tag{5.119}$$

where C_{ijkl} are the creep functions, which in general have 21 constants, see Shu and Onat [5.17], for an anisotropic linear viscoelastic material. For an isotropic material the number of creep functions is reduced to two.

As pointed out before, it is more convenient to separate the stress and strain tensors into deviatoric and dilatational components, respectively. Accordingly (5.85) and (5.86) can be generalized for multiaxial states of stress as follows

$$s_{ij}(t) = 2\int_0^t G(t-\xi)\frac{\partial d_{ij}(\xi)}{\partial \xi}\, d\xi, \tag{5.120}$$

$$\sigma_{ii}(t) = 3\int_0^t K(t-\xi)\frac{\partial \varepsilon_{ii}(\xi)}{\partial \xi}\, d\xi. \tag{5.121}$$

Instead of a material constant G defining elastic shear behavior in (5.85), $G(t)$ in (5.120) is the stress relaxation modulus in shear which is a time function. Similarly $K(t)$ in (5.121) is the hydrostatic (or bulk) stress relaxation modulus. Equations (5.120) and (5.121) have been used by several investigators for viscoelastic analysis, see for example [5.18].

An alternative representation for the stress-strain relationship in three-dimensional elasticity is

$$\sigma_{ij} = \lambda\varepsilon_{kk}\delta_{ij}+2G\varepsilon_{ij}, \tag{5.122}$$

where λ and G are material constants (Lamé's constants). The constant G, the shear modulus, is the same in both (5.85) and (5.122), see Table 4.1. Again, by using the analogy between linear elastic and linear viscoelastic materials, (5.122) may be written in the following form in the case of time-dependent behavior

$$\sigma_{ij}(t) = \delta_{ij}\left[\lambda\varepsilon_{kk}(t)-\int_0^t \psi_1(t-\xi)\,\frac{\partial\varepsilon_{kk}(\xi)}{\partial\xi}\,d\xi\right]$$

$$+2G\varepsilon_{ij}(t)-\int_0^t \psi_2(t-\xi)\,\frac{\partial\varepsilon_{ij}(\xi)}{\partial\xi}\,d\xi, \tag{5.123}$$

where λ and G are constants describing the linear time-independent (elastic) stress-strain relations, ψ_1 and ψ_2 are the stress relaxation functions for ε_{kk} and ε_{ij}, respectively.

Strains can be expressed in terms of stresses by the following inversion of (5.122), compare with (4.15b),

$$\varepsilon_{ij} = -\frac{\lambda}{2G(2G+3\lambda)}\sigma_{kk}\delta_{ij}+\frac{1}{2G}\sigma_{ij}. \tag{5.124}$$

The conversion of (5.124) to a time-dependent state leads to the following form [5.19]

$$\varepsilon_{ij}(t) = \delta_{ij}\left[a_0\sigma_{kk}(t)+\int_0^t \varphi_1(t-\xi)\,\frac{\partial\sigma_{kk}(\xi)}{\partial\xi}\,d\xi\right]$$

$$+b_0\sigma_{ij}(t)+\int_0^t \varphi_2(t-\xi)\,\frac{\partial\sigma_{ij}(\xi)}{\partial\xi}\,d\xi, \tag{5.125}$$

where $a_0 = \dfrac{-\lambda}{2G(2G+3\lambda)}$, $b_0 = 1/2G$ are material constants, φ_1 and φ_2 are the creep functions (or kernel functions). It is noted that this representation does not require separate determination for dilatational behavior. For example a_0, b_0, φ_1, φ_2 can be determined from a uniaxial and a pure shear creep tests. Equation (5.125) can be divided into two equations similar to (5.120 and 5.121). Therefore the constants and the time function of both representations may be expressed in terms of each other.

The creep functions and the material constants for (5.125) can be determined from constant stress creep tests in tension and in torsion as follows: For example the creep data of many materials, especially plastics, under constant stress can be described by equations of the following form

$$\text{For tension } \sigma,\ \varepsilon_{11} = \varepsilon_{11}^{0}\sigma + \varepsilon_{11}^{+}t^{n}\sigma, \tag{5.126}$$

where ε_{11}^{0}, ε_{11}^{+}, n are material constants in tension.

$$\text{For torsion } \tau,\ \varepsilon_{12} = \varepsilon_{12}^{0}\tau + \varepsilon_{12}^{+}t^{q}\tau, \tag{5.127}$$

where ε_{12}^{0}, ε_{12}^{+} and q are material constants in shear.

For constant tensile stress σ_0, $\sigma(t) = \sigma_0 H(t)$, so (5.125) reduces to

$$\varepsilon_{11}(t) = a_0\sigma_0 + \sigma_0\varphi_1(t) + b_0\sigma_0 + \sigma_0\varphi_2(t). \tag{5.128}$$

Similarly, for torsion under constant shear stress τ_0, (5.125) reduces to

$$\varepsilon_{12}(t) = b_0\tau_0 + \varphi_2(t)\tau_0. \tag{5.129}$$

Comparing (5.127) with (5.129) for $\tau = \tau_0$ and taking $\varphi_2(0) = 0$ the following is obtained

$$b_0 = \varepsilon_{12}^{0}, \quad \varphi_2(t) = \varepsilon_{12}^{+}t^{q}. \tag{5.130}$$

Similarly, substituting (5.130) into (5.128), comparing with (5.126) for $\sigma = \sigma_0$, and taking $\varphi_1(0) = 0$ yields

$$a_0 = \varepsilon_{11}^{0} - \varepsilon_{12}^{0}, \quad \varphi_1(t) = \varepsilon_{11}^{+}t^{n} - \varepsilon_{12}^{+}t^{q}. \tag{5.131}$$

Substituting (5.130) and (5.131) into (5.125) yields the following general equation for material whose creep data are describable by (5.126, 5.127)

$$\varepsilon_{ij}(t) = \delta_{ij}\left\{(\varepsilon_{11}^{0} - \varepsilon_{12}^{0})\sigma_{kk}(t) + \int_0^t [\varepsilon_{11}^{+}(t-\xi)^{n} - \varepsilon_{12}^{+}(t-\xi)^{q}]\,\frac{\partial\sigma_{kk}(\xi)}{\partial\xi}\,d\xi\right\}$$

$$+\varepsilon_{12}^{0}\sigma_{ij}(t) + \int_0^t \varepsilon_{12}^{+}(t-\xi)^{q}\,\frac{\partial\sigma_{ij}(\xi)}{\partial\xi}\,d\xi. \tag{5.132}$$

5.17 Behavior of Linear Viscoelastic Material under Oscillating Loading

In the previous sections the representation of the mechanical behavior of linear viscoelastic materials by means of constitutive equations has been discussed. These have considered those cases where creep or stress relaxation experiments are required in order to determine the constitutive equations either in the form of differential operators or integrals.

However, creep and relaxation experiments are not capable of providing complete information concerning the mechanical behavior of viscoelastic solids. In certain cases the response of structures to a loading for times "substantially shorter" than the lower limit of time of usual creep measurements (approximately 10 sec.) may be of practical importance. For example, duration of the impact of a steel ball on a viscoelastic block may be of the order of 10^{-5} sec. In order to be able to calculate the stresses under this condition, the mechanical behavior of the material at high rates of loading or short duration loading must be known. Creep or stress relaxation experiments may provide data from about 10 sec. to about 10 years (8.76×10^4 hr. or 3.15×10^8 sec.) while dynamic experiments may provide data from about 10^{-8} sec. to about 10^3 sec. Thus there is an overlapping region where data can be obtained from both types of experiments.

To investigate the response of viscoelastic materials for very short times of loading it is more convenient to use oscillatory than static loading. The types of experimental apparatus and techniques used are quite different in the two methods. Problems associated with accuracy, sensitivity and stability of instrumentation, and thermal effects in materials are somewhat more difficult to solve in oscillatory than static experiments. In the following the basic concepts involved in determining viscoelastic properties of materials by means of oscillatory loading will be discussed.

5.18 Complex Modulus and Compliance

An external source may apply a constant amplitude of force to a member made of viscoelastic material. If the source contains a single frequency the force may be represented by

$$F = F_0 \cos \omega t,$$

where F_0 is the amplitude of the force and ω is the angular frequency ($f = \omega/2\pi$ is the cyclic frequency). Upon application of such a force transient vibrations are induced in the viscoelastic member. If the member has a single degree of freedom or vibrates only in one mode the transient vibration will consist of a sinusoidal vibration at the natural frequency for that mode $f_m = p/2\pi$, where p is the circular frequency of natural vibration. In the presence of even small damping this transient vibration will disappear with time, but a steady-state vibration at the same frequency ω of the source will continue as long as the source persists. This steady-state vibration will induce a variation in stress at a given point,

$$\sigma = \sigma_0 \cos \omega t, \tag{5.133}$$

where σ_0 is the stress amplitude, ω is the frequency of the forced vibration, and $T = 2\pi/\omega$ is the period of the oscillation.

Since

$$e^{i\omega t} = \cos \omega t + i \sin \omega t, \tag{5.134}$$

where $i = \sqrt{-1}$ represents the imaginary axis of the complex variable, it follows that (5.133) can be expressed by using the real part of (5.134) as follows

$$\sigma = \sigma_0 e^{i\omega t}. \tag{5.135}$$

If the material is linearly viscoelastic and the input is an oscillatory stress (5.135) the strain response will be an oscillation at the same frequency as the stress but lagging behind by a phase angle δ, see Fig. 5.15a. Thus

$$\varepsilon = \varepsilon_0 \cos (\omega t - \delta), \tag{5.136}$$

or

$$\varepsilon = \varepsilon_0 e^{i(\omega t - \delta)}, \tag{5.137}$$

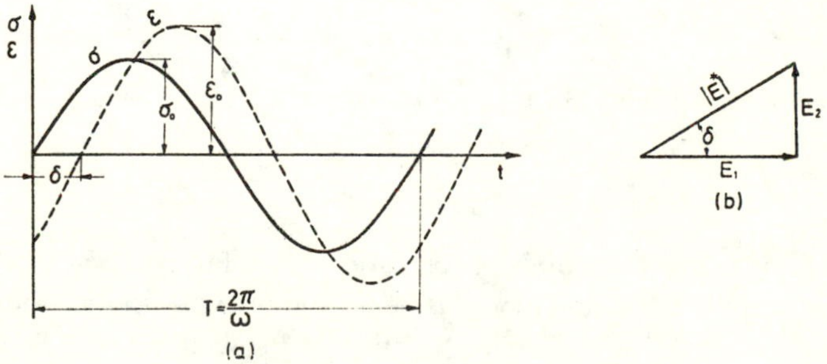

Fig. 5.15. Oscillating Stress σ, Strain ε and Phase lag δ.

where ε_0 is the strain amplitude. The phase angle δ is often called the loss angle and is a function of the internal friction of the material. In order to describe the relation between ε and δ for linear viscoelastic materials it is convenient to write (5.137) as follows

$$\varepsilon = (\varepsilon_0 e^{-i\delta}) e^{i\omega t} = \varepsilon^* e^{i\omega t}, \tag{5.138}$$

where ε^*, the complex strain amplitude,

$$\varepsilon^* = \varepsilon_0 e^{-i\delta} = \varepsilon_0 (\cos \delta - i \sin \delta) \tag{5.139}$$

is a complex number.

Inserting (5.135) and (5.138) into (5.68) for viscoelastic material yields

$$[p_0+i\omega p_1+(i\omega)^2p_2+\dots]\sigma_0 e^{i\omega t} = [q_0+i\omega q_1+(i\omega)^2q_2+\dots]\varepsilon^* e^{i\omega t} \tag{5.140}$$

or

$$\frac{\varepsilon^*}{\sigma_0} = J^*(\omega) = \frac{[p_0+(i\omega)p_1+(i\omega)^2p_2+\dots]}{[q_0+(i\omega)q_1+(i\omega)^2q_2+\dots]}, \tag{5.141}$$

where J^* is called the complex creep compliance or complex compliance, analogous to the creep compliance defined for quasi-static tests, see (5.96). The complex compliance depends on the frequency ω as shown by (5.141).

Similarly, if the input is an oscillatory strain

$$\varepsilon(t) = \varepsilon_0 e^{i\omega t} \tag{5.142}$$

then the stress response will lead the strain by a phase angle δ which can be expressed as follows

$$\sigma(t) = \sigma_0 e^{i(\omega t+\delta)}, \tag{5.143}$$

or

$$\sigma(t) = \sigma^* e^{i\omega t}, \tag{5.144}$$

where σ^* is the complex stress amplitude

$$\sigma^* = \sigma_0 e^{i\delta} = \sigma_0 (\cos\delta + i\sin\delta). \tag{5.145}$$

Inserting (5.142) and (5.144) into (5.68) yields the following

$$\frac{\sigma^*}{\varepsilon_0} = E^*(\omega) = \frac{[q_0+(i\omega)q_1+(i\omega)^2q_2+\dots]}{[p_0+(i\omega)p_1+(i\omega)^2p_2+\dots]}, \tag{5.146}$$

where E^* is called the complex relaxation modulus or complex modulus analogous to the relaxation modulus defined in the quasi-static loading case, see (5.98). Again E^* is a complex function of the frequency ω. Substituting (5.145) in the left-hand side of (5.146) yields

$$E^* = \frac{\sigma_0}{\varepsilon_0} e^{i\delta} = \frac{\sigma_0}{\varepsilon_0} [\cos\delta + i\sin\delta] = E_1 + iE_2 = |E^*|\, e^{i\delta}. \tag{5.147}$$

The first term on the right-hand side of (5.147) is in phase with the strain and is the real part of the complex modulus, often called the storage modulus,

$$E_1 = (\sigma_0/\varepsilon_0)\cos\delta. \tag{5.148}$$

The second term of (5.147) represents the imaginary part of the complex (or dynamic) modulus, often called the loss modulus,

$$E_2 = (\sigma_0/\varepsilon_0)\sin\delta. \tag{5.149}$$

The complex modulus is thus expressed in terms of E_1 and E_2, and the stress-strain relation takes the form

$$\sigma = (E_1 + \mathrm{i}E_2)\varepsilon. \tag{5.150}$$

As shown in Fig. 5.15b

$$\tan\delta = E_2/E_1, \tag{5.151}$$

where δ is the phase angle. Tan δ is called the mechanical loss. The magnitude of the complex modulus is

$$|E^*| = (E_1^2 + E_2^2)^{1/2} = \sigma_0/\varepsilon_0. \tag{5.152}$$

In an analogous manner to the derivation of the storage modulus and loss modulus from the complex relaxation modulus, the complex creep compliance J^*, defined by (5.141) can be expressed as follows by substituting (5.139) into the left-hand side of (5.141)

$$\begin{aligned} J^* &= \varepsilon^*/\sigma_0 = (\varepsilon_0/\sigma_0)e^{-\mathrm{i}\delta} = (\varepsilon_0/\sigma_0)(\cos\delta - \mathrm{i}\sin\delta) \\ &= J_1 - \mathrm{i}J_2 = |J^*|\, e^{-\mathrm{i}\delta}, \end{aligned} \tag{5.153}$$

where J_1 is called the storage compliance, J_2 is the loss compliance, $|J^*|$ is the magnitude of the complex compliance and δ is the phase angle (loss angle) between J_1 and J_2. The magnitude of the complex compliance is

$$|J^*| = (J_1^2 + J_2^2)^{1/2} = \varepsilon_0/\sigma_0. \tag{5.154}$$

The relations among J_1, J_2 and δ are as follows

$$\tan\delta = J_2/J_1, \tag{5.155}$$

$$J_1 = |J^*|\cos\delta = |J^*|(1+\tan^2\delta)^{-1/2}, \tag{5.156}$$

$$J_2 = |J^*|\sin\delta = |J^*|\tan\delta\,(1+\tan^2\delta)^{-1/2}. \tag{5.157}$$

Comparison of (5.141) and (5.146) or (5.147) and (5.153) shows that J^* and E^* are reciprocal. Thus

$$J^*E^* = 1. \tag{5.158}$$

From (5.152) and (5.154) and from (5.151) and (5.155) it can be shown that

$$|J^*| = \frac{1}{|E^*|} \quad \text{and} \quad \tan\delta = \frac{E_2}{E_1} = \frac{J_2}{J_1}. \tag{5.159}$$

In quasi-static experiments such as creep the time t is a variable while it is common to use frequency ω as the corresponding variable in cyclic tests. The period T of a cyclic variation is

$$T = \frac{1}{\omega}, \quad \text{accordingly} \quad \log T = -\log\omega. \tag{5.160}$$

Consequently the time scales of experimental creep curves are usually converted into frequency scales by means of (5.160) considering T to be the elapsed time t.

Figure 5.16 shows a continuous variation of the complex compliance J^*, the storage compliance J_1 and the loss compliance J_2 for a polymer exhibiting no irrecoverable viscous flow (standard solid). For long loading times

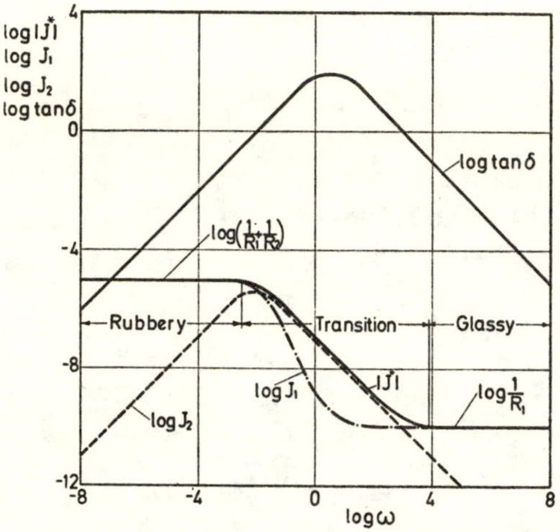

Fig. 5.16. Standard Solid Under Oscillating Loading. The spring constant of the spring in series $R_1 = 10^{10}$ dynes/cm². Spring constant in the Kelvin element $R_2 = 10^5$ dynes/cm². Viscosity of the dashpot $\eta_2 = 10^7$ poises.

and for very low frequencies the loss compliance J_2 approaches zero and the dynamic compliance J^* coincides with the storage compliance J_1. This corresponds to a creep experiment of a material which exhibits a rubbery state. On the other hand, if the frequency is very high, or the loading time is very short, the complex compliance decreases (or rigidity of the material increases) and the loss modulus approaches zero, i.e., negligible energy loss occurs. In this range the material exhibits a glassy state. In the transition region, whose location on the time scale depends on the physical characteristics of the material, the complex compliance J^* is larger than the storage compliance J_1; the loss compliance J_2 is different from zero and shows a maximum. Thus tan δ becomes a maximum in this region also, Fig. 5.16.

Actually the time-dependent characteristics of the dynamic moduli of viscoelastic materials, such as plastics, are strongly related to their internal structure and environmental conditions.

5.19 Dissipation

The storage modulus E_1 as defined previously is associated with energy storage and release during periodic deformation. On the other hand, the loss modulus E_2 is associated with the dissipation of energy and its transformation into heat, therefore called the loss modulus. The ratio $E_2/E_1 = \tan\delta$ is widely used as a measure of the damping capacity of viscoelastic materials.

Damping energy is defined as $\Delta W/W$ where ΔW is the energy loss per cycle of vibration of a given amplitude and W is the maximum energy which the system can store for a given amplitude. There will be no energy loss in one cycle if the stress and the strain are in phase, and hence $\delta = 0$. The amount of energy loss in one unit volume during one complete cycle can be calculated by integrating the increment of work done $\sigma\, d\varepsilon$ over a complete cycle of period T, as follows

$$\Delta W = \int_0^T \sigma \frac{d\varepsilon}{dt}\, dt. \tag{5.161}$$

Inserting $\sigma = \sigma_0 \sin\omega t$ and $d\varepsilon/dt = \omega\varepsilon_0 \cos(\omega t - \delta)$ into (5.161)

$$\Delta W = \int_0^T \varepsilon_0\sigma_0\omega \sin\omega t \cos(\omega t - \delta) dt \tag{5.162}$$

is obtained. Integration of (5.162) yields the following expression for energy loss per unit volume per cycle, which is equivalent to the heat generated per cycle.

$$\Delta W = \pi\sigma_0\varepsilon_0 \sin\delta. \tag{5.163}$$

Substituting $\varepsilon_0 E_2$ for $\sigma_0 \sin\delta$ from (5.149) yields

$$\Delta W = \pi\varepsilon_0^2 E_2. \tag{5.164}$$

The maximum energy W, that the material can store in one cycle may be computed by integrating the increment of work $\sigma\, d\varepsilon$ from zero to a maximum stress, that is over 1/4 the period T of one cycle.

$$W = \int_0^{T/4} \sigma \frac{d\varepsilon}{dt}\, dt = \int_0^{T/4} \sigma_0\varepsilon_0\omega \sin\omega t \cos\omega t\, dt = \sigma_0\varepsilon_0/2. \tag{5.165}$$

In the above calculation the stress and strain are considered to be in phase, i.e., $\delta = 0$, since only in this case will the energy be a maximum.

Thus the damping energy of a viscoelastic material as described by $\Delta W/W$ is found from (5.163) and (5.165) to be

$$\frac{\Delta W}{W} = 2\pi \sin \delta. \tag{5.166}$$

This equation shows that the damping ability of a linearly viscoelastic material is only dependent on the phase angle, which is a function of frequency and is a measure of a physical property of the material, but is independent of the stress and strain amplitudes.

5.20 Complex Compliance and Complex Modulus of Some Viscoelastic Models

Mechanical models with a small number of constants have proven useful in simulating the basic character of the mechanical behavior of real materials. In the following the effect of frequency on the mechanical behavior of some specific viscoelastic models will be discussed in detail.

5.21 Maxwell Model

Equation (5.7) gives the stress-strain rate relation of the Maxwell model, see also Table 5.1,

$$\sigma + (\eta/R)\dot{\sigma} = \eta\dot{\varepsilon}. \tag{5.167}$$

Inserting (5.142) and (5.143) or (5.144), respectively, for strain and stress into (5.167) the following relations are obtained,

$$\sigma_0 e^{i\delta}(1 + i\omega\eta/R) = i\omega\varepsilon_0\eta, \tag{5.168}$$

or

$$\sigma^*(1 + i\omega\eta/R) = i\omega\varepsilon_0\eta. \tag{5.169}$$

From these two equations the complex relaxation modulus, as defined by (5.146) and (5.147) can be obtained,

$$E^* = \sigma^*/\varepsilon_0 = (\sigma_0/\varepsilon_0)e^{i\delta} = i\omega\eta/(1 + i\omega\eta/R). \tag{5.170}$$

By separating the last expression into real and imaginary parts the complex modulus of the Maxwell model takes the form

$$E^* = \frac{\eta^2\omega^2/R}{1 + \eta^2\omega^2/R^2} + i\,\frac{\eta\omega}{1 + \eta^2\omega^2/R^2}\,. \tag{5.171}$$

Substituting the equivalents of E_1 and E_2 from (5.171) into (5.152) and (5.151) yields

$$|E^*| = \sigma_0/\varepsilon_0 = \omega\eta(1+\omega^2\eta^2/R^2)^{-1/2} \tag{5.172}$$

and

$$\tan\delta = E_2/E_1 = R/\omega\eta. \tag{5.173}$$

Figure 5.17 shows the variation of the dynamic modulus $|E^*|$ and the mechanical loss tan δ of the Maxwell material with respect to frequency ω for different values of relaxation time $t_R = \eta/R$. If the frequency ω ap-

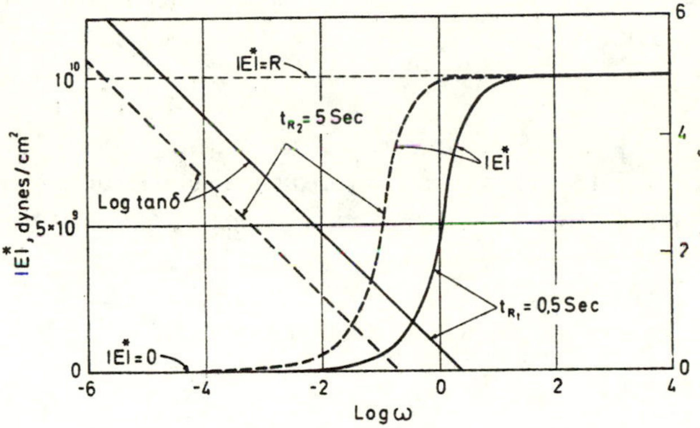

Fig. 5.17. Maxwell Model Under Oscillating Loading. Spring constants $R_1 = R_2 = R = 10^{10}$ dynes/cm². Viscosities $\eta_1 = 5\times10^9$ poises, $\eta_2 = 5\times10^{10}$ poises.

proaches infinity, the dynamic modulus approaches the spring constant R of the Maxwell model. For low frequencies the material exhibits fluid-like behavior, $|E^*|$ approaches zero and flow occurs, while it behaves like an elastic solid body at high frequencies. Log tan δ which is a measure of the energy loss decreases linearly with log frequency. If the relaxation time $t_R = \eta/R$ increases while the spring constant R remains constant, tan δ and the $|E^*|$ transition region will be displaced toward the left.

5.22 Kelvin Model

The constitutive equation of the Kelvin model has the following form as given by (5.15), see also Table 5.1,

$$\sigma = R\varepsilon + \eta\dot{\varepsilon}. \tag{5.174}$$

Inserting (5.142) and (5.143) or (5.144) into (5.174) the following is obtained

$$\sigma^*/\varepsilon_0 = (\sigma_0/\varepsilon_0)e^{i\delta} = R + i\omega\eta, \tag{5.175}$$

which according to the definition of E^* given in (5.146) and (5.147) yields the complex modulus E^* of the Kelvin model,

$$E^* = R + i\omega\eta. \tag{5.176}$$

The magnitude of the complex modulus is

$$|E^*| = (R^2 + \omega^2\eta^2)^{1/2} = R(1 + t_c^2\omega^2)^{1/2}, \tag{5.177}$$

where $t_c = \eta/R$ is the retardation time of the Kelvin model. The mechanical loss for the model is

$$\tan\delta = t_c\omega. \tag{5.178}$$

As shown in Fig. 5.18, for low frequencies the magnitude of the complex modulus $|E^*|$ of the Kelvin model approximately equals the spring constant R; while at high frequencies the dynamic modulus increases very rapidly,

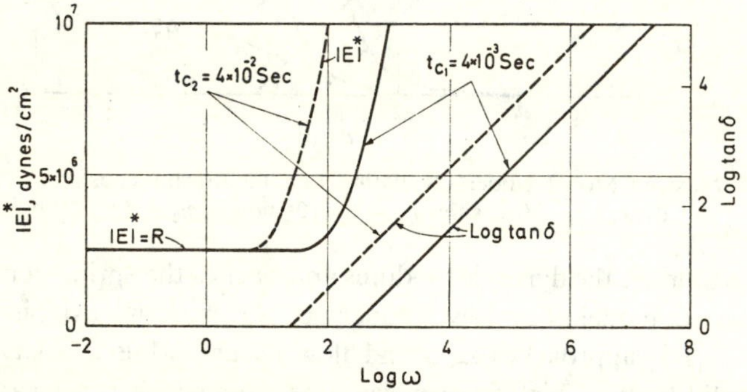

Fig. 5.18. Kelvin Model Under Oscillating Loading. Spring constants $R_1 = R_2 = R = 2.5 \times 10^6$ dynes/cm². Viscosities: $\eta_1 = 10^4$, $\eta_2 = 10^5$ poises.

i.e., the material becomes more and more stiff. If the spring constant R stays fixed while the viscosity of the dashpot increases, the retardation time t_c also increases and the transition frequency of $|E^*|$ and the energy loss, $\tan\delta$ will be displaced toward the left.

The behavior of most real viscoelastic materials cannot be described very well by Maxwell or Kelvin models (which have only two parameters). More complicated models with a larger number of adjustable parameters can be used to approximate more closely the behavior of real materials.

5.23 Burgers Model

The complex modulus for the Burgers model may be determined as follows. The constitutive equation (5.26) for the Burgers model may be written as (see Table 5.1)

$$\sigma+p_1\dot{\sigma}+p_2\ddot{\sigma} = q_1\dot{\varepsilon}+q_2\ddot{\varepsilon}, \tag{5.179}$$

where

$$p_1 = \frac{\eta_1}{R_1}+\frac{\eta_1}{R_2}+\frac{\eta_2}{R_2}, \quad p_2 = \frac{\eta_1\eta_2}{R_1R_2},$$

$$q_1 = \eta_1, \quad q_2 = \frac{\eta_1\eta_2}{R_2}.$$

Substituting (5.142) and (5.144) for ε and σ yields

$$\sigma^*(1+p_1\mathrm{i}\omega-p_2\omega^2) = \varepsilon_0(q_1\mathrm{i}\omega-q_2\omega^2). \tag{5.180}$$

Solving (5.180) for $\sigma^*/\varepsilon_0 = E^*$ from (5.146), rearranging terms and also using (5.147)

$$E^* = \frac{\sigma^*}{\varepsilon_0} = \frac{\sigma_0 e^{\mathrm{i}\delta}}{\varepsilon_0} = \frac{[p_1q_1\omega^2-q_2\omega^2(1-p_2\omega^2)]+\mathrm{i}[p_1q_2\omega^2+q_1(1-p_2\omega^2)]\omega}{p_1^2\omega^2+(1-p_2\omega^2)^2}. \tag{5.181}$$

The magnitude of the complex modulus $|E^*|$ and mechanical loss tan δ can be computed from (5.181) by using (5.152) and (5.151) respectively,

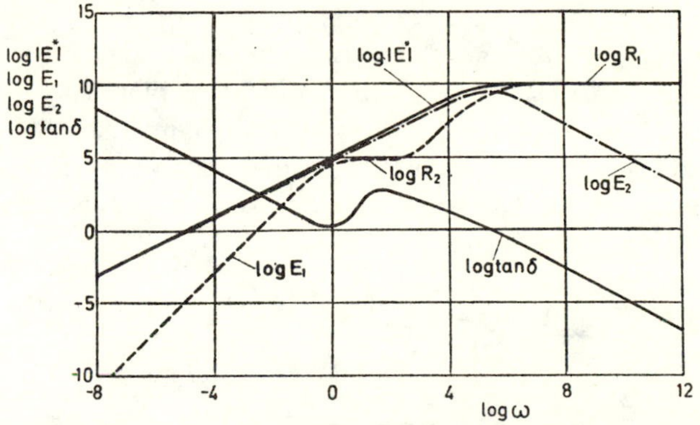

Fig. 5.19. Burgers Model Under Oscillating Loading. The constants of the Maxwell element: $R_1 = 10^{10}$ dynes/cm², $\eta_1 = 10^5$ poises. The constants of the Kelvin element: $R_2 = 10^5$ dynes/cm², $\eta_2 = 10^5$ poises.

since the real part of (5.181) is E_1 and the imaginary part is E_2, (see Table 5.1).

The variation of $|E^*|$ and tan δ with frequency ω are shown in Figure 5.19. At high frequency the Burgers model behaves like a stiff elastic solid (glassy) with an elastic modulus of R_1. For lower frequencies (below a transition region) it becomes flexible but elastic (rubbery state) with a much lower modulus than the glassy state. If the frequency is very low Newtonian flow in the Maxwell model becomes dominant, so fluid-like behavior is observed.

5.24 Relation Between the Relaxation Modulus and the Complex Relaxation Modulus

In previous sections it was shown that the relaxation modulus or creep compliance of materials is sufficient information from which to describe the viscoelastic behavior of linear materials. Therefore, given the stress and strain under oscillating loading described by (5.142), (5.143) or (5.144), together with a given relaxation modulus of the material, the complex modulus E^* and therefore E_1, E_2, tan δ, and $|E^*|$ should be obtainable. Similarly, the relaxation modulus (or creep compliance) should be obtainable if the complex relaxation modulus (or complex compliance) of the material is given.

The actual procedure involves a fairly lengthy mathematical calculation. Usually such mathematical techniques as Fourier transforms, Fourier integrals and contour integrals are required. For more detailed information on this subject, the reader is referred to [5.3] and [5.20].

To illustrate the problem of predicting creep or relaxation from oscillatory data and vice versa, one method will be employed to compute the complex modulus E^* from the stress relaxation modulus $E(t)$ for linear viscoelastic material and vice versa. The stress-strain-time relation for variable strain history is given in terms of the relaxation modulus $E(t)$ by (5.105) for uniaxial stress

$$\sigma(t) = \int_{-\infty}^{t} E(t-\xi)\frac{\partial \varepsilon(\xi)}{\partial \xi}\,d\xi. \tag{5.182}$$

In (5.182) the lower limit of integration is taken at $-\infty$ for the present purposes. Consider that an oscillatory strain has been imposed for a long time so that transient vibrations have disappeared. Then changing the variable of integration from ξ to $\xi' = t-\xi$ equation (5.182) becomes

$$\sigma(t) = -\int_0^\infty E(\xi')\frac{\partial\varepsilon(t-\xi')}{\partial\xi'}\,d\xi'. \tag{5.183}$$

In changing variables in (5.183) the limits of integration are: when $\xi = -\infty$, $\xi' = \infty$; when $\xi = t$, $\xi' = 0$. The minus sign before (5.183) resulted from interchanging upper and lower limits. Inserting an oscillatory strain (5.142),

$$\varepsilon(t) = \varepsilon_0 e^{i\omega t} = \varepsilon_0[\cos\omega t + i\sin\omega t],$$

or

$$\varepsilon(t-\xi') = \varepsilon_0 e^{i\omega t}e^{-i\omega\xi'} = \varepsilon_0 e^{i\omega t}[\cos\omega\xi' - i\sin\omega\xi'], \tag{5.184}$$

into (5.183) yields

$$\sigma(t) = \varepsilon_0\omega e^{i\omega t}\int_0^\infty E(\xi')(\sin\omega\xi' + i\cos\omega\xi')\,d\xi', \tag{5.185}$$

where $\sigma(t)$ is the output of an oscillatory strain input $\varepsilon_0 e^{i\omega t}$. Thus it is equivalent to (5.144),

$$\sigma(t) = \sigma^* e^{i\omega t}.$$

Also using (5.146)

$$\sigma(t) = E^*\varepsilon_0 e^{i\omega t}. \tag{5.186}$$

Inserting the exponential form of (5.184) and (5.186) into (5.183) yields

$$E^*\varepsilon_0 e^{i\omega t} = -\int_0^\infty E(\xi')\varepsilon_0 e^{i\omega t}(-i\omega)e^{-i\omega\xi'}\,d\xi',$$

from which after cancelling $\varepsilon_0 e^{i\omega t}$ from both sides

$$E^*(\omega) = i\omega\int_0^\infty E(\xi')e^{-i\omega\xi'}\,d\xi' = (i\omega)\mathscr{L}[E(\xi')]_{s=i\omega}. \tag{5.187}$$

Equation (5.187) is one form of relationship between the complex modulus $E^*(\omega)$ and the relaxation modulus $E(t)$. The right-hand expression in (5.187) involves the Laplace transform of $E(\xi')$, where $s = i\omega$ is the transform variable and $\xi' = t-\xi$.

Another form relating the complex modulus $E^*(\omega)$ to the relaxation modulus $E(t)$ is found from the sinusoidal form of (5.184) by equating (5.185) and (5.186). After cancelling $\varepsilon_0 e^{i\omega t}$ from both sides

$$E^*(\omega) = \omega \int_0^\infty E(\xi') \sin \omega\xi' \, d\xi' + i\omega \int_0^\infty E(\xi') \cos \omega\xi' \, d\xi'. \quad (5.188)$$

Thus the real E_1 and imaginary E_2 parts of the complex modulus are as follows from (5.147) and (5.188)

$$E_1(\omega) = \omega \int_0^\infty E(\xi') \sin \omega\xi' \, d\xi', \quad (5.189)$$

$$E_2(\omega) = \omega \int_0^\infty E(\xi') \cos \omega\xi' \, d\xi'. \quad (5.190)$$

These integrals are one-sided Fourier transforms and may be used to compute the real and imaginary parts of the complex modulus of a material for different frequencies ω when the relaxation modulus $E(t)$ is known.

By employing the Fourier inversion to (5.189), (5.190), the following expressions for the relaxation modulus $E(t)$ as a function of components of the complex modulus may be obtained

$$E(\xi') = \frac{2}{\pi} \int_0^\infty \frac{E_1(\omega)}{\omega} \sin \omega\xi' \, d\omega, \quad (5.191)$$

$$E(\xi') = \frac{2}{\pi} \int_0^\infty \frac{E_2(\omega)}{\omega} \cos \omega\xi' \, d\omega. \quad (5.192)$$

By using (5.191) or (5.192) the relaxation modulus $E(t)$ can be computed for particular values of t when the functions E_1 or E_2 are known over the complete spectrum of frequency ω.

For example, to compute the complex modulus for a Maxwell model from its relaxation modulus use (5.189) and (5.190). The relaxation modulus from (5.9) or Table 5.1 is

$$E(t) = Re^{-Rt/\eta}. \quad (5.193)$$

Substituting (5.193) into (5.189), (5.190)

$$E_1(\omega) = \omega R \int_0^\infty e^{-R\xi'/\eta} \sin \omega\xi' \, d\xi' = \frac{\omega^2 R}{(R/\eta)^2 + \omega^2}, \tag{5.194}$$

$$E_2(\omega) = \omega R \int_0^\infty e^{-R\xi'/\eta} \cos \omega\xi' \, d\xi' = \frac{\omega R^2/\eta}{(R/\eta)^2 + \omega^2}. \tag{5.195}$$

Equations (5.194), (5.195) are identical to the real and imaginary parts of (5.171) respectively (see Table 5.1).

5.25 Relation Between Creep Compliance and Complex Compliance

Following the same procedure as above for relaxation modulus and complex relaxation modulus, or noting the similarity of (5.103) and (5.105), (5.103) can be written after changing variables from ξ to $\xi' = t-\xi$ as follows

$$\varepsilon(t) = -\int_0^\infty J(\xi') \frac{\partial \sigma(t-\xi')}{\partial \xi'} \, d\xi', \tag{5.196}$$

where $J(\xi')$ is the creep compliance.

Introducing an oscillatory stress $\sigma = \sigma_0 e^{i\omega t}$ (5.135) as input with strain $\varepsilon(t) = \varepsilon^* e^{i\omega t}$ (5.138) as output into (5.196), expressions for the complex compliance $J^*(\omega)$ (5.141) may be obtained. Using an exponential form for stress the following is obtained

$$J^*(\omega) = i\omega \int_0^\infty J(\xi') e^{-i\omega\xi'} \, d\xi' = i\omega \mathcal{L}[J(\xi')]_{s=i\omega}. \tag{5.197}$$

where $s = i\omega$ is the Laplace transform variable and $\xi' = t-\xi$.

Using a sinusoidal form for stress yields

$$J^*(\omega) = \omega \int_0^\infty J(\xi') \sin \omega\xi' \, d\xi' + i\omega \int_0^\infty J(\xi') \cos \omega\xi' \, d\xi'. \tag{5.198}$$

Thus the real J_1 and imaginary J_2 components of the complex compliance are

$$J_1(\omega) = \omega \int_0^\infty J(\xi') \sin \omega\xi' \, d\xi', \tag{5.199}$$

$$J_2(\omega) = \omega \int_0^\infty J(\xi') \cos \omega\xi' \, d\xi'. \tag{5.200}$$

Taking the inverse Fourier transform of (5.199), (5.200) yields expressions for the creep compliance as a function of either component of the complex compliance,

$$J(\xi') = \frac{2}{\pi} \int_0^\infty \frac{J_1(\omega)}{\omega} \sin \omega\xi' \, d\omega, \tag{5.201}$$

$$J(\xi') = \frac{2}{\pi} \int_0^\infty \frac{J_2(\omega)}{\omega} \cos \omega\xi' \, d\omega. \tag{5.202}$$

5.26 Complex Compliance for t^n

If the creep compliance $J(t)$ is expressed as a power function of time

$$J(t) = \varepsilon_0 + mt^n, \tag{5.203}$$

then the corresponding complex compliance can be obtained by inserting the Laplace transform of (5.203) into (5.197) as follows

$$J^*(\omega) = i\omega \mathcal{L} | J(t) |_{s=i\omega} = i\omega \left[\frac{\varepsilon_0}{i\omega} + \frac{m\Gamma(n+1)}{(i\omega)^{n+1}} \right],$$

or

$$J^*(\omega) = J_1(\omega) + iJ_2(\omega) = \varepsilon_0 + m\Gamma(n+1)(i\omega)^{-n}. \tag{5.204}$$

Using the standard formula

$$(\cos\theta + i \sin\theta)^{-n} = \cos n\theta - i \sin n\theta \tag{5.205}$$

and letting $\theta = (\pi/2) + 2\pi k$, $k = 0, \pm 1, \pm 2, \ldots$, so that $\cos\theta = 0$, $\sin\theta = 1$ it follows that (5.205) becomes

$$i^{-n} = \cos n\pi(2k + \tfrac{1}{2}) - i \sin n\pi(2k + \tfrac{1}{2}), \quad k = 0, \pm 1, \pm 2, \ldots . \tag{5.206}$$

Inserting (5.206) into (5.204)

$$\mathrm{J}_1(\omega)+i\mathrm{J}_2(\omega) = \varepsilon_0+m\Gamma(n+1)\omega^{-n}[\cos n\pi(2k+\tfrac{1}{2})-i\sin n\pi(2k+\tfrac{1}{2})] \tag{5.207}$$

Thus the real $\mathrm{J}_1(\omega)$ and imaginary $\mathrm{J}_2(\omega)$ parts of the complex compliance are given by

$$\begin{aligned} \mathrm{J}_1(\omega) &= \varepsilon_0+m\Gamma(n+1)\omega^{-n}\cos n\pi(2k+\tfrac{1}{2}),\\ \mathrm{J}_2(\omega) &= -m\Gamma(n+1)\omega^{-n}\sin n\pi(2k+\tfrac{1}{2}),\\ k &= 0, \quad \pm 1, \quad \pm 2, \ldots. \end{aligned} \tag{5.208}$$

5.27 Temperature Effect and Time-Temperature Superposition Principle

In the discussion so far, the effect of temperature on the constitutive equations has not been taken into account. In general, it is possible to consider the effect of temperature on the constitutive equations by assuming that the coefficients of each term of (5.68) in the differential operator form or the creep compliance and relaxation modulus of (5.96) and (5.98) in the integral representation to be functions of temperature and time, such as

$$E = E(T, t), \tag{5.209}$$

where E is the relaxation modulus as defined by (5.98), T is the temperature and t is the time. The unknown function could be determined in principle, from a proper set of tests which would require at least several creep or relaxation tests under different isothermal temperature levels. This could pose a problem if the number of tests required are very large. Furthermore, even if this approach is workable the analysis of, say, a transient thermal mechanical problem becomes very involved.

Fortunately, theoretical and experimental results (see, for example, [5.9, 5.21]) indicate that for a certain class of material the effect due to time and temperature can be combined into a single parameter through the concept of the "time-temperature superposition principle" which implies that the following relations exist

$$E(T, t) = E(T_0, \zeta), \tag{5.210}$$

$$\zeta = t/a_T(T), \tag{5.211}$$

where t is the actual time of observation measured from first application of load, T is the temperature and ζ is the "reduced time". The reduced time is related to the real time t by the temperature shift factor $a_T(T)$, and T_0 is the reference temperature. The time-temperature superposition principle cited above states that the effect of temperature on the time dependent mechanical behavior is equivalent to a stretching (or shrinking) of the real time for temperatures above (or below) the reference temperature. In other words, the behavior of materials at high temperature and high strain rate is similar to that at low temperature and low strain rate. Materials exhibiting this property are called "thermorheologically simple" after Schwarzl and Staverman [5.22]. Thus, determination of the temperature shift factor $a_T(T)$ as a function of temperature will provide the necessary information for determination of the reduced time ζ.

One of the most common functions relating the shift factor and temperature has been proposed by Williams, Landel and Ferry [5.23] as follows

$$\log_{10} a_T(t) \equiv \log \frac{t}{\zeta} = \frac{-k_1(T-T_0)}{k_2+(T-T_0)}, \tag{5.212}$$

where k_1 and k_2 are material constants and T_0 is the reference temperature. This form is referred to as the WLF equation. This equation has been used to describe the temperature effect on the relaxation behavior of many polymers with fairly satisfactory results (see for example [5.21]).

For transient temperature conditions where temperatures vary with time, Morland and Lee [5.24] have introduced the following reduced time

$$\zeta(t) = \int_0^t \frac{dt'}{a_T[T(t')]}, \tag{5.213}$$

where t' is an arbitrary time prior to t.

The time-temperature superposition principle can be explained more clearly by the following examples [5.25]. Figure 5.20 shows a log-log plot of relaxation modulus versus time at different temperature levels obtained from a series of uniaxial tension stress-relaxation tests on a polybutadiene acrylic acid (PBAA) propellant under various isothermal temperature levels as shown in the figure. If all of these curves except that at 70°F are shifted sideways parallel to the time axis an appropriate distance they will be found to lie along a single line as shown in Fig. 5.21. The magnitude of the total shift is shown in the insert on Fig. 5.21.

The insert in Fig. 5.21 gives the relationship of the shift factor a_T as a function of temperature T. This relationship can be expressed either in

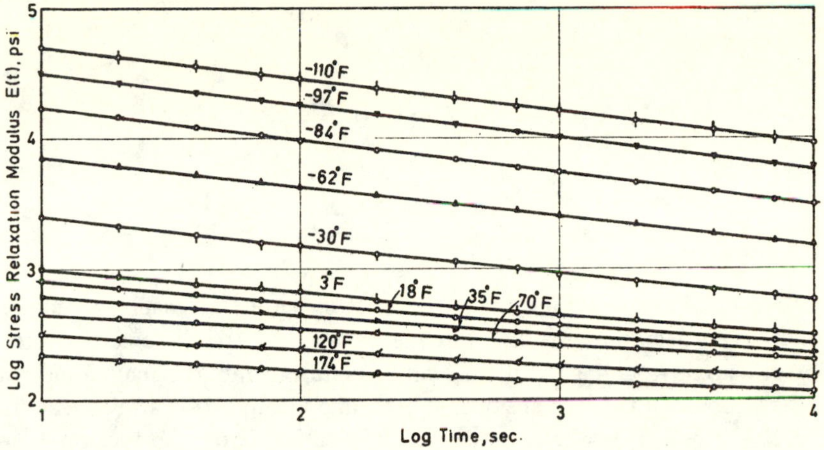

Fig. 5.20. Stress Relaxation Data for Temperatures Between −110°F and +174°F for PBAA Propellant. From [5.25], courtesy Chemical Propellant Information Agency.

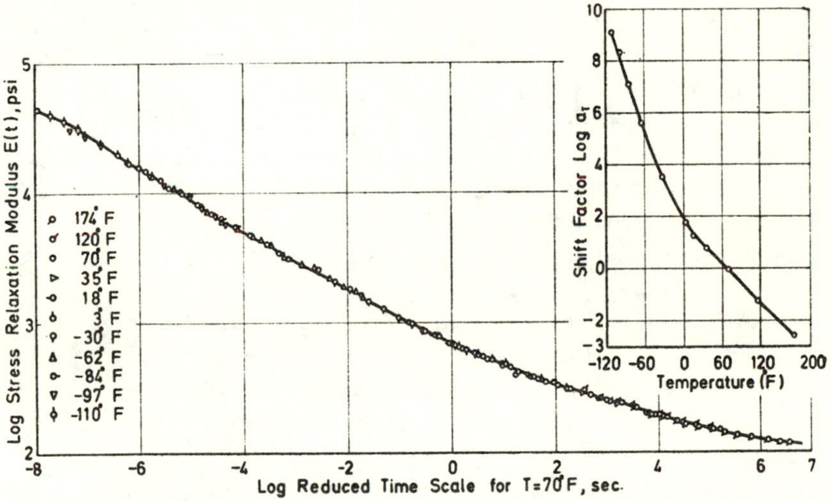

Fig. 5.21. Master Stress Relaxation Modulus at T = 70°F and Shift Factor for PBAA Propellant. From [5.25]), courtesy Chemical Propellant Information Agency.

analytical or in numerical form. The resultant smooth curve obtained by horizontally shifting these stress relaxation curves at different temperature levels into a single smooth curve is called a "master stress relaxation curve" as shown in Figure 5.21.

CHAPTER 6

LINEAR VISCOELASTIC STRESS ANALYSIS

6.1 Introduction

Design of load bearing members of a structure or machine requires analysis to determine the stress in the member due to the external forces or displacement acting on it. In many situations, only the magnitude and distribution of the stress are of interest while in others the strains also are needed. This is true where stiffness and deflections are important or where deformation affects the stress distribution.

In any solid body subjected to external forces and/or displacements the stress and strain at every point due to this loading must simultaneously satisfy three basic equations, the equilibrium equations, the kinematic equations, and the constitutive equations. In addition the stresses and displacements at the boundary must agree with the external forces and displacements imposed by the prescribed boundary conditions. In Chapter 4, it has been pointed out that the first two equations are independent of the type of the material or whether it is linear or not while the third represents the mechanical behavior of the material. The rate effects (or time factor) which are usually neglected in elasticity and plasticity must be considered in viscoelastic materials.

In Chapter 5 various suggested representations for linear viscoelastic behavior have been discussed and some appropriate constitutive equations have been presented. In this chapter the means by which those constitutive relations may be used to solve stress analysis problems in viscoelastic materials will be considered.

Generally, viscoelastic stress analysis problems are more involved than elasticity problems due to inclusion of the time variable in the differential equations in addition to the space variables. However, in many problems where the types of boundary conditions and temperature remain constant in time, the time variable can be removed by employing the Laplace transform.* The problem is thus converted to an equivalent elastic problem.

---

* See Appendix A4.

When a solution for the desired variable has been found in terms of the Laplace transform variable s the inverse Laplace transformation yields the desired solution in the time variable t for the time dependent behavior in the viscoelastic problem. In Sections 6.3 and 6.4 this systematic method, called the elastic-viscoelastic analogy or the "correspondence principle" will be explained and some examples will be given. But before that, in Section 6.2, a beam problem, which is essentially a one-dimensional problem, will be solved by using both the direct solution of the viscoelastic problem, and the correspondence principle. A comparison of the two methods in this relatively simple problem, will provide some feeling for the correspondence principle.

Although there is a large class of problems which can be readily solved by using the correspondence principle, and there are a number of examples given in the literature, there are few viscoelastic problems where the correspondence principle is not applicable that have been examined. In Section 6.6 some of them will be referenced.

6.2 Beam Problems

In order to keep the problem simple the elementary beam theory is used in this section to analyze various beam problems. The assumptions made in the elementary beam theory include, (a) plane sections before bending remain plane after bending, (b) linear viscoelastic stress-strain relations hold, (c) tensile and compression properties are the same, (d) the effect due to shear is negligible.

Direct Solution of a Beam in Pure Bending. A linearly viscoelastic beam of length l as shown in Figure 6.1 is subjected to a known pure bending moment $M(t)$ and the stress and strain distributions and the displacement are sought. Assumption (a) that plane sections before bending remain

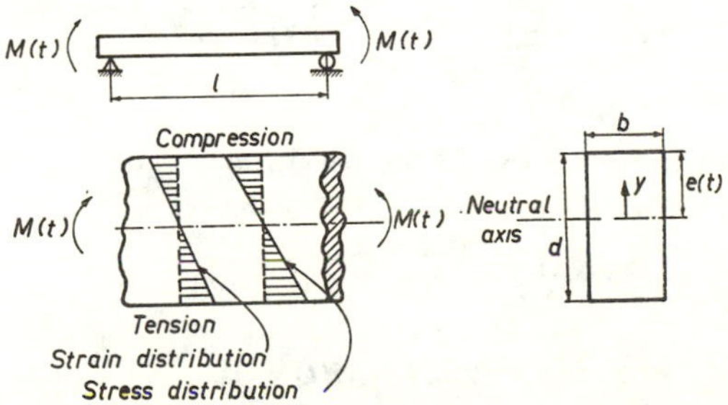

Fig. 6.1. A Linearly Viscoelastic Beam Under Pure Bending.

plane after bending implies that strains are linearly proportional to the distance from the neutral axis whose location is to be determined. Therefore,

$$\varepsilon(t, y) = -\Omega(t)y, \tag{6.1}$$

where ε represents the axial strain, positive for tension and negative for compression, y is the distance measured positively upward from the neutral axis and Ω is the curvature (one/radius of curvature). The negative sign is required so that a positive moment M corresponding to a positive curvature Ω (concave upward is positive) gives compression (negative) strains for positive values of y. Assumptions (b) and (c) imply that

$$\sigma_T(t) = \int_0^t E(t-\xi)\dot{\varepsilon}_T(\xi)\,d\xi \quad \text{or} \quad \varepsilon_T(t) = \int_0^t J(t-\xi)\dot{\sigma}_T(\xi)\,d\xi, \tag{6.2a}$$

$$\sigma_c(t) = \int_0^t E(t-\xi)\dot{\varepsilon}_c(\xi)\,d\xi \quad \text{or} \quad \varepsilon_c(t) = \int_0^t J(t-\xi)\dot{\sigma}_c(\xi)\,d\xi, \tag{6.2b}$$

and

$$E(t) = E_c(t) = E_T(t) \quad \text{or} \quad J(t) = J_c(t) = J_T(t), \tag{6.3}$$

where $E(t)$ and $J(t)$ are the relaxation modulus and creep compliance respectively and subscripts T and c denote tension and compression respectively. In (6.2) integral representations such as (5.103) and (5.105) are used for the linear viscoelastic stress-strain relation. However, the differential operator form such as (5.68) can be used as well. Inserting (6.1) into (6.2) yields

$$\sigma_T(t) = -y\int_0^t E(t-\xi)\dot{\Omega}(\xi)\,d\xi, \quad y < 0, \tag{6.4a}$$

$$\sigma_c(t) = -y\int_0^t E(t-\xi)\dot{\Omega}(\xi)\,d\xi, \quad y > 0. \tag{6.4b}$$

The equilibrium conditions yield the following relations:

$$\iint_A \sigma\,dA = 0, \tag{6.5}$$

$$\iint_A \sigma y\,dA + M(t) = 0. \tag{6.6}$$

Equation (6.5) is the balance of the resultant force of the axial stress on the cross section of the beam and (6.6) is the balance of the internal moment of the axial stress on the cross section of a portion of the beam with the external applied moment $M(t)$ on the same portion. In (6.5), (6.6), A signifies the area of the cross-section. Inserting (6.4a), (6.4b) into (6.5) and (6.6), yields two simultaneous equations for two unknowns, curvature $\Omega(t)$, and position of the neutral axis $e(t)$. For a rectangular beam of width b, depth d, relaxation modulus in compression E_c and in tension E_T these equations are

$$b\int_0^{e(t)'} \sigma_c(t,y)\,dy - b\int_0^{-[d-e(t)]} \sigma_T(t,y)\,dy = 0$$

or

$$b\int_0^{e(t)}\left[-\int_0^t E_c(t-\xi)\dot{\Omega}(\xi)\,d\xi\right]y\,dy - b\int_0^{-[d-e(t)]}\left[-\int_0^t E_T(t-\xi)\dot{\Omega}(\xi)\,d\xi\right]y\,dy = 0 \tag{6.7}$$

and

$$b\int_0^{e(t)}\left[-\int_0^t E_c(t-\xi)\dot{\Omega}(\xi)\,d\xi\right]y^2\,dy - b\int_0^{-[d-e(t)]}\left[-\int_0^t E_T(t-\xi)\dot{\Omega}(\xi)\,d\xi\right]y^2\,dy + M(t) = 0, \tag{6.8}$$

where $e(t)$ represents the position of the neutral axis as measured from the top surface. The assumption (c) given by (6.3) simplifies the solution* since the brackets in the two integrals in (6.7) become identical and can be removed from the integral sign. After integration over y the following relation can be obtained from (6.7)

$$b\left\{-\int_0^t \{E(t-\xi)\dot{\Omega}(\xi)\,d\xi\right\}\left\{\frac{[e(t)]^2}{2}-\frac{[d-e(t)]^2}{2}\right\} = 0, \tag{6.9}$$

which implies that

$$e(t) = \frac{d}{2}. \tag{6.10}$$

* In general for the bilinear material where E_T and E_c are not the same, (6.7) and (6.8) become two coupled differential-integral equations. Thus the solution is very involved.

Thus the position of the neutral axis does not change with time and is at the center (coincides with the axis of the centroid). This would not be true if the tension and compression properties were unequal. Inserting (6.10) into (6.8) with $E_c = E_T = E$ and performing the integration with respect to y yields the following result

$$-\frac{bd^3}{12}\int_0^t E(t-\xi)\dot{\Omega}(\xi)\,\mathrm{d}\xi + M(t) = 0,$$

or

$$\int_0^t E(t-\xi)\dot{\Omega}(\xi)\,\mathrm{d}\xi = \frac{M(t)}{I}, \tag{6.11}$$

where I is the moment of inertia of the cross section. Equation (6.11) is a Volterra integral equation of $\Omega(t)$, which can be readily solved by using the Laplace transform.

The Laplace transform of (6.11) yields

$$\hat{E}(s)s\hat{\Omega}(s) = \hat{M}/I,$$

or

$$\hat{\Omega}(s) = \frac{1}{I}\frac{1}{s\hat{E}(s)}\hat{M}(s). \tag{6.12}$$

Since the Laplace transform of the relaxation modulus $\hat{E}(s)$ and creep compliance $\hat{J}(s)$ are related by (5.110), (6.12) can be rewritten as follows:

$$\hat{\Omega}(s) = \frac{1}{I}\hat{J}(s)s\hat{M}(s). \tag{6.13}$$

Then the convolution theorem, number 29 Table A4.1, yields the inverse Laplace transform of (6.13),

$$\Omega(t) = \frac{1}{I}\int_0^t J(t-\xi)\frac{\partial M(\xi)}{\partial \xi}\,\mathrm{d}\xi. \tag{6.14}$$

For given I and $J(t)$, $\Omega(t)$ can be calculated as a function of the applied external moment $M(t)$.

The strain distribution can be obtained by inserting (6.14) into (6.1)

$$\varepsilon(t, y) = -\frac{y}{I}\int_0^t J(t-\xi)\dot{M}(\xi)\,\mathrm{d}\xi, \tag{6.15}$$

The stress can be obtained by taking the Laplace transform of (6.4) and using No. 29 in Table A4.1. Then substituting (6.13) and applying the inverse Laplace transform yields the following after making use of (5.110),

$$\sigma(t) = -\frac{M(t)y}{I}. \tag{6.16}$$

For small deflections the strain, ε, and deflection, w, are related by the following equation

$$\varepsilon(t, y) = -y\frac{d^2w(t, x)}{dx^2}, \tag{6.17}$$

(see any book on strength of materials).

Inserting (6.15) into (6.17) yields

$$\frac{d^2w(x, t)}{dx^2} = \frac{1}{I}\int_0^t J(t-\xi)\dot{M}(\xi)\, d\xi. \tag{6.18}$$

Thus, the deflection can be calculated from (6.18) for a given set of boundary conditions and given I and $J(t)$. For example, if the beam of Fig. 6.1 is made of Maxwell material (see Table 5.1),

$$J(t) = \frac{1}{R}+\frac{t}{\eta}, \tag{6.19}$$

where R is the spring constant and η is the viscosity of the dashpot in the Maxwell model, and if the external moment is a constant M_0,

$$M(t) = M_0H(t), \tag{6.20}$$

then inserting (6.19) and (6.20) into (6.15), (6.16) and (6.18) yields the following results for strain, stress and deflection,

$$\varepsilon(t, y) = -\frac{M_0}{I}\left(\frac{1}{R}+\frac{t}{\eta}\right)y \tag{6.21}$$

$$\sigma(t, y) = -\frac{M_0 y}{I}, \tag{6.22}$$

and

$$\frac{d^2w(x, t)}{dx^2} = \frac{M_0}{I}\left(\frac{1}{R}+\frac{t}{\eta}\right). \tag{6.23}$$

Integrating (6.23) twice and applying the boundary conditions that $\mathrm{d}w/\mathrm{d}x = 0$ at $x = l/2$ and $w = 0$ at $x = 0$ yields

$$w(x, t) = \frac{M_0}{I}\left(\frac{1}{R}+\frac{t}{\eta}\right)\left[\frac{x^2}{2}-\frac{l}{2}x\right], \tag{6.24}$$

which is the equation for the deflection curve of the beam shown in Fig. 6.1. Equations (6.21) and (6.24) show that the strain and deflection increase continuously with time under a constant bending moment, though the stress distribution remains constant.

Solution of a Beam in Pure Bending Using the Elastic-Viscoelastic Analogy. The equations for stress, strain and deflection of a linearly elastic beam under pure bending can be obtained from any elementary strength of material book as follows:

$$\varepsilon(t, y) = -\frac{M(t)y}{E_e I}, \tag{6.25}$$

$$\sigma(t, y) = -\frac{M(t)y}{I}, \tag{6.26}$$

$$\frac{\mathrm{d}^2 w(x, t)}{\mathrm{d}x^2} = \frac{M(t)}{E_e I}, \tag{6.27}$$

where E_e is the Young's modulus of the material and I is the moment of inertia of the cross-section of the beam with respect to the neutral axis. Note that by letting $\eta = \infty$, $R = E_e$ and $M(t) = M_0$, (6.21, 6.22, 6.23) become identical to (6.25, 6.26, 6.27). Applying the Laplace transform to the elastic solution, (6.25, 6.26, 6:27), first and then replacing E_e (which equals σ/ε) by $s\hat{E}(s)$* or $\hat{\mathrm{Q}}(s)/\hat{\mathrm{P}}(s)$† (which equals $\hat{\sigma}/\hat{\varepsilon}$) yields the following relations

$$\hat{\varepsilon}(s) = -\frac{y}{I}\frac{\hat{M}(s)}{s\hat{E}(s)} = -\frac{y}{I}\hat{\mathrm{J}}(s)s\hat{M}(s), \tag{6.28a}$$

or

$$\hat{\varepsilon}(s) = -\frac{y}{I}\frac{\hat{\mathrm{P}}(s)}{\hat{\mathrm{Q}}(s)}\hat{M}(s), \tag{6.28b}$$

$$\hat{\sigma}(s) = -\frac{y}{I}\hat{M}(s), \tag{6.29}$$

---

* $E(s)$ is the Laplace transform of the relaxation modulus $E(t)$.

† $\hat{\mathrm{Q}}(s)$, $\hat{\mathrm{P}}(s)$ are the Laplace transforms of the operators in (5.68).

$$\frac{d^2\hat{w}(x,s)}{dx^2} = \frac{1}{I}\frac{\hat{M}(s)}{s\hat{E}(s)} = \frac{1}{I}\hat{J}(s)s\hat{M}(s), \tag{6.30a}$$

or

$$\frac{d^2\hat{w}(x,s)}{dx^2} = \frac{1}{I}\frac{\hat{P}(s)}{\hat{Q}(s)}\hat{M}(s). \tag{6.30b}$$

The second form of (6.28a, 6.30a) result from use of (5.110). Inversion of (6.28a, 6.29, 6.30a) yields the following results by using the convolution transform

$$\varepsilon(t) = -\frac{y}{I}\int_0^t J(t-\xi)\dot{M}(\xi)\,d\xi, \tag{6.31}$$

$$\sigma(t) = -\frac{M(t)y}{I}, \tag{6.32}$$

$$\frac{d^2w(x,t)}{dx^2} = \frac{1}{I}\int_0^t J(t-\xi)\dot{M}(\xi)\,d\xi. \tag{6.33}$$

These results, obtained from the elastic-viscoelastic analogy, are identical to (6.15), (6.16), and (6.18) resulting from the direct viscoelastic solution. Although both approaches yield the same result, it is clear that the elastic-viscoelastic (or correspondence) approach is much simpler.

Elastic-Viscoelastic Correspondence Principle (*for beam problems*). Considering the above, the following recipe can be written for viscoelastic beams. The stresses, strains and deflections of a linearly viscoelastic beam, may be found as follows: (a) Solve the corresponding elastic problem for the desired quantities, (b) take the Laplace transform of the elastic solution, (c) replace E_e, the elastic modulus, by $s\hat{E}(s)$ or $\hat{Q}(s)/\hat{P}(s)$, the Laplace transform of the viscoelastic relaxation modulus of the integral representation or of the differential operator representation, respectively, (d) solve for the desired quantities as functions of the Laplace variables and (e) take the inverse Laplace transform. The result will be the solution of the viscoelastic beam problem as a function of time t.

Elastic Beam on Viscoelastic Foundation. In the following, the correspondence principle cited above will be utilized to solve a somewhat more complicated problem, an elastic beam of infinite length on a viscoelastic foundation as shown in Fig. 6.2. This solution is based on Winkler-Zimmerman's

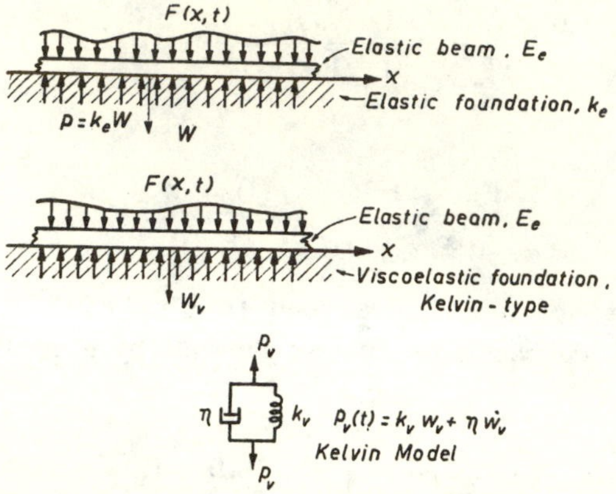

Fig. 6.2. Infinite Elastic Beam on a Viscoelastic Foundation.

assumption that the deflection w at any point is proportional to the foundation pressure p at that point, and does not depend on the pressure at any other point of the foundation. The foundation pressure is the normal force per unit length transmitted across the interface between the beam and supporting material.

(a) *The Associated Elastic Beam on an Elastic Foundation.* Consider an infinite linearly elastic beam of constant cross-section supported on a Winkler-Zimmerman elastic foundation and subjected to a space-periodic load $F(x, t)$. The differential equation for the deflection curve can be obtained from Timoshenko [6.1] as follows:

$$E_eI\frac{\partial^4 w(x, t)}{\partial x^4} = F(x, t)-p(x, t), \tag{6.34}$$

where E_e is the Young's modulus of the beam, I is the moment of inertia of the cross-section of the beam, p is the foundation pressure. The foundation pressure p is related to the deflection w by the resilience modulus k_e of the foundation material, as follows

$$p(x, t) = k_e w(x, t). \tag{6.35}$$

Inserting (6.35) into (6.34), after rearranging, yields

$$\frac{\partial^4 w(x, t)}{\partial x^4}+\frac{k_e}{E_eI}w(x, t) = \frac{A_0(t)}{2E_eI}+\frac{1}{E_eI}\sum_{m=1}^{\infty}\left[A_m(t)\cos mx+B_m(t)\sin mx\right], \tag{6.36}$$

where the space-periodic loading $F(x, t)$ has been represented by a Fourier series

$$F(x, t) = \frac{A_0(t)}{2} + \sum_{m=1}^{\infty} [A_m(t) \cos mx + B_m(t) \sin mx]. \qquad (6.37)$$

The coefficients $A_m(t)$ and $B_m(t)$ can be determined from the known loading by the Fourier coefficients

$$A_m(t) = \frac{1}{\pi} \int_{-\pi}^{\pi} F(x, t) \cos mx \, dx; \quad B_m(t) = \frac{1}{\pi} \int_{-\pi}^{\pi} F(x, t) \sin mx \, dx. \qquad (6.38)$$

The particular solution of (6.36) yields

$$w(x, t) = \frac{A_0(t)}{2k_e} + \sum_{m=1}^{\infty} \left(\frac{1}{E_e I m^4 + k_e} \right) [A_m(t) \cos mx + B_m(t) \sin mx] \qquad (6.39)$$

and $p(x, t) = k_e w(x, t)$

$$= \frac{A_0(t)}{2} + \sum_{m=1}^{\infty} \left(\frac{k_e}{(E_e I m^4 + k_e)} \right) [A_m(t) \cos mx + B_m(t) \sin mx]. \qquad (6.40)$$

(b) *Solution of Beam on Viscoelastic Foundation.* The Laplace transform of (6.39) with respect to the time variable t, is

$$\hat{w}(x, s) = \frac{\hat{A}_0(s)}{2k_e} + \sum_{m=1}^{\infty} \left(\frac{1}{E_e I m^4 + k_e} \right) [\hat{A}_m(s) \cos mx + \hat{B}_m(s) \sin mx.] \qquad (6.41)$$

In (6.41) replace k_e by the corresponding Laplace transformed viscoelastic resilience modulus of the foundation $s\hat{K}_v(s)$. This yields the Laplace transformed deflection curve $\hat{w}_v(s)$ of the elastic beam on a viscoelastic foundation as follows:

$$\hat{w}_v(x, s) = \frac{\hat{A}_0(s)}{2s\hat{K}_v(s)} + \sum_{m=1}^{\infty} \frac{1}{E_e I m^4 + s\hat{K}_v(s)} [\hat{A}_m(s) \cos mx + \hat{B}_m(s) \sin mx]. \qquad (6.42)$$

The actual deflection curve $w_v(x, t)$ is found from the inverse Laplace transform of (6.42). Before this can be determined the viscoelastic nature of the foundation $\hat{K}_v(s)$ and the values of $A_m(t)$ and $B_m(t)$ which describe the nature of the loading must be inserted. For example, if the loading is stationary with respect to time, then A_0, A_m and B_m are constants, so $\hat{A}_0(s) = A_0/s$, for example. If the foundation is a Kelvin type material as shown in the insert of Fig. 6.2, then $s\hat{K}_v(s)$ in (6.42) becomes $s\hat{K}_v(s) = \hat{p}_v(s)/\hat{w}_v(s) = k_v+\eta s$, see Fig. 6.2. In this case, (6.42) becomes

$$\hat{w}_v(x, s) = \frac{A_0/2\eta}{s(s+k_v/\eta)} + \sum_{m=1}^{\infty} \frac{1/\eta}{s[s+(E_eIm^4+k_v)/\eta]}(A_m \cos mx + B_m \sin mx), \tag{6.43}$$

where A_0, A_m, B_m are constants. The inverse Laplace transform of (6.43) yields the following for the deflection curve

$$w_v(x, t) = \frac{A_0}{2k_v}[1-e^{-k_v t/\eta}]$$
$$+\sum_{m=1}^{\infty}\left(\frac{1}{k_v+E_eIm^4}\right)(A_m \cos mx + B_m \sin mx)[1-e^{-(E_eIm^4+k_v)t/\eta}]. \tag{6.44}$$

The foundation pressure, p_v, can be obtained as follows. For a Kelvin material as shown in Fig. 6.2

$$p_v(x, t) = k_v w_v + \eta \dot{w}_v. \tag{6.45}$$

Substituting (6.44) and its time derivative into (6.45) yields

$$p_v(x, t) = \frac{A_0}{2} + \sum_{m=1}^{\infty}\left(\frac{k_v}{k_v+E_eIm^4}\right)(A_m \cos mx + B_m \sin mx) \tag{6.46}$$
$$\left[1+\frac{E_eIm^4}{k_v}e^{-(E_eIm^4+k_v)t/\eta}\right].$$

If the loading varies with time, then the solutions are similar to (6.44) and (6.46) except that the right-hand side of these equations will include convolution integrals involving $A_0(t)$, $A_m(t)$, $B_m(t)$ and the time t; for example, the first term of (6.44) becomes

$$\frac{1}{2k_v}\int_0^t A_0(t-\xi)[e^{-k_v\xi/\eta}]\,d\xi, \tag{6.47}$$

etc., as found by introducing $k_v+\eta s$ for $s\hat{K}_v(s)$ in (6.42) and inverting the first term by using the convolution theorem, Table A4.1.

Direct solutions for an elastic beam on various kinds of viscoelastic foundations have been described by Freudenthal and Lorsch [6.2].

6.3 Stress Analysis of Quasi-static Viscoelastic Problems Using the Elastic-Viscoelastic Correspondence Principle

In the analysis of stress and deformation, three kinds of equations are needed: equilibrium equations, kinematic equations and constitutive equations of the material, plus the boundary conditions. These equations have been discussed in Chapter 4. The first two kinds of equations are the same for elastic as well as for viscoelastic stress analysis problems, provided that in the latter case, the variables are dependent explicitly on time. The main feature which differentiates the viscoelastic from elastic analysis is in the constitutive equations of the material. In the following, the equations used in the solution of stress analysis problems for linear viscoelastic material are formulated briefly. The detailed derivations are omitted as most of them are similar to those described in Chapter 4.

Equations for Linear Viscoelastic Stress Analysis

A) The equations of equilibrium are, from Chapter 4,

$$\frac{\partial \sigma_{ij}(x, t)}{\partial x_i} + F_j(x, t) = 0, \tag{6.48}$$

where $x = x_k$ refers to the location of the element of material considered and t is time. In these equations all the stress components are generally time dependent.

B) The linear viscoelastic stress-strain-time relations are, from (5.120, 5.121),

$$s_{ij}(t) = 2\int_0^t G(t-\xi)\dot{d}_{ij}(\xi)\, \mathrm{d}\xi$$

and

$$\sigma_{ii}(t) = 3\int_0^t K(t-\xi)\dot{\varepsilon}_{ii}(\xi)\, \mathrm{d}\xi; \tag{6.49}$$

or the equivalent forms in the integral representation for creep; or the following for the differential operator form

$$\mathrm{P}_1 s_{ij}(t) = \mathrm{Q}_1 d_{ij}(t),$$

and

$$\mathrm{P}_2 \sigma_{ii}(t) = \mathrm{Q}_2 \varepsilon_{ii}(t). \tag{6.50}$$

The first of equations (6.49) and (6.50) are the deviatoric viscoelastic stress-strain-time relations while the second equations are the dilatational stress-strain-time relations.

C) The strain-displacement relations are, from (3.67)

$$\varepsilon_{ij}(t) = \frac{1}{2}\left[\frac{\partial u_i(t)}{\partial x_j} + \frac{\partial u_j(t)}{\partial x_i}\right]. \tag{6.51}$$

D) The compatibility relations are as given by equation (4.9), or as follows for two dimensions

$$2\frac{\partial^2 \varepsilon_{xy}(t)}{\partial x\,\partial y} = \frac{\partial^2 \varepsilon_{xx}(t)}{\partial y^2} + \frac{\partial^2 \varepsilon_{yy}(t)}{\partial x^2}. \tag{6.52}$$

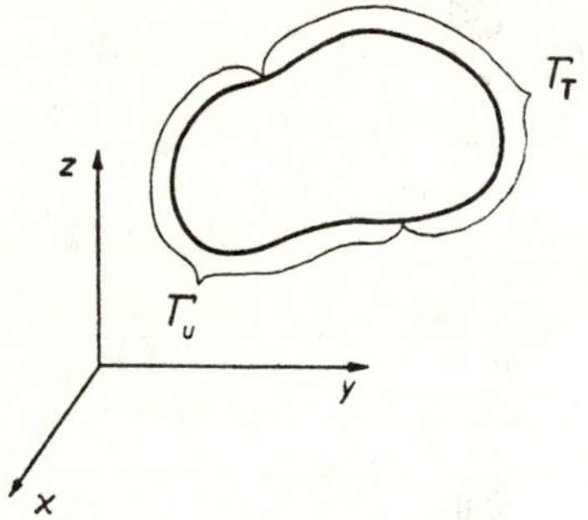

Fig. 6.3. Boundary Conditions.

E) Boundary conditions:

The boundary Γ, as shown in Figure 6.3, is divided into a region Γ_T and a complementary region $\Gamma_u = \Gamma - \Gamma_T$. On Γ_T the external loading (tractions) T_j are prescribed as

$$T_j(x, t) = \sigma_{ij}(x, t)n_i, \tag{6.53}$$

where x denotes the position on Γ_T. On Γ_u, the surface displacement U_j are prescribed as

$$U_j(x, t) = u_j(x, t), \tag{6.54}$$

where x indicates position on Γ_u. Equations (6.48) through (6.54) form a complete set for the solution of linear viscoelastic stress and deformation analysis.

Solution Using the Elastic-Viscoelastic Correspondence Principle. The Laplace transforms of (6.48) through (6.54) yield the following results:

A) $$\frac{\partial \hat{\sigma}_{ij}(x, s)}{\partial x_i} + \hat{F}_j(x, s) = 0, \tag{6.55}$$

B) $$\hat{s}_{ij}(s) = 2s\hat{G}(s)\hat{d}_{ij}(s), \tag{6.56}$$

and

$$\hat{\sigma}_{ii}(s) = 3s\hat{K}(s)\hat{\varepsilon}_{ii}(s), \tag{6.57}$$

or

$$\hat{s}_{ij}(s) = \frac{\hat{Q}_1(s)}{\hat{P}_1(s)}\, \hat{d}_{ij}(s), \tag{6.56a}$$

and

$$\hat{\sigma}_{ii}(s) = \frac{\hat{Q}_2(s)}{\hat{P}_2(s)}\, \hat{\varepsilon}_{ii}(s), \tag{6.57a}$$

C) $$\hat{\varepsilon}_{ij}(s) = \frac{1}{2}\left[\frac{\partial \hat{u}_i(s)}{\partial x_j} + \frac{\partial \hat{u}_j(s)}{\partial x_i}\right], \tag{6.58}$$

D) $$2\frac{\partial^2 \hat{\varepsilon}_{xy}(s)}{\partial x\, \partial y} = \frac{\partial^2 \hat{\varepsilon}_{xx}(s)}{\partial y^2} + \frac{\partial^2 \hat{\varepsilon}_{yy}(s)}{\partial x^2}, \tag{6.59}$$

E) $$\hat{T}_j(x, s) = \hat{\sigma}_{ij}(x, s)n_i \quad \text{on } \Gamma_T, \tag{6.60}$$

$$\hat{U}_j(x, s) = \hat{u}_j(x, s) \quad \text{on } \Gamma_u. \tag{6.61}$$

Comparison of (6.55) through (6.61) with the corresponding linear elasticity equations, (4.3), (4.11), (3.67), (4.8), (4.16), and (4.17), shows that (6.55) through (6.61) describe a fictitious elastic problem in which the "elastic constants" $s\hat{G}(s)$, $s\hat{K}(s)$, body forces $\hat{F}_j(s)$, external forces $\hat{T}_j(s)$ applied on the boundary Γ_T, and the external displacement, $\hat{U}_j(s)$, applied on the boundary Γ_u are functions of the transform parameter s. If this associated elastic problem can be solved to yield $\hat{\sigma}_{ij}(x, s)$ and or $\hat{u}_j(x, s)$ throughout the body, then a subsequent inverse Laplace transformation gives $\sigma_{ij}(x, t)$ and or $u_j(x, t)$ for the viscoelastic stress analysis problem. This is called the "Elastic-Viscoelastic Correspondence Principle."

The limitation of this approach lies in the boundary conditions (6.53, 6.54) or (6.60, 6.61). The correspondence principle is applicable only to those boundary conditions where the interface between the boundary Γ_T on which stress is prescribed and the boundary Γ_u on which displacements are prescribed is independent of time. However, the conditions prescribed on Γ_T and Γ_u themselves can be time dependent. Figure 6.4a, 6.4b, illustrate these two different types of boundary conditions. Figure 6.4a shows a rigid cylinder indenting a semi-infinite viscoelastic body where Γ_T varies but

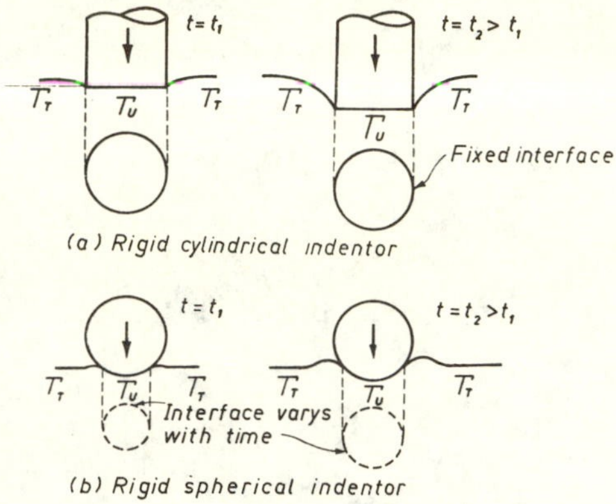

Fig. 6.4. Illustration of Fixed and Variable Interface Between Regions of Prescribed Tractions versus Prescribed Displacements.

where the interface between Γ_T and Γ_u is fixed at all times. Figure 6.4b is a rigid spherical indenter on a semi-infinite viscoelastic body. Here the interface varies as the indentation proceeds.

The use of the correspondence principle is illustrated in the following examples.

6.4 Thick-walled Viscoelastic Tube*

For a linear elastic material it can be shown, see [6.1] for example, that the radial σ_r and tangential σ_θ stresses and radial displacement u for an axially symmetric open-ended tube subjected to internal pressure p_i are as follows:

$$\sigma_r = \frac{p_i a^2}{b^2 - a^2}\left(1 - \frac{b^2}{r^2}\right), \tag{6.62}$$

$$\sigma_\theta = \frac{p_i a^2}{b^2 - a^2}\left(1 + \frac{b^2}{r^2}\right), \tag{6.63}$$

$$u = \frac{p_i a^2}{b^2 - a^2}\,\frac{1+\nu}{E}\left(\frac{b^2}{r} + \frac{1-\nu}{1+\nu}\,r\right), \tag{6.64}$$

* An erroneous solution of this problem has been reported elsewhere.

where a and b are the inside and outside radii respectively of the tube. As shown by (6.62) and (6.63) the stresses in the elastic body are independent of the material constants. Therefore in accordance with the elastic-viscoelastic correspondence principle the stresses will be the same as (6.62, 6.63) in a linearly viscoelastic tube of the same size.

Since the displacement u is a function of material constants E and ν it will be time dependent if the material is viscoelastic. The time dependence may be found by employing the correspondence principle as follows. Taking the Laplace transform of (6.64) and substituting the transformed equivalents of E and ν from (5.91) and (5.92) into (6.64) yields

$$\hat{u}(s) = \frac{a^2}{b^2-a^2}\hat{p}_i\frac{\hat{P}_1}{\hat{Q}_1}\left[\frac{b^2}{r}+\frac{r}{3}\left(\frac{2\hat{P}_2\hat{Q}_1}{\hat{P}_1\hat{Q}_2}+1\right)\right], \tag{6.65}$$

where $\hat{P}_1$, $\hat{Q}_1$; $\hat{P}_2$, $\hat{Q}_2$ are the Laplace transforms of the time operators for deviatoric (5.83) and dilatational (5.84) constitutive equations respectively.

Consider, for example, that the tube is made of material whose behavior is elastic under hydrostatic pressure and Maxwell-type viscoelastic in distortion, namely,

$$\hat{P}_1(s) = 1+\frac{\eta'}{R'}s, \quad \hat{Q}_1(s) = \eta' s, \tag{6.66}$$

$$\hat{P}_2(s) = 1, \quad \hat{Q}_2(s) = 3K, \tag{6.67}$$

where η' and R' are the coefficients of viscosity of the dashpot and the spring constant respectively of the Maxwell model in shear, and K is the bulk modulus; see Table 5.1.

For a suddenly applied internal pressure p, $\hat{p}_i(s) = p/s$. Inserting this and (6.66), (6.67) into (6.65) and rearranging yields the displacement as a function of the transform variable s,

$$\hat{u}(s) = \frac{pa^2}{b^2-a^2}\left(\frac{\mathsf{M}}{s^2}+\frac{\mathsf{N}}{s}\right), \tag{6.68}$$

where

$$\mathsf{M} = \frac{1}{\eta'}\left(\frac{b^2}{r}+\frac{r}{3}\right),$$

$$\mathsf{N} = \frac{1}{R'}\left(\frac{b^2}{r}+\frac{r}{3}\right)+\frac{2r}{9k}.$$

Employing the inverse Laplace transform for each term in the parentheses in (6.68) gives the displacement as a function of time

$$u(t) = \frac{pa^2}{b^2-a^2}(\mathsf{M}t+\mathsf{N}). \tag{6.69}$$

The N-term in (6.69) describes the "instantaneous" displacement produced by the suddenly applied pressure and the M-term describes a constant rate of flow resulting from the fluid nature of the Maxwell model. This method of analysis does not have sufficient accuracy if the deformations become large, due to the assumption of small deformation made in the analysis of the elastic problem.

The radial ε_r and tangential ε_θ strains in a thick tube under internal or external pressure are given by [6.1]

$$\varepsilon_r = \frac{\partial u}{\partial r}, \qquad \varepsilon_\theta = \frac{u}{r}. \tag{6.70}$$

Substituting (6.69) in (6.70) yields the strains as a function of time

$$\varepsilon_r = \frac{pa^2}{b^2-a^2}\left[\left(\frac{t}{\eta'}+\frac{1}{R'}\right)\left(\frac{1}{3}-\frac{b^2}{r^2}\right)+\frac{2}{9K}\right], \tag{6.71}$$

$$\varepsilon_\theta = \frac{pa^2}{b^2-a^2}\left[\left(\frac{t}{\eta'}+\frac{1}{R'}\right)\left(\frac{1}{3}+\frac{b^2}{r^2}\right)+\frac{2}{9K}\right]. \tag{6.72}$$

The stress distribution in the examples given above are independent of time. However, in the following problems it will be observed that the stress distribution varies with time.

Viscoelastic Hollow Cylinder Reinforced with an Elastic Sheath. Woodward and Radok [6.3] analyzed the stress distribution of a viscoelastic hollow cylinder reinforced on the outside by a thin elastic sheath and subjected to constant pressure on the inside. For a linear elastic material (the associated elastic problem) this problem can be solved by applying the following boundary conditions

$$\sigma_r = -p_i, \; r = a, \tag{6.73}$$

$$\sigma_r = \alpha\sigma_\theta, \quad r = b, \tag{6.74}$$

where,

$$\alpha = \frac{(1-\nu^2)}{\nu(1+\nu)-(1-\nu_R^2)\dfrac{bE}{hE_R}}, \tag{6.74a}$$

to the general solution for the axially symmetric thick tube under plane strain

$$\sigma_r = A - \frac{B}{r^2}, \tag{6.75}$$

$$\sigma_\theta = A + \frac{B}{r^2}, \tag{6.76}$$

$$\sigma_z = \nu(\sigma_r + \sigma_\theta), \tag{6.77}$$

$$u = r\varepsilon_\theta = \frac{r}{E}\left[\sigma_\theta - \nu(\sigma_r + \sigma_z)\right] = \frac{r(1-\nu^2)}{E}\left[\sigma_\theta - \frac{\nu}{1-\nu}\sigma_r\right], \tag{6.78}$$

to solve for A and B.

In (6.74a) E and ν are the Young's modulus and Poisson's ratio of the cylinder, E_R and ν_R are the Young's modulus and Poisson's ratio of the elastic sheath and h is the thickness of the sheath, which is assumed to be very small in comparison with the outside radius of the cylinder b.

Equation (6.74) was derived from compatibility of strains by equating the circumferential strain ε_θ^R in the sheath under plane strain,

$$\varepsilon_\theta^R = (1-\nu_R^2)F_\theta/hE_R,$$

to the circumferential strain ε_θ in the cylinder, under plane strain, at $r = b$

$$\varepsilon_\theta = \frac{1-\nu^2}{E}\left[\sigma_\theta - \frac{\nu}{1-\nu}\sigma_r\right],$$

where F_θ and h are the tangential force per unit length and thickness of the sheath, respectively. The derivation also required applying the condition of equilibrium to the sheath. From elementary theory of thin tubes subjected to internal pressure, $-\sigma_r$,

$$F_\theta = -b\sigma_r.$$

Employing (6.73), (6.74) and (6.74a) in (6.75) and (6.76) the radial (and tangential) stresses for elastic materials are found to be

$$\sigma_r \text{ (and } \sigma_\theta) = \mp p_i \left[\frac{\alpha\left(\frac{b^2}{r^2} \pm 1\right) - \left(\frac{b^2}{r^2} \mp 1\right)}{\alpha\left(\frac{b^2}{a^2} + 1\right) - \left(\frac{b^2}{a^2} - 1\right)}\right] \tag{6.79}$$

In these two equations, stresses are dependent on the material properties of both the cylinder and the sheath since α as given by (6.74a) is dependent on those properties.

To obtain the solution of the viscoelastic cylinder bounded by a thin elastic sheath, the Laplace transform is first applied to the elastic solution. In the following, only σ_θ will be considered as an example. The transform of σ_θ as given by (6.79) is

$$\hat{\sigma}_\theta = \hat{p}_i(s) \left[\frac{\hat{\alpha}(s)\left(\frac{b^2}{r^2}-1\right)-\left(\frac{b^2}{r^2}+1\right)}{\hat{\alpha}(s)\left(\frac{b^2}{a^2}+1\right)-\left(\frac{b^2}{a^2}-1\right)} \right], \tag{6.80}$$

where $\hat{p}_i(s)$ is the Laplace transform of p_i which in general is time dependent, and $\hat{\alpha}$ is the Laplace transform of α, (6.74a). $\hat{\alpha}(s)$ in (6.80) can be obtained by replacing E and ν in (6.74a) by (5.91) and (5.92) respectively and rearranging as follows

$$\hat{\alpha}(s) = \left(\frac{1}{\hat{P}_2}+\frac{2}{\hat{P}_1}\frac{\hat{Q}_1}{\hat{Q}_2}\right) \Big/ \left[\frac{1}{\hat{P}_2}-\frac{1}{\hat{P}_1}\frac{\hat{Q}_1}{\hat{Q}_2}-\frac{\hat{Q}_1}{\hat{P}_1}\left(\frac{2}{\hat{P}_2}+\frac{1}{\hat{P}_1}\frac{\hat{Q}_1}{\hat{Q}_2}\right)\frac{(1-\nu_R^2)b}{hE_R}\right], \tag{6.81}$$

where $\hat{P}_1$, $\hat{P}_2$, $\hat{Q}_1$, $\hat{Q}_2$, are functions of s.

The inverse Laplace transform of (6.80) yields the tangential stress σ_θ. In general, the inversion is very cumbersome and only in certain restricted cases is the exact inversion convenient. For example, if the material of the tube is assumed to have incompressible behavior under hydrostatic pressure and Maxwell-like behavior in distorsion, the transformed P and Q operators have the same form as (6.66, 6.67) except that for incompressible material $\hat{Q}_2 = \infty$. Substituting these values into (6.81) yields

$$\hat{\alpha}(s) = 1 \Big/ \left\{1-\frac{2\eta' s}{(1+\eta' s/R')}\,\frac{(1-\nu_R^2)b}{hE_R}\right\}. \tag{6.82}$$

For an abruptly applied constant internal pressure p, $\hat{p}_i(s) = p/s$. Introducing p/s and (6.82) into (6.80) and rearranging terms

$$\hat{\sigma}_\theta(s) = -p\,\frac{B}{A}\left[\frac{A}{s(s+B)}+\frac{1}{s+B}\right], \tag{6.83}$$

where

$$A = \eta'\left[\frac{1}{R'}-\frac{(1-\nu_R^2)}{E_R}\,\frac{b}{h}\left(\frac{b^2}{r^2}+1\right)\right]^{-1},$$

$$B = \eta'\left[\frac{1}{R'}+\frac{(1-\nu_R^2)}{E_R}\,\frac{h}{b}\left(\frac{b^2}{a^2}-1\right)\right]^{-1}.$$

Taking the inverse Laplace transform of the two terms in (6.83) and collecting terms yields the stress σ_θ as a function of time

$$\sigma_\theta(t) = -p\left[1+\left(\frac{B}{A}-1\right)e^{-Bt}\right]. \tag{6.84}$$

Equation (6.84) says that immediately after loading $\sigma_\theta(0+)$ is given by

$$\sigma_\theta(0+) = -p\frac{B}{A} = -p\,\frac{C-\left(\dfrac{b^2}{r^2}+1\right)}{C+\left(\dfrac{b^2}{a^2}-1\right)}, \tag{6.85}$$

where

$$C = E_R h/R'b(1-\nu_R^2).$$

As expected the immediate response to an abrupt load is independent of η' but dependent on both R' and E_R. If E_R is taken to be zero then C is zero and (6.85) reduces to (6.63). The response given by (6.85) is followed by a stress distribution which changes at each material point in an exponential manner. After a sufficiently long time, $\sigma_\theta(\infty) = -p$ everywhere, independent of r, and the full pressure is transmitted to the reinforcing sheath. This characteristic results from the fluid nature of the Maxwell model. Equation (6.84) is plotted in Fig. 6.5 for a particular set of the constants for the cylinder and reinforcement.

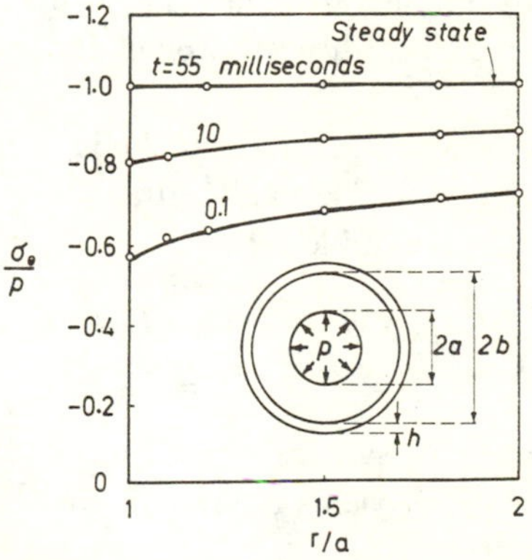

Fig. 6.5. Circumferential Stress Distribution in a Hollow Viscoelastic Cylinder with an Elastic Sheath Subject to a Suddenly Applied Constant Pressure. The cylinder was made of incompressible Maxwell material for which the relaxation time was 10 milliseconds and the elastic constant was 10^5 psi. The sheath was 1/8 in. steel and the inner and outer diameters of the cylinder were 4 and 8 in. respectively. From [6.3], courtesy Brown University.

6.5 Point Force Acting on the Surface of a Semi-infinite Viscoelastic Solid

In this example, due to Lee [6.4] point force F applied on the surface of a semi-infinite viscoelastic body will be considered (Fig. 6.6). The elastic solution of this problem yields the following for the radial stress σ_r in

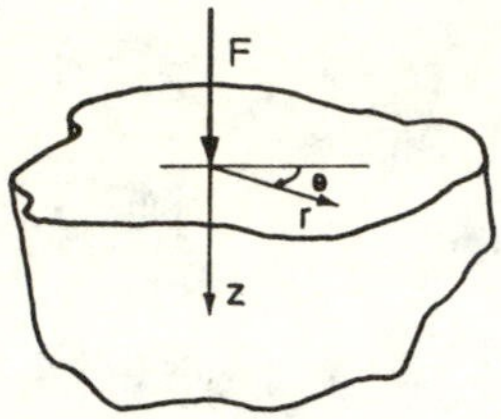

Fig. 6.6. Point Force on the Surface of a Semi-infinite Viscoelastic Body.

cylindrical coordinates r, θ, z

$$\sigma_r = \frac{F}{2\pi}[(1-2\nu)A-B], \tag{6.86}$$

where

$$A = \frac{1}{r^2}-\frac{z}{r^2}(r^2+z^2)^{-1/2}, \quad B = 3r^2z(r^2+z^2)^{-5/2}.$$

The viscoelastic body in this example will be taken to have Kelvin-like behavior in distorsion and elastic behavior under hydrostatic compression, (see Table 5.1).

The only elastic constant appearing in (6.86) is ν for which the viscoelastic equivalent is given by (5.92). The Laplace transforms of P_1, Q_1, P_2, Q_2, for the chosen material are, see Table 5.1,

$$\hat{P}_1(s) = 1, \quad \hat{Q}_1(s) = R'+\eta's,$$
$$\hat{P}_2(s) = 1, \quad \hat{Q}_2(s) = 3K.$$

Introducing these into the Laplace transform of (5.92) and then into the Laplace transform of (6.86) yields the radial stress in terms of the Laplace transform variable s

$$\hat{\sigma}_r = \frac{F}{2\pi s}\left[\frac{3(R'+\eta's)}{6K+R'+\eta's}A-B\right], \tag{6.87}$$

where it has been assumed that a constant force F was applied for which $\hat{F}(s) = F/s$.

Equation (6.87) may be rearranged as follows

$$\hat{\sigma}_r = \frac{F}{2\pi}\left[\frac{3R'A/\eta'}{s\left(s+\dfrac{6K+R'}{\eta'}\right)}+\frac{3A}{s+\dfrac{6K+R'}{\eta'}}-\frac{B}{s}\right], \tag{6.88}$$

and then the inverse transformation may be applied term by term. After rearranging following inverse transformation the radial stress as a function of time is found to be

$$\sigma_r(t) = \frac{F}{2\pi}\left[\left(\frac{3R'A}{6K+R'}-B\right)+\left(\frac{18KA}{6K+R'}\right)\exp\left(-\frac{6K+R'}{\eta'}t\right)\right]. \tag{6.89}$$

Equation (6.89) indicates that immediately after loading $\sigma_r(0+)$ is given by

$$\sigma_r(0+) = (F/2\pi)(3A-B). \tag{6.90}$$

Comparing (6.90) with (6.86) implies that Poisson's ratio $\nu(0+) = -1$. Poisson's ratio is given by, see Table 4.1,

$$\nu = \frac{(3K/G)-2}{(6K/G)+2}\,. \tag{6.91}$$

For a load applied abruptly at $t = 0+$ the shear modulus G (relation between deviatoric stress and deviatoric strain) of a Kelvin model approaches infinity while the elastic volumetric modulus K is unchanged. Thus, allowing G to approach infinity in (6.91) yields $\nu = -1$ in agreement with (6.90).

The second term of (6.89) describes a gradual decay which approaches zero as t approaches infinity. Hence the stress $\sigma_r(\infty)$ after sufficient time is given by

$$\sigma_r(\infty) = \frac{F}{2\pi}\left[\frac{3R'}{6K+R'}A-B\right]. \tag{6.92}$$

This problem has been extended by Lee [6.4] to a situation in which the point force moves along a straight line. The superposition principle was employed in this solution.

All the above problems have been treated as quasi-static problems by ignoring inertia terms even though loads or pressures have been applied abruptly. Thus, in real situations where the time span of interest is short, the neglected transient response may be very significant.

6.6 Concluding Remarks

Although general principles for solving linear viscoelastic stress analysis problems have been formulated, the detailed solution technique has been focused on using the elastic-viscoelastic correspondence principle. This method is by far the simplest method though, as mentioned before, not all linear viscoelastic problems can be solved by this method.

There is a large class of problems where the interface of the stress and displacement boundaries in the mixed boundary value problem is not constant; thus the elastic-viscoelastic correspondence principle is not applicable. In this case, the stress analysis problem becomes more involved; and, in general, there is no unified approach to this type of problem. The solution of this type of problem is not considered here. For further study of this subject, references [6.5, 6.6, 6.7, 6.8] may be consulted.

The transitory thermo-viscoelastic problem is not considered in this chapter. In general, the response of viscoelastic materials is very sensitive to the temperature variation encountered in many engineering applications. Thermal effects interact with the stress distribution whether the stress distribution is due to external force or residual stress. Solution of some thermo-viscoelastic stress analysis problems which are of practical importance may be found in references [6.9, 6.10, 6.11, 6.12].

CHAPTER 7

MULTIPLE INTEGRAL REPRESENTATION

7.1 Introduction

In Chapter 5, the Boltzmann superposition principle was employed to derive a constitutive equation for linear viscoelastic materials under uniaxial stressing and later extended to the case of multi-axial stressing condition. However, when the stress-strain relation of a given material is nonlinear, the Boltzmann superposition principle is not applicable; so a constitutive equation to describe the nonlinear viscoelastic behavior has to be sought by other means. A formal approach to the subject of nonlinear viscoelastic constitutive equations may be derived from continuum mechanics, though it is somewhat involved (see Appendix A2). Instead, a different approach is employed in this chapter to derive a nonlinear viscoelastic constitutive equation which in a sense is an extension of the linear superposition to the nonlinear range. It is hoped that this approach will allow the reader without a background in continuum mechanics to gain an insight into nonlinear viscoelastic constitutive relations without being frustrated by the mathematical complexity of the formal derivation. For the sake of simplifying the discussion, the results obtained in the rest of this chapter are based on an assumption of small deformation; that is, the dependence of strain on the stress history, and vice versa, is nonlinear even for small deformation. A nonlinear viscoelastic constitutive relation under uniaxial and multi-axial loadings will be derived. The derivation though less rigorous, has results essentially identical to that obtained from the formal rigorous continuum mechanics approach included in Appendix A2.

The reader may find that the complete formulation is too complicated for use in many stress analysis situations. Also the determination of the material parameters is involved. Therefore, some simplification of the constitutive equation is desirable in order to obtain a practically useful theory. At the present time, it seems that there are two possible approaches. One is to simplify the constitutive equation by restricting the types of situations to which they are applicable, such as restricting consideration to incompressibility, restricted stress or strain states or restricted loading conditions. The other approach is simplification of the material functions.

In this chapter, certain simplifications in the first category will be discussed including linearly compressible and incompressible behavior, short time approximations and the superposition of small time dependent deformation on large deformation. In subsequent chapters, constitutive relations under restricted stress and strain states, and simplifications in the material functions will be discussed.

Other means of introducing nonlinearity so as to retain only single integral forms are also considered. Three types of nonlinear single integral constitutive equations are described at the end of this chapter.

7.2 Nonlinear Viscoelastic Behavior under Uniaxial Loading

Consider a nonlinear viscoelastic material subjected to a constant stress applied at time $t = 0$ as shown in Figure 7.1a. The time-dependent strain

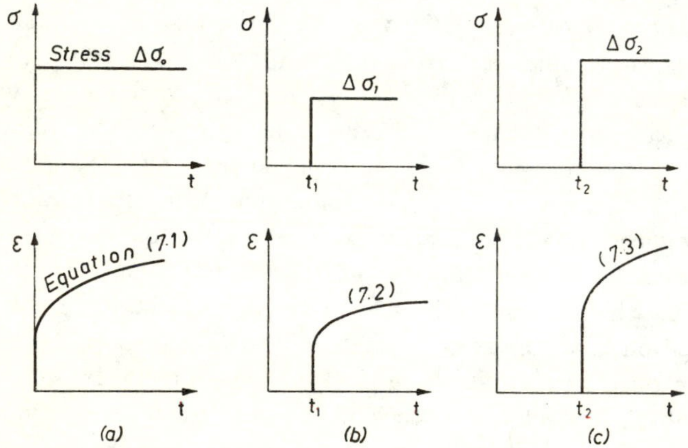

Fig. 7.1. Creep Response to Stresses Applied at Different Times.

resulting from this nonlinear constant stress $\Delta\sigma_0$ may be represented by a polynomial such as the following:

$$\varepsilon_0(t) = (\Delta\sigma_0)\,\varphi_1(t)+(\Delta\sigma_0)^2\,\varphi_2(t)+(\Delta\sigma_0)^3\,\varphi_3(t)+\ \ldots\ . \tag{7.1}$$

In (7.1), φ_1, φ_2 and φ_3 are time-dependent material functions. The accuracy of description of nonlinear behavior can be improved, in principle, by including more terms in (7.1). However, for the sake of clarity in demonstrating the basic concept, the remainder of this section will consider only the first three terms of the polynomial in (7.1). There is no difficulty in including higher order terms, if needed. Also, experiments have shown three terms to be adequate for plastics investigated to date [7.1, 7.2, 7.3] for example.

In the cases shown in Figure 7.1b and 7.1c, where the stresses $\Delta\sigma_1$ and $\Delta\sigma_2$ are applied at the instance $t = t_1$ and $t = t_2$ respectively, the corresponding time-dependent strain response in both cases can be obtained, according to (7.1) as follows:

$$\varepsilon_1(t) = (\Delta\sigma_1)\,\varphi_1(t-t_1)+(\Delta\sigma_1)^2\,\varphi_2(t-t_1)+(\Delta\sigma_1)^3\,\varphi_3(t-t_1),\quad t \geqslant t_1, \tag{7.2}$$

$$\varepsilon_1(t) = 0 \quad \text{for} \quad t < t_1, \tag{7.2a}$$

$$\varepsilon_2(t) = (\Delta\sigma_2)\,\varphi_1(t-t_2)+(\Delta\sigma_2)^2\,\varphi_2(t-t_2)+(\Delta\sigma_2)^3\,\varphi_3(t-t_2),\quad t \geqslant t_2, \tag{7.3}$$

$$\varepsilon_2(t) = 0 \quad \text{for} \quad t < t_2. \tag{7.3a}$$

In (7.2) and (7.3) the time-dependent functions φ_1, φ_2 and φ_3 are identical to those in (7.1) provided the time argument t is replaced by $(t-t_1)$ and $(t-t_2)$ respectively.

In other words, it has been assumed that the material response to the application of stress is exactly the same at any instant of time. This assumption is applicable to most engineering materials at least for short time spans. However, there are many exceptions, Portland cement concrete responds to external stress differently at different curing periods. Materials of this kind are called "aging materials". Most biological materials belong to this category.

Now consider a multi-step stressing in the nonlinear range as shown in Figure 7.2. This can be considered as the superposition of the stresses of Figure 7.1a, b, and c. The time-dependent strain at the first step, between $t = 0$ and $t = t_1$, is the same as (7.1). However, at the second step, although the stress can be considered as the sum of that shown in Figure 7.1a and 7.1b, the strain is composed not only of the sum of all terms in (7.1) and (7.2), but also must include all the possible cross products to account for nonlinearity, as follows:

$$\begin{aligned}\varepsilon(t) = {} & (\Delta\sigma_0)\varphi_1\,(t)+(\Delta\sigma_0)^2\,\varphi_2(t,\,t)+(\Delta\sigma_0)^3\,\varphi_3(t,\,t,\,t)\\ & +(\Delta\sigma_1)\,\varphi_1(t-t_1)+(\Delta\sigma_1)^2\,\varphi_2(t-t_1,\,t-t_1)+(\Delta\sigma_1)^3\,\varphi_3(t-t_1,\,t-t_1,\\ & \quad t-t_1)+2(\Delta\sigma_0)\,(\Delta\sigma_1)\,\varphi_2(t,\,t-t_1)+3(\Delta\sigma_0)^2\,(\Delta\sigma_1)\,\varphi_3(t,\,t,\,t-t_1)\\ & +3(\Delta\sigma_0)\,(\Delta\sigma_1)^2\,\varphi_3(t,\,t-t_1,\,t-t_1),\quad t_1 < t \leqslant t_2.\end{aligned} \tag{7.4}$$

This equation needs further clarification. The time function $\varphi_2(t)$ in (7.1) has become $\varphi_2(t, t)$ in (7.4). This can be interpreted as follows. The term $(\Delta\sigma_0)^2$ may be considered as the separate action of $(\Delta\sigma_0)$ and $(\Delta\sigma_0)$, each term having its effect upon φ_2. Therefore two time parameters, one for each stress, are utilized in the time function $\varphi_2(t, t)$.

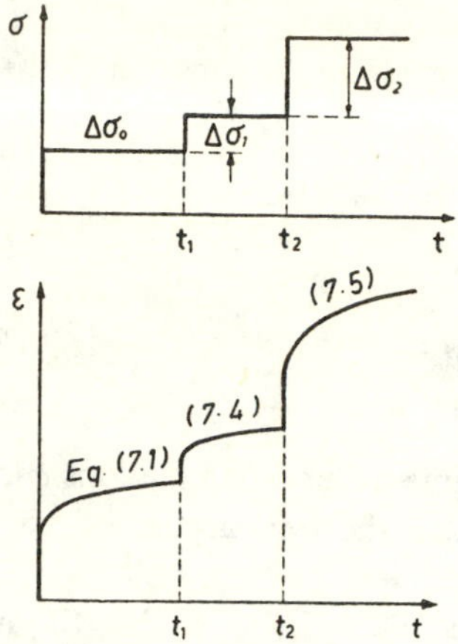

Fig. 7.2. Creep Response to Multiple Steps of Stress.

A similar explanation applies to $\varphi_3(t, t, t)$, $\varphi_2(t-t_1, t-t_1)$ and $\varphi_3(t-t_1, t-t_1, t-t_1)$. The concept becomes especially important in terms involving products of unlike stress increments such as $(\Delta\sigma_0)(\Delta\sigma_1)\varphi_2(t, t-t_1)$. This term $(\Delta\sigma_0)(\Delta\sigma_1)\varphi_2(t, t-t_1)$ can be interpreted as the additional contribution to the time dependent strain in the nonlinear range due to the cross effect of $\Delta\sigma_0$ applied at $t = 0$ and $\Delta\sigma_1$ applied at $t = t_1$. Also, in this term, it has been assumed that the cross effect of $\Delta\sigma_0$ upon $\Delta\sigma_1$ is identical to the cross effect of $\Delta\sigma_1$ upon $\Delta\sigma_0$; that is, $\varphi_2(t, t-t_1) = \varphi_2(t-t_1, t)$. Hence, these two terms have been summed resulting in the factor 2, and φ_2 is called symmetric with respect to its time arguments. A similar explanation applies to the last two terms of (7.4), where there are now three cross effect terms which are assumed identical and summed. At the third step of loading the strain response can be obtained as follows in a similar manner to (7.4):

$$
\begin{aligned}
\varepsilon(t) = {} & (\Delta\sigma_0)\,\varphi_1(t)+(\Delta\sigma_0)^2\,\varphi_2(t, t)+(\Delta\sigma_0)^3\,\varphi_3(t, t, t)+(\Delta\sigma_1)\,\varphi_1(t-t_1) \\
& +(\Delta\sigma_1)^2\,\varphi_2(t-t_1, t-t_1)+(\Delta\sigma_1)^3\,\varphi_3(t-t_1, t-t_1, t-t_1) \\
& +(\Delta\sigma_2)\,\varphi_1(t-t_2)+(\Delta\sigma_2)^2\,\varphi_2(t-t_2, t-t_2)+(\Delta\sigma_2)^3\,\varphi_3(t-t_2, t-t_2, \\
& \quad t-t_2)+2(\Delta\sigma_0)(\Delta\sigma_1)\,\varphi_2(t, t-t_1)+2(\Delta\sigma_0)(\Delta\sigma_2)\,\varphi_2(t, t-t_2)
\end{aligned}
$$

$$+2(\Delta\sigma_1)(\Delta\sigma_2)\varphi_2(t-t_1, t-t_2)+3(\Delta\sigma_0)^2(\Delta\sigma_1)\varphi_3(t, t, t-t_1)+3(\Delta\sigma_0)(\Delta\sigma_1)^2\varphi_3(t, t-t_1, t-t_1)+3(\Delta\sigma_0)^2(\Delta\sigma_2)\varphi_3(t, t, t-t_2)+3(\Delta\sigma_0)(\Delta\sigma_2)^2\varphi_3(t, t-t_2, t-t_2)+3(\Delta\sigma_1)^2(\Delta\sigma_2)\varphi_3(t-t_1, t-t_1, t-t_2)+3(\Delta\sigma_1)(\Delta\sigma_2)^2\varphi_3(t-t_1, t-t_2, t-t_2)+6(\Delta\sigma_0)(\Delta\sigma_1)(\Delta\sigma_2)\varphi_3(t, t-t_1, t-t_2), \quad t \geqslant t_2. \tag{7.5}$$

The factor 6 for $\varphi_3(t, t-t_1, t-t_2)$ results from the fact that there are six different possible cross effects resulting from the three different time parameters, i.e., $\varphi_3^1(t, t-t_1, t-t_2)$, $\varphi_3^2(t-t_1, t, t-t_2)$, $\ldots$, $\varphi_3^6(t-t_2, t-t_1, t)$. Assuming φ_3 to be symmetric means that all 6 are identical and hence, may be summed.

The results obtained so far can be expanded in a straightforward manner to N steps of stress as follows:

$$\begin{aligned}\varepsilon(t) &= \sum_{i=0}^{N}(\Delta\sigma_i)\varphi_1(t-t_i)\\ &+\sum_{i=0}^{N}\sum_{j=0}^{N}(\Delta\sigma_i)(\Delta\sigma_j)\varphi_2(t-t_i, t-t_j)\\ &+\sum_{i=0}^{N}\sum_{j=0}^{N}\sum_{k=0}^{N}(\Delta\sigma_i)(\Delta\sigma_j)(\Delta\sigma_k)\varphi_3(t-t_i, t-t_j, t-t_k).\end{aligned} \tag{7.6}$$

If φ_2 and φ_3 vanish for all time arguments, (7.6) becomes the Boltzmann linear superposition principle. Equation (7.6) can be considered, in a sense, as an extension of the linear superposition principle by adding the nonlinear effect. It is also to be noted from (7.1), (7.4) and (7.5) that a linear response of a material under a constant stress (namely $\varphi_2(t, t)$ and $\varphi_3(t, t, t)$ equal zero) is not a sufficient proof that the material is a linear viscoelastic material. It still may be a nonlinear viscoelastic material if terms like $\varphi_2(t, t-t_1)$ etc., still exist.

Arbitrarily varying stress history can be considered as a limiting case of (7.6) consisting of an infinite number of infinitesimal stepwise stress inputs. Thus the sums of (7.6) become integrals as follows:

$$\begin{aligned}\varepsilon(t) &= \int_0^t \varphi_1(t-\xi_1)\dot{\sigma}(\xi_1)\,d\xi_1\\ &+\int_0^t\int_0^t \varphi_2(t-\xi_1, t-\xi_2)\dot{\sigma}(\xi_1)\dot{\sigma}(\xi_2)\,d\xi_1\,d\xi_2\\ &+\int_0^t\int_0^t\int_0^t \varphi_3(t-\xi_1, t-\xi_2, t-\xi_3)\dot{\sigma}(\xi_1)\dot{\sigma}(\xi_2)\dot{\sigma}(\xi_3)\,d\xi_1\,d\xi_2\,d\xi_3,\end{aligned} \tag{7.7}$$

where $\dot{\sigma}(\xi)\,d\xi = [\partial\sigma(\xi)/\partial\xi]\,d\xi$ is introduced instead of $d\sigma(\xi)$ so that time t becomes the independent variable.

The linear kernel function $\varphi_1(t)$ of (7.7) is described in terms of a single time parameter t. Thus $\varphi_1(t)$ versus t may be illustrated by a curve such as shown in Figure 7.3. However, the nonlinear kernel functions such as φ_2 and φ_3 of (7.7) require more than one time parameter for their description.

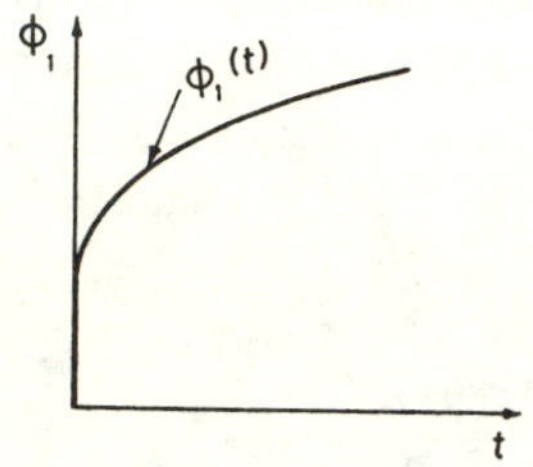

Fig. 7.3. Linear Kernel Function vs. Time.

For example, the second order kernel function $\varphi_2(t, t-\xi_1)$ is described in terms of two time parameters t and $t-\xi_1$. The variation of $\varphi_2(t, t-\xi_1)$ with these two time parameters may be illustrated by a surface A as shown in Figure 7.4 with coordinates φ_2, t and t.

In Figure 7.4 the line B in surface A describes the ordinate $\varphi_2(t, t)$ versus time t when the two time parameters are equal. Hence, line B lies in a plane which bisects the angle between the coordinates t and t. Line C in surface A of Figure 7.4 represents $\varphi_2(t, t-\xi_1)$ versus time t when the two time

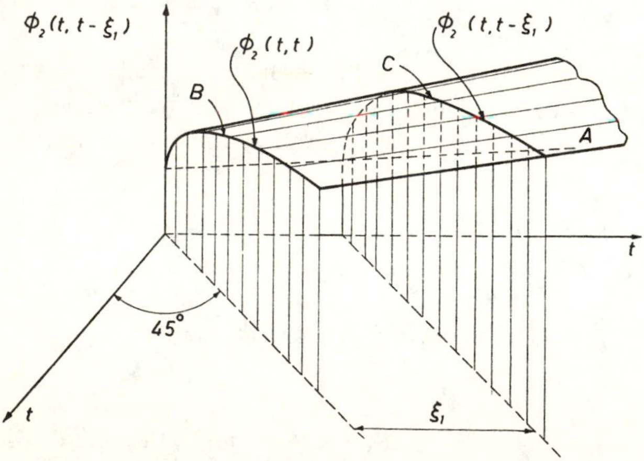

Fig. 7.4. Second Order Kernel Functions vs. Two Time Parameters.

parameters differ by a particular value ξ_1. Thus line C lies in a plane parallel to the plane of line B. A set of lines like C in parallel planes may be found by using a set of values of ξ_1, thus defining surface A. This suggests an experimental method of determining $\varphi_2(t, t-\xi_1)$. A set of creep experiments may be performed using two stresses applied at different times, one at $t = 0$ and the other at $t = t-\xi_1$, where ξ_1 is different for each experiment of the set. Thus the results of a set of such experiments all using the same stresses but suitably chosen values of ξ_1 will yield information from which $\varphi_2(t, t-\xi_1)$ may be obtained. This method will be further discussed in Chapter 9.

Since the kernel functions are considered symmetrical with respect to their time parameters,

$$\varphi_2(t, t-\xi_1) = \varphi_2(t-\xi_1, t).$$

Thus, surface A is symmetrical with respect to the plane defined by line B.

A third order kernel function does not yield to meaningful pictorial representation.

7.3 Nonlinear Viscoelastic Behavior under a Multiaxial Stress State

The representation of the uniaxial nonlinear constitutive relation derived in the previous section can be extended to a multi-axial stress state by using similar concepts to those employed in deriving the multi-axial constitutive relation for linear viscoelastic theory. A similar approach has been used elsewhere as discussed in Chapter 2. By considering all the possible combinations of stress tensors and stress invariants (written in terms of traces of stress tensors, see Chapter 3), all possible terms through the third order are shown as follows for isotropic material:

First order terms:

$$\mathbf{I}\int_0^t Q_1(t-\xi)\,\mathrm{tr}\dot{\boldsymbol{\sigma}}(\xi)\,d\xi, \quad \int_0^t S_1(t-\xi)\,\dot{\boldsymbol{\sigma}}(\xi)\,d\xi; \tag{7.8a}$$

Second order terms:

$$\mathbf{I}\int_0^t\int_0^t Q_2(t-\xi_1, t-\xi_2)\,\mathrm{tr}\dot{\boldsymbol{\sigma}}(\xi_1)\,\mathrm{tr}\dot{\boldsymbol{\sigma}}(\xi_2)\,d\xi_1\,d\xi_2,$$

$$\mathbf{I}\int_0^t\int_0^t Q_3(t-\xi_1, t-\xi_2)\,\mathrm{tr}[\dot{\boldsymbol{\sigma}}(\xi_1)\,\dot{\boldsymbol{\sigma}}(\xi_2)]\,d\xi_1\,d\xi_2,$$

$$\int_0^t \int_0^t S_2(t-\xi_1, t-\xi_2) \operatorname{tr}\dot{\boldsymbol{\sigma}}(\xi_1)\, \dot{\boldsymbol{\sigma}}(\xi_2)\, d\xi_1\, d\xi_2,$$

$$\int_0^t \int_0^t S_3(t-\xi_1, t-\xi_2)\, \dot{\boldsymbol{\sigma}}(\xi_1)\, \dot{\boldsymbol{\sigma}}(\xi_2)\, d\xi_1\, d\xi_2; \tag{7.8b}$$

Third order terms:

$$\mathbf{I} \int_0^t \int_0^t \int_0^t Q_4(t-\xi_1, t-\xi_2, t-\xi_3) \operatorname{tr}\dot{\boldsymbol{\sigma}}(\xi_1) \operatorname{tr}\dot{\boldsymbol{\sigma}}(\xi_2) \operatorname{tr}\dot{\boldsymbol{\sigma}}(\xi_3)\, d\xi_1\, d\xi_2\, d\xi_3,$$

$$\mathbf{I} \int_0^t \int_0^t \int_0^t Q_5(t-\xi_1, t-\xi_2, t-\xi_3) \operatorname{tr}[\dot{\boldsymbol{\sigma}}(\xi_1)\dot{\boldsymbol{\sigma}}(\xi_2)] \operatorname{tr}\dot{\boldsymbol{\sigma}}(\xi_3)\, d\xi_1\, d\xi_2\, d\xi_3,$$

$$\mathbf{I} \int_0^t \int_0^t \int_0^t Q_6(t-\xi_1, t-\xi_2, t-\xi_3) \operatorname{tr}[\dot{\boldsymbol{\sigma}}(\xi_1)\, \dot{\boldsymbol{\sigma}}(\xi_2)\, \dot{\boldsymbol{\sigma}}(\xi_3)]\, d\xi_1\, d\xi_2\, d\xi_3,$$

$$\int_0^t \int_0^t \int_0^t S_4(t-\xi_1, t-\xi_2, t-\xi_3) \operatorname{tr}\dot{\boldsymbol{\sigma}}(\xi_1) \operatorname{tr}\dot{\boldsymbol{\sigma}}(\xi_2)\, \dot{\boldsymbol{\sigma}}(\xi_3)\, d\xi_1\, d\xi_2\, d\xi_3,$$

$$\int_0^t \int_0^t \int_0^t S_5(t-\xi_1, t-\xi_2, t-\xi_3) \operatorname{tr}[\dot{\boldsymbol{\sigma}}(\xi_1)\, \dot{\boldsymbol{\sigma}}(\xi_2)]\, \dot{\boldsymbol{\sigma}}(\xi_3)\, d\xi_1\, d\xi_2\, d\xi_3,$$

$$\int_0^t \int_0^t \int_0^t S_6(t-\xi_1, t-\xi_2, t-\xi_3) \operatorname{tr}\dot{\boldsymbol{\sigma}}(\xi_1)\, \dot{\boldsymbol{\sigma}}(\xi_2)\, \dot{\boldsymbol{\sigma}}(\xi_3)\, d\xi_1\, d\xi_2\, d\xi_3,$$

$$\int_0^t \int_0^t \int_0^t S_7(t-\xi_1, t-\xi_2, t-\xi_3)\, \dot{\boldsymbol{\sigma}}(\xi_1)\, \dot{\boldsymbol{\sigma}}(\xi_2)\, \dot{\boldsymbol{\sigma}}(\xi_3)\, d\xi_1\, d\xi_2\, d\xi_3. \tag{7.8c}$$

However, the kernel functions S_4, S_5, S_6, and S_7, Q_4, Q_5, and Q_6 in (7.8c) are related to each other through the Hamilton-Cayley equation (see Chapter 3), since each of their stress functions is found in (3.39). Therefore, any one of them, say Q_4 may be eliminated. Omitting Q_4, combining (7.8a), (7.8b), (7.8c) and renumbering the kernel functions yields the following equation for the third order nonlinear constitutive relation for creep strain in terms of stress history.

Creep

$$\boldsymbol{\epsilon}(t) = \int_0^t [\mathbf{I}K_1\bar{\dot{\boldsymbol{\sigma}}}(\xi_1) + K_2\,\dot{\boldsymbol{\sigma}}(\xi_1)]\,\mathrm{d}\xi_1 + \int_0^t\int_0^t \Big\{\mathbf{I}\Big[K_3\bar{\dot{\boldsymbol{\sigma}}}(\xi_1)\,\bar{\dot{\boldsymbol{\sigma}}}(\xi_2) + K_4\,\overline{\dot{\boldsymbol{\sigma}}(\xi_1)\dot{\boldsymbol{\sigma}}(\xi_2)}\Big]$$

$$+K_5\,\bar{\dot{\boldsymbol{\sigma}}}(\xi_1)\,\dot{\boldsymbol{\sigma}}(\xi_2) + K_6\,\dot{\boldsymbol{\sigma}}(\xi_1)\,\dot{\boldsymbol{\sigma}}(\xi_2)\Big\}\,\mathrm{d}\xi_1\,\mathrm{d}\xi_2$$

$$+\int_0^t\int_0^t\int_0^t \Big\{\mathbf{I}\Big[K_7\,\overline{\dot{\boldsymbol{\sigma}}(\xi_1)\,\dot{\boldsymbol{\sigma}}(\xi_2)\,\dot{\boldsymbol{\sigma}}(\xi_3)} + K_8\,\overline{\dot{\boldsymbol{\sigma}}(\xi_1)\,\dot{\boldsymbol{\sigma}}(\xi_2)}\,\bar{\dot{\boldsymbol{\sigma}}}(\xi_3)\Big]$$

$$+K_9\bar{\dot{\boldsymbol{\sigma}}}(\xi_1)\bar{\dot{\boldsymbol{\sigma}}}(\xi_2)\,\dot{\boldsymbol{\sigma}}(\xi_3) + K_{10}\,\overline{\dot{\boldsymbol{\sigma}}(\xi_1)\,\dot{\boldsymbol{\sigma}}(\xi_2)}\,\dot{\boldsymbol{\sigma}}(\xi_3) + K_{11}\,\bar{\dot{\boldsymbol{\sigma}}}(\xi_1)\,\dot{\boldsymbol{\sigma}}(\xi_2)\,\dot{\boldsymbol{\sigma}}(\xi_3)$$

$$+K_{12}\,\dot{\boldsymbol{\sigma}}(\xi_1)\,\dot{\boldsymbol{\sigma}}(\xi_2)\,\dot{\boldsymbol{\sigma}}(\xi_3)\Big\}\,\mathrm{d}\xi_1\,\mathrm{d}\xi_2\,\mathrm{d}\xi_3, \tag{7.9}$$

where K_1 to K_{12} are kernel functions which must be determined from experiments, see Chapters 8 and 9, and are functions of the time variables as follows:

$$\begin{aligned}
K_\alpha &= K_\alpha(t-\xi) & \alpha &= 1, 2,\\
K_\alpha &= K_\alpha(t-\xi_1, t-\xi_2) & \alpha &= 3, 4, 5, 6,\\
K_\alpha &= K_\alpha(t-\xi_1, t-\xi_2, t-\xi_3) & \alpha &= 7, 8, 9, 10, 11, 12.
\end{aligned} \tag{7.10}$$

K_α are considered to be symmetric with respect to their time variables, so that the order of listing the time variables is immaterial. Also,

$$\begin{aligned}
\bar{\dot{\boldsymbol{\sigma}}} &= \mathrm{tr}\,\dot{\boldsymbol{\sigma}} = \dot{\sigma}_{ii}, && \text{is the trace of the stress rate tensor,}\\
\dot{\boldsymbol{\sigma}}\dot{\boldsymbol{\sigma}} &= \dot{\sigma}_{ip}\dot{\sigma}_{pj}, && i, p, j = 1, 2, 3,\\
\dot{\boldsymbol{\sigma}}\dot{\boldsymbol{\sigma}}\dot{\boldsymbol{\sigma}} &= \dot{\sigma}_{ip}\dot{\sigma}_{pq}\dot{\sigma}_{qj}, && i, p, q, j = 1, 2, 3,\\
\bar{\dot{\boldsymbol{\sigma}}}\bar{\dot{\boldsymbol{\sigma}}} &= \mathrm{tr}(\dot{\sigma}_{ij})\,\mathrm{tr}(\dot{\sigma}_{ij}) = \dot{\sigma}_{ii}\dot{\sigma}_{jj}, && i, j = 1, 2, 3,\\
\overline{\dot{\boldsymbol{\sigma}}\dot{\boldsymbol{\sigma}}} &= \mathrm{tr}(\dot{\sigma}_{ip}\dot{\sigma}_{pj}) = \dot{\sigma}_{ij}\dot{\sigma}_{ji}, && i, p, j = 1, 2, 3,\\
\overline{\dot{\boldsymbol{\sigma}}\dot{\boldsymbol{\sigma}}\dot{\boldsymbol{\sigma}}} &= \mathrm{tr}(\dot{\sigma}_{ip}\dot{\sigma}_{pq}\dot{\sigma}_{qj}) = \dot{\sigma}_{ip}\dot{\sigma}_{pq}\dot{\sigma}_{qi}, && i, p, q, j = 1, 2, 3.
\end{aligned} \tag{7.11}$$

In (7.9) the material is taken to be isotropic and stress and strain free prior to time zero.

Multiple integrals higher than the third order shown in (7.9) may be included and their kernel functions may be readily listed. However, as in the case of one of the kernel functions in the third order, many are redundant and should be omitted. Spencer and Rivlin [7.4] show that in integrals higher than the fifth order no term need be retained which includes a product of more than five stress tensors or a trace of a product of more than six stress tensors.

Stress Relaxation:

The inverse of creep (stress relaxation) which expresses stress in terms of strain history, has the same form as (7.9) except ε replaces σ and vice versa as follows:

$$\boldsymbol{\sigma}(t) = \int_0^t [\mathbf{I}\psi_1 \bar{\dot{\boldsymbol{\epsilon}}}(\xi_1) + \psi_2 \dot{\boldsymbol{\epsilon}}(\xi_1)]\, d\xi_1 + \int_0^t \int_0^t \{\mathbf{I}[\psi_3 \bar{\dot{\boldsymbol{\epsilon}}}(\xi_1) \bar{\dot{\boldsymbol{\epsilon}}}(\xi_2) + \psi_4 \overline{\dot{\boldsymbol{\epsilon}}(\xi_1) \dot{\boldsymbol{\epsilon}}(\xi_2)}]$$
$$+ \psi_5 \bar{\dot{\boldsymbol{\epsilon}}}(\xi_1) \dot{\boldsymbol{\epsilon}}(\xi_2) + \psi_6 \dot{\boldsymbol{\epsilon}}(\xi_1) \dot{\boldsymbol{\epsilon}}(\xi_2)\}\, d\xi_1\, d\xi_2 + \int_0^t \int_0^t \int_0^t \{\mathbf{I}[\psi_7 \overline{\dot{\boldsymbol{\epsilon}}(\xi_1) \dot{\boldsymbol{\epsilon}}\xi_2) \dot{\boldsymbol{\epsilon}}(\xi_3)}$$
$$+ \psi_8 \overline{\dot{\boldsymbol{\epsilon}}(\xi_1) \dot{\boldsymbol{\epsilon}}(\xi_2)}\, \bar{\dot{\boldsymbol{\epsilon}}}(\xi_3)] + \psi_9 \bar{\dot{\boldsymbol{\epsilon}}}(\xi_1) \bar{\dot{\boldsymbol{\epsilon}}}(\xi_2) \dot{\boldsymbol{\epsilon}}(\xi_3) + \psi_{10} \overline{\dot{\boldsymbol{\epsilon}}(\xi_1) \dot{\boldsymbol{\epsilon}}(\xi_2)}\, \dot{\boldsymbol{\epsilon}}(\xi_3)$$
$$+ \psi_{11} \bar{\dot{\boldsymbol{\epsilon}}}(\xi_1) \dot{\boldsymbol{\epsilon}}(\xi_2) \dot{\boldsymbol{\epsilon}}(\xi_3) + \psi_{12} \dot{\boldsymbol{\epsilon}}(\xi_1) \dot{\boldsymbol{\epsilon}}(\xi_2) \dot{\boldsymbol{\epsilon}}(\xi_3)\}\, d\xi_1\, d\xi_2\, d\xi_3, \tag{7.12}$$

where the relaxation kernel functions, ψ_1, ψ_2, are functions of $(t-\xi_1)$; $\psi_3, \psi_4, \psi_5, \psi_6$, are functions of $(t-\xi_1, t-\xi_2)$; $\psi_7, \ldots, \psi_{12}$ are functions of $(t-\xi_1, t-\xi_2, t-\xi_3)$; and the traces of strain tensors and products of strain tensors have the same form as given for stresses in (7.11).

7.4 A Linearly Compressible Material

Some materials are linearly compressible in the range of stress permitted in tension even though the stress-strain relation, say, in tension, is nonlinear [7.5, 7.6]. For example as shown in [7.6] a stress-strain plot for polyurethane, shown in Fig. 7.5A for a test performed at a constant strain rate of 0.05 in. per in. per min., is nearly linear. However, at constant stress, nonlinear creep resulted. A plot of the one hour tensile strain divided by the applied constant stress (the one hour value of the tensile compliance) versus stress is shown in Fig. 7.5B along with the 0.1 hour value of the tensile compliance. This figure shows that the one hour value of the compliance at 6000 psi is about 8 per cent greater than at 1000 psi.

In Fig. 7.5C the difference in the values of the compliance at 1.0 and 0.1 hour plotted versus stress shows that the time-dependent strain is actually much more nonlinear than suggested by Fig. 7.5A and 7.5B. This difference is about 250 percent greater at 6000 psi than at 1000 psi. Figure 7.5D shows

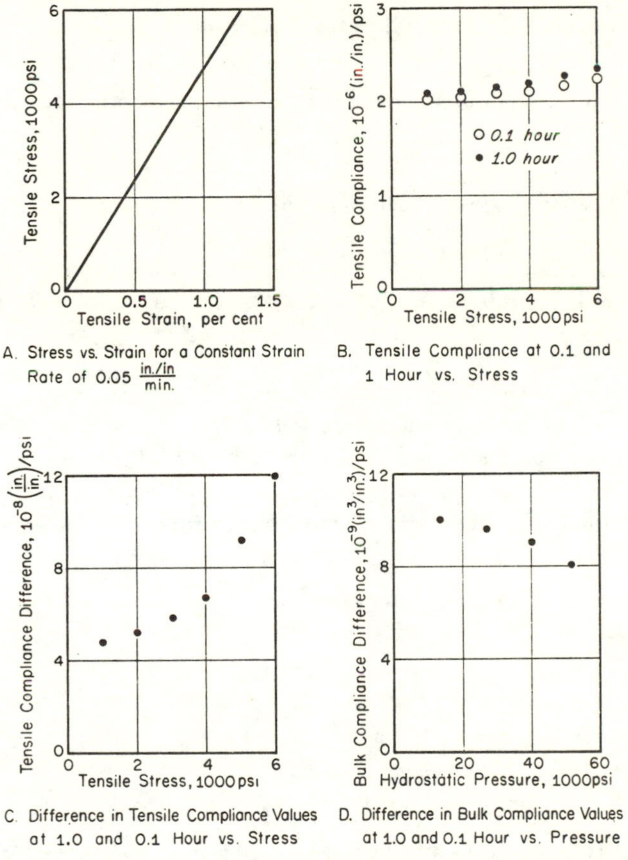

A. Stress vs. Strain for a Constant Strain Rate of 0.05 $\frac{\text{in./in}}{\text{min.}}$

B. Tensile Compliance at 0.1 and 1 Hour vs. Stress

C. Difference in Tensile Compliance Values at 1.0 and 0.1 Hour vs. Stress

D. Difference in Bulk Compliance Values at 1.0 and 0.1 Hour vs. Pressure

Fig. 7.5. Time-dependence of Tensile and Bulk Compliance of Polyurethane. From [7.6], courtesy ASME.

that the difference in the bulk compliance under constant hydrostatic pressure at one hour minus that at 0.1 hour decreased with increase in hydrostatic pressure. This difference in the bulk compliance is one order of magnitude smaller than the corresponding quantity for the tensile compliance and decreases by only 20 percent for values of pressure between 13,300 psi and 50,000 psi. This implies that the time-dependent portion of the volume strain is considerably smaller and much more linear than the time-dependent portion of the tensile strain. This result is the basis for the assumption of linear compressibility employed in this section.

Creep. The creep formulation for a linearly compressible material assumption as described by Nolte and Findley [7.6] is formed as follows. Taking the trace of (7.9) yields the following expression

$$\bar{\boldsymbol{\epsilon}}(t) = \int_0^t [3K_1+K_2]\bar{\dot{\boldsymbol{\sigma}}}(\xi_1)\,d\xi_1 + \int_0^t\int_0^t [(3K_3+K_5)\,\bar{\dot{\boldsymbol{\sigma}}}(\xi_1)\,\bar{\dot{\boldsymbol{\sigma}}}(\xi_2)$$

$$+(3K_4+K_6)\,\overline{\dot{\boldsymbol{\sigma}}(\xi_1)\,\dot{\boldsymbol{\sigma}}}(\xi_2)]\,d\xi_1\,d\xi_2$$

$$+\int_0^t\int_0^t\int_0^t [(3K_7+K_{12})\,\overline{\dot{\boldsymbol{\sigma}}(\xi_1)\,\dot{\boldsymbol{\sigma}}(\xi_2)\,\dot{\boldsymbol{\sigma}}}(\xi_3)+(3K_8+K_{10}+K_{11})$$

$$\overline{\dot{\boldsymbol{\sigma}}(\xi_1)\,\dot{\boldsymbol{\sigma}}}(\xi_2)\,\bar{\dot{\boldsymbol{\sigma}}}(\xi_3)+K_9\,\bar{\dot{\boldsymbol{\sigma}}}(\xi_1)\bar{\dot{\boldsymbol{\sigma}}}(\xi_2)\,\bar{\dot{\boldsymbol{\sigma}}}(\xi_3)]\,d\xi_1\,d\xi_2\,d\xi_3, \qquad (7.13)$$

where $\bar{\boldsymbol{\epsilon}} = \varepsilon_{kk} = 3\varepsilon_v$ is the change in volume per unit volume. The linearly compressible assumption implies that the sum of all the terms on the right-hand side of (7.13) except the first term (which is linear) vanish. To satisfy this condition for arbitrary stress histories each function of K's in parentheses in (7.13) except the first one, must vanish; therefore,

$$\begin{aligned} K_3 &= -\tfrac{1}{3}K_5, \\ K_4 &= -\tfrac{1}{3}K_6, \\ K_7 &= -\tfrac{1}{3}K_{12}, \\ K_8 &= -\tfrac{1}{3}(K_{10}+K_{11}), \\ K_9 &= 0. \end{aligned} \qquad (7.14)$$

Inserting (7.14) into (7.9) yields

$$\boldsymbol{\epsilon}(t) = \int_0^t [\mathbf{I}K_1\bar{\dot{\boldsymbol{\sigma}}}(\xi_1)+K_2\,\dot{\boldsymbol{\sigma}}(\xi_1)]\,d\xi_1 + \int_0^t\int_0^t \{K_5[\dot{\boldsymbol{\sigma}}(\xi_1)-\tfrac{1}{3}\mathbf{I}\,\bar{\dot{\boldsymbol{\sigma}}}(\xi_1)]\,\bar{\dot{\boldsymbol{\sigma}}}(\xi_2)$$

$$+K_6[\dot{\boldsymbol{\sigma}}(\xi_1)\,\dot{\boldsymbol{\sigma}}(\xi_2)-\tfrac{1}{3}\mathbf{I}\,\overline{\dot{\boldsymbol{\sigma}}(\xi_1)\,\dot{\boldsymbol{\sigma}}}(\xi_2)]\}\,d\xi_1\,d\xi_2 + \int_0^t\int_0^t\int_0^t \{K_{10}[\dot{\boldsymbol{\sigma}}(\xi_1)$$

$$-\tfrac{1}{3}\mathbf{I}\,\bar{\dot{\boldsymbol{\sigma}}}(\xi_1)]\,\overline{\dot{\boldsymbol{\sigma}}(\xi_2)\dot{\boldsymbol{\sigma}}}(\xi_3)+K_{11}[\dot{\boldsymbol{\sigma}}(\xi_1)\dot{\boldsymbol{\sigma}}(\xi_2)-\tfrac{1}{3}\mathbf{I}\,\overline{\dot{\boldsymbol{\sigma}}(\xi_1)\,\dot{\boldsymbol{\sigma}}}(\xi_2)]\,\bar{\dot{\boldsymbol{\sigma}}}(\xi_3)$$

$$+K_{12}[\dot{\boldsymbol{\sigma}}(\xi_1)\,\dot{\boldsymbol{\sigma}}(\xi_2)\dot{\boldsymbol{\sigma}}(\xi_3)-\tfrac{1}{3}\mathbf{I}\,\overline{\dot{\boldsymbol{\sigma}}(\xi_1)\,\dot{\boldsymbol{\sigma}}(\xi_2)\,\dot{\boldsymbol{\sigma}}}(\xi_3)]\}\,d\xi_1\,d\xi_2\,d\xi_3. \qquad (7.15)$$

Therefore, only seven independent material functions remain.

Stress Relaxation. The relaxation formulation for a linearly compressible assumption is formed by using the same method as described in the creep formulation. Taking the trace of (7.12) yields the following expression:

$$\bar{\sigma}(t) = \int_0^t [3\psi_1+\psi_2]\,\bar{\dot{\epsilon}}(\xi_1)\,d\xi_1 + \int_0^t\int_0^t [(3\psi_3+\psi_5)\,\bar{\dot{\epsilon}}(\xi_1)\,\bar{\dot{\epsilon}}(\xi_2)$$

$$+(3\psi_4+\psi_6)\,\overline{\dot{\epsilon}(\xi_1)\,\dot{\epsilon}(\xi_2)}]\,d\xi_1 d\xi_2$$

$$+\int_0^t\int_0^t\int_0^t [(3\psi_7+\psi_{12})\,\overline{\dot{\epsilon}(\xi_1)\,\dot{\epsilon}(\xi_2)\,\dot{\epsilon}(\xi_3)}$$

$$+(3\psi_8+\psi_{10}+\psi_{11})\,\overline{\dot{\epsilon}(\xi_1)\,\dot{\epsilon}(\xi_2)}\bar{\dot{\epsilon}}(\xi_3)+\psi_9\,\bar{\dot{\epsilon}}(\xi_1)\,\bar{\dot{\epsilon}}(\xi_2)\,\bar{\dot{\epsilon}}(\xi_3)]\,d\xi_1\,d\xi_2\,d\xi_3, \tag{7.16}$$

where $\boldsymbol{\sigma} = \sigma_{kk} = 3\sigma_v = -3p$ is proportional to the pressure p.

The linearly compressible assumption in this case implies that the induced pressure, proportional to tr$\boldsymbol{\sigma}$, is linearly dependent on the change in volume per unit volume, tr$\boldsymbol{\epsilon}$. Thus the second and third order terms in (7.16) must be equal to zero. Hence,

$$\begin{aligned}
\psi_3 &= -\tfrac{1}{3}\psi_5,\\
\psi_4 &= -\tfrac{1}{3}\psi_6,\\
\psi_7 &= -\tfrac{1}{3}\psi_{12},\\
\psi_8 &= -\tfrac{1}{3}(\psi_{10}+\psi_{11}),\\
\psi_9 &= 0.
\end{aligned} \tag{7.17}$$

Inserting (7.17) into (7.12) yields

$$\boldsymbol{\sigma}(t) = \int_0^t [\mathbf{I}\psi_1\,\bar{\dot{\epsilon}}(\xi_1)+\psi_2\,\dot{\boldsymbol{\epsilon}}(\xi_1)]\,d\xi_1 + \int_0^t\int_0^t \{\psi_5[\dot{\boldsymbol{\epsilon}}(\xi_1)-\tfrac{1}{3}\mathbf{I}\,\bar{\dot{\epsilon}}(\xi_1)]\,\bar{\dot{\epsilon}}(\xi_2)$$

$$+\psi_6[\dot{\boldsymbol{\epsilon}}(\xi_1)\,\dot{\boldsymbol{\epsilon}}(\xi_2)-\tfrac{1}{3}\mathbf{I}\,\overline{\dot{\epsilon}(\xi_1)\,\dot{\epsilon}(\xi_2)}]\}\,d\xi_1\,d\xi_2$$

$$+\int_0^t\int_0^t\int_0^t \{\psi_{10}[\dot{\boldsymbol{\epsilon}}(\xi_1)-\tfrac{1}{3}\mathbf{I}\,\bar{\dot{\epsilon}}(\xi_1)]\,\overline{\dot{\epsilon}(\xi_2)\,\dot{\epsilon}(\xi_3)}+\psi_{11}[\dot{\boldsymbol{\epsilon}}(\xi_1)\,\dot{\boldsymbol{\epsilon}}(\xi_2)$$

$$-\tfrac{1}{3}\mathbf{I}\,\overline{\dot{\epsilon}(\xi_1)\,\dot{\epsilon}(\xi_2)}]\,\bar{\dot{\epsilon}}(\xi_3)+\psi_{12}[\dot{\boldsymbol{\epsilon}}(\xi_1)\,\dot{\boldsymbol{\epsilon}}(\xi_2)\,\dot{\boldsymbol{\epsilon}}(\xi_3)$$

$$-\tfrac{1}{3}\mathbf{I}\,\overline{\dot{\epsilon}(\xi_1)\,\dot{\epsilon}(\xi_2)\,\dot{\epsilon}(\xi_3)}]\}\,d\xi_1\,d\xi_2\,d\xi_3, \tag{7.18}$$

which is the same form as the corresponding creep function.

7.5 Incompressible Material Assumption

A more drastic assumption which sometimes may be convenient and sufficiently accurate is to treat the analysis as though the material was incompressible. Since materials clearly are never completely incompressible this assumption is artificial and less accurate than linear compressibility. The behavior of a material may be taken as that of an incompressible material if the effect of compressibility is unimportant in the application considered. Whether or not the effect of compressibility is negligible may be difficult to establish, however. It should not be expected that an incompressible model will describe real material behavior as accurately as a compressible model.

In this section an incompressible material is defined as one whose response to stressing or straining is insensitive to volumetric-type changes in strain or stress. That is, the relation between volumetric strain and stress has been uncoupled and the relation between volumetric stress and strain has been uncoupled. This definition of incompressibility is used in order to yield a dual restriction in the creep and relaxation formulations. The resulting relations given by Findley and Onaran [7.7] for creep and relaxation of an incompressible and a linearly compressible isotropic material are derived as follows for small strains.

Creep. In the creep formulation incompressibility implies that (A) a volumetric stress σ_v (mean normal stress or negative pressure) has no effect on the strain, (B) the stress has no effect on volumetric-type strain such as ε_v.

The stress tensor $\boldsymbol{\sigma}(t)$ may be written in terms of the deviatoric stress tensor $\mathbf{s}(t)$ and the volumetric stress $\sigma_v(t)$ and inserted into (7.9). According to (3.11)

$$\boldsymbol{\sigma} = \mathbf{s} + \mathbf{I}\sigma_v = \begin{vmatrix} s_{11}+\sigma_v & s_{12} & s_{13} \\ s_{21} & s_{22}+\sigma_v & s_{23} \\ s_{31} & s_{32} & s_{33}+\sigma_v \end{vmatrix}, \tag{7.19}$$

where

$$\sigma_v = \tfrac{1}{3}\sigma_{kk} = -p, \tag{7.20}$$

and p is the hydrostatic pressure. The trace of (7.19) is

$$\bar{\boldsymbol{\sigma}} = \bar{\mathbf{s}} + 3\sigma_v = 3\sigma_v, \tag{7.21}$$

since by definition of $\mathbf{s}$, $\operatorname{tr}\mathbf{s} = 0$. The tensor products and their traces required in (7.9) may be found by forming the indicated products of (7.19), taking the traces of the products and noting that by definition of $\mathbf{s}$, $\mathrm{tr}\,\mathbf{s} = \bar{\mathbf{s}} = 0$. In this manner the following relations are found

$$\begin{aligned}
\boldsymbol{\sigma\sigma} &= \mathbf{ss} + 2\mathbf{s}\sigma_v + \mathbf{I}\sigma_v^2, \\
\overline{\boldsymbol{\sigma\sigma}} &= \overline{\mathbf{ss}} + 3\sigma_v^2, \\
\boldsymbol{\sigma\sigma\sigma} &= \mathbf{sss} + 3\mathbf{ss}\sigma_v + 3\mathbf{s}\sigma_v^2 + \mathbf{I}\sigma_v^3, \\
\overline{\boldsymbol{\sigma\sigma\sigma}} &= \overline{\mathbf{sss}} + 3\overline{\mathbf{ss}}\sigma_v + 3\sigma_v^3.
\end{aligned} \tag{7.22}$$

Inserting (7.19), (7.21) and (7.22) into (7.9) and rearranging terms yields

$$\begin{aligned}
\boldsymbol{\epsilon}(t) = \mathbf{I}\hat{\varepsilon}(t) &+ \int_0^t [\mathbf{I}(3K_1+K_2)\dot{\sigma}_v + K_2\dot{\mathbf{s}}]\, d\xi_1 + \int_0^t\int_0^t [\mathbf{I}(9K_3+3K_4+3K_5+K_6)\dot{\sigma}_v\dot{\sigma}_v \\
&+ (3K_5+2K_6)\dot{\sigma}_v\dot{\mathbf{s}} + \mathbf{I}K_4\overline{\dot{\mathbf{s}}\dot{\mathbf{s}}} + K_6\,\dot{\mathbf{s}}\dot{\mathbf{s}}]\, d\xi_1\, d\xi_2 + \int_0^t\int_0^t\int_0^t [\mathbf{I}(3K_7+9K_8+9K_9 \\
&+ 3K_{10}+3K_{11}+K_{12})\dot{\sigma}_v\dot{\sigma}_v\dot{\sigma}_v + (9K_9+3K_{10}+6K_{11}+3K_{12})\dot{\sigma}_v\dot{\sigma}_v\,\dot{\mathbf{s}} \\
&+ \mathbf{I}(3K_7+3K_8+K_{10})\dot{\sigma}_v\overline{\dot{\mathbf{s}}\dot{\mathbf{s}}} + (3K_{11}+3K_{12})\dot{\sigma}_v\dot{\mathbf{s}}\dot{\mathbf{s}} \\
&+ \mathbf{I}K_7\,\overline{\dot{\mathbf{s}}\dot{\mathbf{s}}\dot{\mathbf{s}}} + K_{10}\,\overline{\dot{\mathbf{s}}\dot{\mathbf{s}}}\;\dot{\mathbf{s}} + K_{12}\,\dot{\mathbf{s}}\dot{\mathbf{s}}\dot{\mathbf{s}}]\, d\xi_1\, d\xi_2\, d\xi_3,
\end{aligned} \tag{7.23}$$

where $\hat{\varepsilon}$, representing any non-stress induced volumetric strain, has been inserted, $\mathbf{s}$ and σ_v are functions of ξ_1, ξ_2, ξ_3 as in (7.9) and the K_j terms are functions of time variables as in (7.10). Incompressibility assumption (A) requires the coefficients of terms containing σ_v to vanish. Therefore

$$\begin{aligned}
&3K_1+K_2 = 0, \\
&9K_3+3K_4+3K_5+K_6 = 0, \\
&3K_5+2K_6 = 0, \\
&3K_7+9K_8+9K_9+3K_{10}+3K_{11}+K_{12} = 0, \\
&3K_9+K_{10}+2K_{11}+K_{12} = 0, \\
&3K_7+3K_8+K_{10} = 0, \\
&K_{11}+K_{12} = 0.
\end{aligned} \tag{7.24}$$

Employing (7.24) in (7.23) and expressing strain in terms of deviatoric strain $\mathbf{d}$ and volumetric strain ε_v yields

$$\begin{aligned}
\boldsymbol{\epsilon}(t) = \mathbf{d}(t) + \mathbf{I}\varepsilon_v(t) = \mathbf{I}\hat{\varepsilon}(t) &+ \int_0^t K_2\dot{\mathbf{s}}\, d\xi_1 + \int_0^t\int_0^t [\mathbf{I}K_4\overline{\dot{\mathbf{s}}\dot{\mathbf{s}}} + K_6\dot{\mathbf{s}}\dot{\mathbf{s}}]\, d\xi_1\, d\xi_2 \\
&+ \int_0^t\int_0^t\int_0^t [\mathbf{I}K_7\overline{\dot{\mathbf{s}}\dot{\mathbf{s}}\dot{\mathbf{s}}} + K_{10}\,\overline{\dot{\mathbf{s}}\dot{\mathbf{s}}}\;\dot{\mathbf{s}} + K_{12}\dot{\mathbf{s}}\dot{\mathbf{s}}\dot{\mathbf{s}}]\, d\xi_1\, d\xi_2\, d\xi_3.
\end{aligned} \tag{7.25}$$

Taking the trace of (7.25), and noting that tr**d** = 0 yields

$$3\varepsilon_v(t) = 3\hat{\varepsilon}(t) + \int_0^t K_2\bar{\dot{s}}\, d\xi_1 + \int_0^t\int_0^t [3K_4+K_6]\overline{\dot{s}\dot{s}}\, d\xi_1\, d\xi_2$$

$$+\int_0^t\int_0^t\int_0^t \{[3K_7+K_{12}]\,\overline{\dot{s}\dot{s}\dot{s}}+K_{10}\overline{\dot{s}\dot{s}}\,\bar{\dot{s}}\}\, d\xi_1\, d\xi_2\, d\xi_3. \tag{7.26}$$

The first and last integral terms of (7.26) vanish because by definition of **s** tr**s** = 0. The incompressibility condition (B) says that ε_v on the left-hand side of (7.26) is unaffected by stress. Since any stress deviator **s** may be prescribed in the right-hand side of (7.26) the only possibility is that the remaining integrals on the right-hand side vanish. Thus

$$\begin{aligned} 3K_4+K_6 &= 0, \\ 3K_7+K_{12} &= 0, \\ \varepsilon_v(t) &= \hat{\varepsilon}(t). \end{aligned} \tag{7.27}$$

Introducing (7.27) into (7.25) yields the following expression containing only four kernel functions

$$\boldsymbol{\epsilon}(t) = \mathbf{d}(t) = \int_0^t K_2\dot{\mathbf{s}}\, d\xi_1 + \int_0^t\int_0^t K_6(\dot{\mathbf{s}}\dot{\mathbf{s}}-\tfrac{1}{3}\mathbf{I}\overline{\dot{\mathbf{s}}\dot{\mathbf{s}}})\, d\xi_1\, d\xi_2$$

$$+\int_0^t\int_0^t\int_0^t [K_{12}(\dot{\mathbf{s}}\dot{\mathbf{s}}\dot{\mathbf{s}}-\tfrac{1}{3}\mathbf{I}\overline{\dot{\mathbf{s}}\dot{\mathbf{s}}\dot{\mathbf{s}}})+K_{10}\overline{\dot{\mathbf{s}}\dot{\mathbf{s}}}\,\dot{\mathbf{s}}]\, d\xi_1\, d\xi_2\, d\xi_3. \tag{7.28}$$

However, a further relation among the kernel functions may be obtained from the last four equations of (7.24). By substitution of one equation in another it may be shown from (7.24) that

$$K_{10} = \tfrac{1}{3}K_{12}-2K_7. \tag{7.29}$$

Then by substitution of the second expression in (7.27) it follows that

$$K_{10} = K_{12}. \tag{7.30}$$

Making use of (7.30) the creep form of constitutive relation for an incompressible isotropic material is expressed in terms of only three kernel functions as follows

$$\boldsymbol{\epsilon}(t) = \mathbf{d}(t) = \int_0^t K_2\dot{\mathbf{s}}\, d\xi_1 + \int_0^t\int_0^t K_6(\dot{\mathbf{s}}\dot{\mathbf{s}} - \tfrac{1}{3}\mathbf{I}\overline{\dot{\mathbf{s}}\dot{\mathbf{s}}})\, d\xi_1\, d\xi_2$$

$$+ \int_0^t\int_0^t\int_0^t K_{12}(\dot{\mathbf{s}}\dot{\mathbf{s}}\dot{\mathbf{s}} - \tfrac{1}{3}\mathbf{I}\overline{\dot{\mathbf{s}}\dot{\mathbf{s}}\dot{\mathbf{s}}} + \overline{\dot{\mathbf{s}}\dot{\mathbf{s}}}\;\dot{\mathbf{s}})\, d\xi_1\, d\xi_2\, d\xi_3. \tag{7.31}$$

It is of interest to note that (7.31) is a function of the stress deviator s and the deviatoric parts of ss and sss as it should. The form given by (7.31) is the same as that derived by a different method by Lockett and Stafford [7.8] when restricted to small strains. Lockett and Stafford found three independent kernel functions required for finite strain as well.

As a check on (7.31) consider the effect predicted by (7.31) when a hydrostatic pressure

$$\boldsymbol{\sigma} = \begin{vmatrix} \sigma & 0 & 0 \\ 0 & \sigma & 0 \\ 0 & 0 & \sigma \end{vmatrix}$$

is applied. According to (7.19) and (7.20) $s_{11} = s_{22} = s_{33} = 0$. Thus $\mathbf{s} = 0$ for this state of stress. Hence the integrals on the right-hand side of (7.31) are zero which means that the hydrostatic pressure has no effect, as required by incompressibility. It may also be shown that the trace of (7.31) is zero for any state of stress.

Stress Relaxation. In the stress relaxation formulation for an incompressible material, incompressibility implies that (a) the volumetric stress σ_v (mean normal stress or negative pressure) is arbitrary (independent of deformation) and has no effect on the volume and (b) a volumetric strain ε_v has no effect on the stress.

The strain rate tensor $\dot{\boldsymbol{\epsilon}}(t)$ may be written in terms of the deviatoric strain rate tensor $\dot{\mathbf{d}}(t)$ and the volumetric strain rate $\dot{\epsilon}_v(t)$ and inserted into (7.12). According to (3.73)

$$\dot{\boldsymbol{\epsilon}} = \dot{\mathbf{d}} + \mathbf{I}\dot{\varepsilon}_v = \begin{vmatrix} \dot{d}_{11}+\dot{\varepsilon}_v & \dot{d}_{12} & \dot{d}_{13} \\ \dot{d}_{21} & \dot{d}_{22}+\dot{\varepsilon}_v & \dot{d}_{23} \\ \dot{d}_{31} & \dot{d}_{32} & \dot{d}_{33}+\dot{\varepsilon}_v \end{vmatrix}, \tag{7.32}$$

where

$$\dot{\varepsilon}_v = \tfrac{1}{3}\dot{\varepsilon}_{kk}. \tag{7.33}$$

The trace of (7.32) is

$$\bar{\dot{\boldsymbol{\epsilon}}} = \bar{\dot{\mathbf{d}}} + 3\dot{\varepsilon}_v = 3\dot{\varepsilon}_v, \tag{7.34}$$

since by definition of $\dot{\mathbf{d}}$, $\operatorname{tr}\dot{\mathbf{d}} = 0$. The tensor products and their traces required in (7.12) may be found by forming the indicated products of (7.32), taking the trace of the products and noting that $\operatorname{tr}\dot{\mathbf{d}} = \bar{\dot{\mathbf{d}}} = 0$ with the following results

$$\begin{aligned}
\dot{\boldsymbol{\epsilon}}\dot{\boldsymbol{\epsilon}} &= \dot{\mathbf{d}}\dot{\mathbf{d}} + 2\dot{\mathbf{d}}\dot{\varepsilon}_v + \mathbf{I}\dot{\varepsilon}_v^2, \\
\overline{\dot{\boldsymbol{\epsilon}}\dot{\boldsymbol{\epsilon}}} &= \overline{\dot{\mathbf{d}}\dot{\mathbf{d}}} + 3\dot{\varepsilon}_v^2, \\
\dot{\boldsymbol{\epsilon}}\dot{\boldsymbol{\epsilon}}\dot{\boldsymbol{\epsilon}} &= \dot{\mathbf{d}}\dot{\mathbf{d}}\dot{\mathbf{d}} + 3\dot{\mathbf{d}}\dot{\mathbf{d}}\dot{\varepsilon}_v 3\dot{\mathbf{d}}\dot{\varepsilon}_v^2 + \mathbf{I}\dot{\varepsilon}_v^3, \\
\overline{\dot{\boldsymbol{\epsilon}}\dot{\boldsymbol{\epsilon}}\dot{\boldsymbol{\epsilon}}} &= \overline{\dot{\mathbf{d}}\dot{\mathbf{d}}\dot{\mathbf{d}}} + 3\overline{\dot{\mathbf{d}}\dot{\mathbf{d}}}\dot{\varepsilon}_v + 3\dot{\varepsilon}_v^3.
\end{aligned} \tag{7.35}$$

Since (7.32) through (7.35) are exactly the same as (7.19) through (7.22) and (7.12) is exactly the same as (7.9) except interchanging stress and strain symbols it follows that insertion of (7.32) through (7.35) into (7.12) yields the same expression as (7.23) except replace $\mathbf{I}\hat{\varepsilon}(t)$ by $\mathbf{I}\hat{\sigma}(t)$, K by ψ, $\boldsymbol{\epsilon}$ by $\boldsymbol{\sigma}$, σ_v by ε_v and $\mathbf{s}$ by $\mathbf{d}$, where $\hat{\sigma}$ is any non-strain induced negative pressure.

Incompressibility condition (b) requires that coefficients of terms containing ε_v vanish. This condition leads to the same set of equations as (7.24) except substituting ψ for K. Introducing the equivalent of (7.24) in the equivalent of (7.23) for relaxation and resolving the stress tensor into the stress deviator $\mathbf{s}$ and volumetric stress σ_v yields

$$\begin{aligned}
\boldsymbol{\sigma}(t) = \mathbf{s}(t) + \mathbf{I}\sigma_v(t) = \mathbf{I}\hat{\sigma}(t) &+ \int_0^t \psi_2\dot{\mathbf{d}}\, d\xi_1 + \int_0^t\int_0^t \left[\mathbf{I}\psi_4\overline{\dot{\mathbf{d}}\dot{\mathbf{d}}} + \psi_6\dot{\mathbf{d}}\dot{\mathbf{d}}\right] d\xi_1\, d\xi_2 \\
&+ \int_0^t\int_0^t\int_0^t \left[\mathbf{I}\psi_7\,\overline{\dot{\mathbf{d}}\dot{\mathbf{d}}\dot{\mathbf{d}}} + \psi_{10}\,\overline{\dot{\mathbf{d}}\dot{\mathbf{d}}}\dot{\mathbf{d}} + \psi_{12}\,\dot{\mathbf{d}}\dot{\mathbf{d}}\dot{\mathbf{d}}\right] d\xi_1\, d\xi_2\, d\xi_3,
\end{aligned} \tag{7.36}$$

where in view of incompressibility condition (a) σ_v represents an arbitrary volumetric stress (or negative pressure).

Taking the trace of (7.36) and noting (7.21)

$$\begin{aligned}
\bar{\boldsymbol{\sigma}} = 3\sigma_v = 3\hat{\sigma}(t) &+ \int_0^t \psi_2\bar{\dot{\mathbf{d}}}\, d\xi_1 + \int_0^t\int_0^t [3\psi_4 + \psi_6]\, \overline{\dot{\mathbf{d}}\dot{\mathbf{d}}}\, d\xi_1\, d\xi_2 \\
&+ \int_0^t\int_0^t\int_0^t \left\{[3\psi_7 + \psi_{12}]\, \overline{\dot{\mathbf{d}}\dot{\mathbf{d}}\dot{\mathbf{d}}} + \psi_{10}\, \overline{\dot{\mathbf{d}}\dot{\mathbf{d}}}\, \bar{\dot{\mathbf{d}}}\right\} d\xi_1\, d\xi_2\, d\xi_3.
\end{aligned} \tag{7.37}$$

By definition of $\dot{\mathbf{d}}$, $\operatorname{tr}\dot{\mathbf{d}} = 0$; thus the first and last integral terms of (7.37) vanish. Incompressibility condition (a) says that the volumetric stress σ_v is arbitrary (independent of deformation), so the left-hand side of (7.37) is arbitrary. But any strain rate deviator $\dot{\mathbf{d}}$ may be prescribed on the right-hand side of (7.37). Therefore the only possibility is that terms on the right-hand side of (7.37) containing $\dot{\mathbf{d}}$ vanish. Hence

$$\begin{aligned} 3\psi_4+\psi_6 &= 0, \\ 3\psi_7+\psi_{12} &= 0, \\ \sigma_v(t) &= \hat{\sigma}(t). \end{aligned} \tag{7.38}$$

Introducing (7.38) into (7.36) yields an expression containing four kernel functions

$$\boldsymbol{\sigma}(t) = \mathbf{s}(t) = \int_0^t \psi_2 \dot{\mathbf{d}}\, d\xi_1 + \int_0^t\int_0^t \psi_6\left(\dot{\mathbf{d}}\dot{\mathbf{d}} - \tfrac{1}{3}\mathbf{I}\overline{\dot{\mathbf{d}}\dot{\mathbf{d}}}\right) d\xi_1\, d\xi_2$$
$$+ \int_0^t\int_0^t\int_0^t \left[\psi_{12}\left(\dot{\mathbf{d}}\dot{\mathbf{d}}\dot{\mathbf{d}} - \tfrac{1}{3}\mathbf{I}\,\overline{\dot{\mathbf{d}}\dot{\mathbf{d}}\dot{\mathbf{d}}}\right) + \psi_{10}\,\overline{\dot{\mathbf{d}}\dot{\mathbf{d}}\dot{\mathbf{d}}}\right] d\xi_1\, d\xi_2\, d\xi_3. \tag{7.39}$$

However, as in the creep formulation a further relation among the kernel functions may be obtained by combining the second expression of (7.38) with the equivalent of the last four of equations (7.24) expressed in terms of ψ's with the result

$$\psi_{10} = \psi_{12}. \tag{7.40}$$

Thus employing (7.40) in (7.39) an expression for the relaxation formulation of an incompressible isotropic material is found which has only three kernel functions, the same number as the corresponding creep formulation

$$\boldsymbol{\sigma}(t) = \mathbf{s}(t) = \int_0^t \psi_2\, \dot{\mathbf{d}}\, d\xi_1 + \int_0^t\int_0^t \psi_6\left(\dot{\mathbf{d}}\dot{\mathbf{d}} - \tfrac{1}{3}\mathbf{I}\,\overline{\dot{\mathbf{d}}\dot{\mathbf{d}}}\right) d\xi_1\, d\xi_2$$
$$+ \int_0^t\int_0^t\int_0^t \psi_{12}\left(\dot{\mathbf{d}}\dot{\mathbf{d}}\dot{\mathbf{d}} - \tfrac{1}{3}\mathbf{I}\overline{\dot{\mathbf{d}}\dot{\mathbf{d}}\dot{\mathbf{d}}} + \overline{\dot{\mathbf{d}}\dot{\mathbf{d}}\dot{\mathbf{d}}}\right) d\xi_1\, d\xi_2\, d\xi_3, \tag{7.41}$$

where $\dot{\mathbf{d}}$ are strain rate deviators. Note that (7.41) is a function of the strain rate deviator $\dot{\mathbf{d}}$ and the deviatoric parts of $\dot{\mathbf{d}}\dot{\mathbf{d}}$ and $\dot{\mathbf{d}}\dot{\mathbf{d}}\dot{\mathbf{d}}$.

As a check on (7.41) consider a volumetric state of strain:

$$\dot{\boldsymbol{\epsilon}} = \begin{vmatrix} \dot{\varepsilon} & 0 & 0 \\ 0 & \dot{\varepsilon} & 0 \\ 0 & 0 & \dot{\varepsilon} \end{vmatrix}.$$

According to (7.32) and (7.33) $\dot{d}_{11} = \dot{d}_{22} = \dot{d}_{33} = 0$. Hence $\dot{\mathbf{d}} = 0$ and the right-hand side of (7.41) is zero. So the volumetric-type strain has no effect on the stress as required by the incompressibility assumption. It may also be shown that the trace of the right-hand side of (7.41) is zero for any state of strain.

Equation (7.39) is essentially the same as that derived by a different method by Pipkin [7.9] and Lockett and Stafford [7.8] where the strain tensors shown in (7.39) are strain deviators. Lockett and Stafford [7.8] obtained a creep formulation containing only three kernel functions as in (7.31) but four kernel functions for the relaxation formulation similar to (7.39). This resulted from the fact that (7.40) did not arise in the method they employed.

7.6 Linearly Compressible, II

Creep. In article 7.4 a linearly compressible creep formulation was derived which placed a restriction only on tr$\boldsymbol{\epsilon}$, namely that tr$\boldsymbol{\epsilon}$ should be linear. This resulted in a relation containing seven kernel functions. A more restricted linearly compressible formulation is obtained by adding a linear compressibility term $K_1\dot{\bar{\boldsymbol{\sigma}}}$ to the incompressible formulation (7.31) as follows

$$\boldsymbol{\epsilon}(t) = \int_0^t [\mathbf{I}K_1\dot{\bar{\boldsymbol{\sigma}}} + K_2\dot{\mathbf{s}}]\, d\xi_1 + \int_0^t\int_0^t K_6(\dot{\mathbf{s}}\dot{\mathbf{s}} - \tfrac{1}{3}\mathbf{I}\,\overline{\dot{\mathbf{s}}\dot{\mathbf{s}}})\, d\xi_1\, d\xi_2$$

$$+ \int_0^t\int_0^t\int_0^t K_{12}(\dot{\mathbf{s}}\dot{\mathbf{s}}\dot{\mathbf{s}} - \tfrac{1}{3}\mathbf{I}\overline{\dot{\mathbf{s}}\dot{\mathbf{s}}\dot{\mathbf{s}}} + \overline{\dot{\mathbf{s}}\dot{\mathbf{s}}}\,\dot{\mathbf{s}})\, d\xi_1\, d\xi_2\, d\xi_3, \quad (7.42)$$

Equation (7.42) contains only four kernel functions. It is less general than (7.15) but should be more representative of actual material behavior than (7.31).

Relaxation. The linearly compressible stress relaxation formulation (7.18) was derived from the assumption that the induced volumetric stress tr$\boldsymbol{\sigma}$ was linearly dependent on the volumetric strain. The resulting formulation contains seven kernel functions. A linearly compressible formulation which is more restricted results from (7.41) by adding a linearly compressible term $\psi_1\bar{\dot{\epsilon}}$ as follows

$$\boldsymbol{\sigma}(t) = \int_0^t [\mathbf{I}\psi_1\bar{\dot{\boldsymbol{\epsilon}}}+\psi_2\dot{\mathbf{d}}]\, d\xi_1 + \int_0^t\int_0^t \psi_6\left(\dot{\mathbf{d}}\dot{\mathbf{d}} - \tfrac{1}{3}\mathbf{I}\overline{\dot{\mathbf{d}}\dot{\mathbf{d}}}\right) d\xi_1\, d\xi_2$$

$$+ \int_0^t\int_0^t\int_0^t \psi_{12}\left(\dot{\mathbf{d}}\dot{\mathbf{d}}\dot{\mathbf{d}} - \tfrac{1}{3}\mathbf{I}\,\overline{\dot{\mathbf{d}}\dot{\mathbf{d}}\dot{\mathbf{d}}} + \overline{\dot{\mathbf{d}}\dot{\mathbf{d}}}\,\dot{\mathbf{d}}\right) d\xi_1\, d\xi_2\, d\xi_3, \tag{7.43}$$

where only four kernel functions remain.

7.7 Constant Volume

For small finite deformations constant volume implies that the determinant of $\mathbf{I}+2\boldsymbol{\epsilon}$ equal one, i.e.,

$$|\mathbf{I}+2\boldsymbol{\epsilon}| = 1. \tag{7.44}$$

Expanding (7.44) yields

$$\bar{\boldsymbol{\epsilon}}+\bar{\boldsymbol{\epsilon}}\,\bar{\boldsymbol{\epsilon}}-\overline{\boldsymbol{\epsilon}\boldsymbol{\epsilon}}+\tfrac{2}{3}[2\overline{\boldsymbol{\epsilon}\boldsymbol{\epsilon}\boldsymbol{\epsilon}}-3\overline{\boldsymbol{\epsilon}\boldsymbol{\epsilon}}\,\bar{\boldsymbol{\epsilon}}+\bar{\boldsymbol{\epsilon}}\,\bar{\boldsymbol{\epsilon}}\,\bar{\boldsymbol{\epsilon}}] = 0. \tag{7.45}$$

For infinitesimal deformation higher order terms vanish so that

$$\bar{\boldsymbol{\epsilon}} = 0. \tag{7.46}$$

It may be shown by solving (7.45) for $\bar{\boldsymbol{\epsilon}}$ and substituting this expression for $\bar{\boldsymbol{\epsilon}}$ in other terms in (7.45) that the following remains when terms higher than the third order are neglected

$$\bar{\boldsymbol{\epsilon}} = \overline{\boldsymbol{\epsilon}\boldsymbol{\epsilon}}-\tfrac{4}{3}\overline{\boldsymbol{\epsilon}\boldsymbol{\epsilon}\boldsymbol{\epsilon}} \tag{7.47}$$

It should be noted that the derivation of the incompressibility equations (7.31) and (7.41) did not involve the expression for constant volume in any of the forms given by (7.44, 7.45 or 7.46).

7.8 Incompressible and Linearly Compressible Creep in Terms of $\boldsymbol{\sigma}$

It is often more convenient to work with $\boldsymbol{\sigma}$ than $\mathbf{s}$ since, for example, in a biaxial state of stress $\boldsymbol{\sigma}$ has four stress components where $\mathbf{s}$ has five.* Equations (7.31) and (7.42) may be expressed in terms of $\boldsymbol{\sigma}$ instead of $\mathbf{s}$ as follows.

* Since $\mathbf{s}$ is obtained from σ by subtracting $\mathbf{I}\sigma_v$ a fifth component $\mathbf{s}_{33} = -\sigma_v$ appears in the $i = 3, j = 3$ position for a biaxial state of stress $\sigma = \sigma_{ij}$.

From (7.19) and (7.21)

$$\mathbf{s} = \boldsymbol{\sigma} - \mathbf{I}\sigma_v, \tag{7.48}$$

$$\bar{\mathbf{s}} = 0, \tag{7.49}$$

$$\bar{\boldsymbol{\sigma}} = 3\sigma_v. \tag{7.50}$$

Forming the square and cube of (7.48), and taking the trace of each making use of (7.50) and (3.53), gives the following results

$$\mathbf{ss} = \boldsymbol{\sigma\sigma} - 2\boldsymbol{\sigma}\sigma_v + \mathbf{I}\sigma_v^2 \tag{7.51}$$

$$\overline{\mathbf{ss}} = \overline{\boldsymbol{\sigma\sigma}} - 3\sigma_v^2, \tag{7.52}$$

$$\mathbf{sss} = \boldsymbol{\sigma\sigma\sigma} - 3\boldsymbol{\sigma\sigma}\sigma_v + 3\boldsymbol{\sigma}\sigma_v^2 - \mathbf{I}\sigma_v^3, \tag{7.53}$$

$$\overline{\mathbf{sss}} = \overline{\boldsymbol{\sigma\sigma\sigma}} - 3\overline{\boldsymbol{\sigma\sigma}}\sigma_v + 6\sigma_v^3. \tag{7.54}$$

Substituting (7.48), (7.51), (7.52), (7.53), (7.54) into (7.31) yields the following expression for incompressible creep in terms of $\boldsymbol{\sigma}$

$$\boldsymbol{\epsilon}(t) = \mathbf{d}(t) = \int_0^t K_2(\dot{\boldsymbol{\sigma}} - \mathbf{I}\dot{\sigma}_v)\, d\xi_1 + \int_0^t\int_0^t K_6[\dot{\boldsymbol{\sigma}}\dot{\boldsymbol{\sigma}} - 2\dot{\boldsymbol{\sigma}}\dot{\sigma}_v - \tfrac{1}{3}\mathbf{I}$$

$$(\overline{\dot{\boldsymbol{\sigma}}\dot{\boldsymbol{\sigma}}} - 6\dot{\sigma}_v^2)]\, d\xi_1\, d\xi_2 + \int_0^t\int_0^t\int_0^t K_{12}(\dot{\boldsymbol{\sigma}}\dot{\boldsymbol{\sigma}}\dot{\boldsymbol{\sigma}} + \overline{\dot{\boldsymbol{\sigma}}\dot{\boldsymbol{\sigma}}}\dot{\boldsymbol{\sigma}} - 3\dot{\boldsymbol{\sigma}}\dot{\boldsymbol{\sigma}}\dot{\sigma}_v - \tfrac{1}{3}\mathbf{I}\,\overline{\dot{\boldsymbol{\sigma}}\dot{\boldsymbol{\sigma}}\dot{\boldsymbol{\sigma}}})\, d\xi_1\, d\xi_2\, d\xi_3. \tag{7.55}$$

Similarly the expression for linearly compressible creep may be obtained by substituting (7.48), (7.51), (7.52), (7.53) and (7.54) into (7.42)

$$\boldsymbol{\epsilon}(t) = \int_0^t [\mathbf{I}K_1\bar{\dot{\boldsymbol{\sigma}}} + K_2(\dot{\boldsymbol{\sigma}} - \mathbf{I}\dot{\sigma}_v)]\, d\xi_1$$

$$+ \int_0^t\int_0^t K_6[\dot{\boldsymbol{\sigma}}\dot{\boldsymbol{\sigma}} - 2\dot{\boldsymbol{\sigma}}\dot{\sigma}_v - \tfrac{1}{3}\mathbf{I}(\overline{\dot{\boldsymbol{\sigma}}\dot{\boldsymbol{\sigma}}} - 6\dot{\sigma}_v^2)]\, d\xi_1\, d\xi_2$$

$$+ \int_0^t\int_0^t\int_0^t K_{12}(\dot{\boldsymbol{\sigma}}\dot{\boldsymbol{\sigma}}\dot{\boldsymbol{\sigma}} + \overline{\dot{\boldsymbol{\sigma}}\dot{\boldsymbol{\sigma}}}\dot{\boldsymbol{\sigma}} - 3\dot{\boldsymbol{\sigma}}\dot{\boldsymbol{\sigma}}\dot{\sigma}_v - \tfrac{1}{3}\mathbf{I}\,\overline{\dot{\boldsymbol{\sigma}}\dot{\boldsymbol{\sigma}}\dot{\boldsymbol{\sigma}}})\, d\xi_1\, d\xi_2\, d\xi_3. \tag{7.56}$$

Taking the trace of (7.55) and (7.56) and introducing a hydrostatic stress, for which $\bar{\boldsymbol{\sigma}} = 3\sigma_v$, the right-hand side of (7.55) reduces to zero and the right-hand side of (7.56) reduces to a linear term as they should.

7.9 Incompressible and Linearly Compressible Relaxation in Terms of $\boldsymbol{\epsilon}$

Equations (7.41) and (7.43) may be expressed in terms of the strain rate tensor $\dot{\boldsymbol{\epsilon}}$ instead of $\dot{\mathbf{d}}$ by introducing

$$\dot{\mathbf{d}} = \dot{\boldsymbol{\epsilon}} - \mathbf{I}\dot{\varepsilon}_v$$

from (7.32) together with the required products and traces. These result in exactly the same expressions as (7.48) through (7.54) except replacing $\mathbf{s}$ by $\dot{\mathbf{d}}$, $\boldsymbol{\sigma}$ by $\dot{\boldsymbol{\epsilon}}$ and σ_v by $\dot{\varepsilon}_v$. Since (7.41) and (7.43) have the same form as (7.31) and (7.42), respectively, except replacing stress terms by corresponding strain terms, it follows that the expression for incompressible stress relaxation in terms of $\boldsymbol{\epsilon}$ is

$$\begin{aligned}\boldsymbol{\sigma}(t) = \mathbf{s}(t) = &\int_0^t \psi_2(\dot{\boldsymbol{\epsilon}} - \mathbf{I}\dot{\varepsilon}_v)\, d\xi_1 + \int_0^t\int_0^t \psi_6[\dot{\boldsymbol{\epsilon}}\dot{\boldsymbol{\epsilon}} - 2\dot{\boldsymbol{\epsilon}}\dot{\varepsilon}_v - \tfrac{1}{3}\mathbf{I}(\overline{\dot{\boldsymbol{\epsilon}}\dot{\boldsymbol{\epsilon}}} - 6\dot{\varepsilon}_v^2)]\, d\xi_1\, d\xi_2 \\ &+ \int_0^t\int_0^t\int_0^t \psi_{12}(\dot{\boldsymbol{\epsilon}}\dot{\boldsymbol{\epsilon}}\dot{\boldsymbol{\epsilon}} + \overline{\dot{\boldsymbol{\epsilon}}\dot{\boldsymbol{\epsilon}}}\dot{\boldsymbol{\epsilon}} - 3\dot{\boldsymbol{\epsilon}}\dot{\boldsymbol{\epsilon}}\dot{\varepsilon}_v - \tfrac{1}{3}\mathbf{I}\overline{\dot{\boldsymbol{\epsilon}}\dot{\boldsymbol{\epsilon}}\dot{\boldsymbol{\epsilon}}})\, d\xi_1\, d\xi_2\, d\xi_3.\end{aligned} \tag{7.57}$$

Similarly the expression for linearly compressible stress relaxation in terms of $\boldsymbol{\epsilon}$ is

$$\begin{aligned}\boldsymbol{\sigma}(t) = &\int_0^t [\mathbf{I}\psi_1\bar{\dot{\boldsymbol{\epsilon}}} + \psi_2(\dot{\boldsymbol{\epsilon}} - \mathbf{I}\dot{\varepsilon}_v)]\, d\xi_1 + \int_0^t\int_0^t \psi_6[\dot{\boldsymbol{\epsilon}}\dot{\boldsymbol{\epsilon}} - 2\dot{\boldsymbol{\epsilon}}\dot{\varepsilon}_v - \tfrac{1}{3}\mathbf{I}(\overline{\dot{\boldsymbol{\epsilon}}\dot{\boldsymbol{\epsilon}}} - 6\dot{\varepsilon}_v^2)]\, d\xi_1\, d\xi_2 \\ &+ \int_0^t\int_0^t\int_0^t \psi_{12}(\dot{\boldsymbol{\epsilon}}\dot{\boldsymbol{\epsilon}}\dot{\boldsymbol{\epsilon}} + \overline{\dot{\boldsymbol{\epsilon}}\dot{\boldsymbol{\epsilon}}}\,\dot{\boldsymbol{\epsilon}} - 3\dot{\boldsymbol{\epsilon}}\dot{\boldsymbol{\epsilon}}\dot{\varepsilon}_v - \tfrac{1}{3}\mathbf{I}\overline{\dot{\boldsymbol{\epsilon}}\dot{\boldsymbol{\epsilon}}\dot{\boldsymbol{\epsilon}}})\, d\xi_1\, d\xi_2\, d\xi_3.\end{aligned} \tag{7.58}$$

7.10 Constitutive Relations under Biaxial Stress and Strain

If (7.9) is restricted to biaxial states of stress then a further reduction in the number of material functions may be effected by employing the Hamilton-Cayley equation for biaxial stress (3.50) and (3.54) and (3.35) for the

2×2 matrix. Care must be exercized in using a 2×2 matrix since its use precludes the possibility of describing those stress or strain components associated with the 3×3 matrix which do not appear in the 2×2 matrix. For example in a state of plane stress a 2×2 matrix describes all the non-zero stress components but does not describe the lateral strain.

Creep. Substituting (3.50) in the K_6 and K_{11} terms of (7.9) and (3.54) in the K_{12} term and making use of (3.35) in K_{11} yields expressions which can be combined with other K-terms so as to eliminate kernel functions K_6, K_{11} and K_{12} in (7.9).

Thus for biaxial stress,

$$\begin{aligned}\boldsymbol{\epsilon}(t) = &\int_0^t [\mathbf{I}K_1\bar{\dot{\boldsymbol{\sigma}}}(\xi_1) + K_2\dot{\boldsymbol{\sigma}}(\xi_1)]\,d\xi_1 + \int_0^t\int_0^t \{\mathbf{I}[K_3'\bar{\dot{\boldsymbol{\sigma}}}(\xi_1)\bar{\dot{\boldsymbol{\sigma}}}(\xi_2) + K_4'\overline{\dot{\boldsymbol{\sigma}}(\xi_1)\dot{\boldsymbol{\sigma}}(\xi_2)}]\\ &+ K_5'\bar{\dot{\boldsymbol{\sigma}}}(\xi_1)\dot{\boldsymbol{\sigma}}(\xi_2)\}\,d\xi_1\,d\xi_2\\ &+ \int_0^t\int_0^t\int_0^t \{\mathbf{I}[K_7'\overline{\dot{\boldsymbol{\sigma}}(\xi_1)\,\dot{\boldsymbol{\sigma}}(\xi_2)\,\dot{\boldsymbol{\sigma}}}(\xi_3) + K_8'\overline{\dot{\boldsymbol{\sigma}}(\xi_1)\dot{\boldsymbol{\sigma}}}(\xi_2)\,\bar{\dot{\boldsymbol{\sigma}}}(\xi_3)]\\ &+ K_9'\bar{\dot{\boldsymbol{\sigma}}}(\xi_1)\,\bar{\dot{\boldsymbol{\sigma}}}(\xi_2)\,\dot{\boldsymbol{\sigma}}(\xi_3) + K_{10}'\overline{\dot{\boldsymbol{\sigma}}(\xi_1)\dot{\boldsymbol{\sigma}}}\,(\xi_2)\,\dot{\boldsymbol{\sigma}}(\xi_3)\}\,d\xi_1\,d\xi_2\,d\xi_3,\end{aligned} \tag{7.59}$$

where

$$\begin{aligned} K_3' &= K_3 - \tfrac{1}{2}K_6,\\ K_4' &= K_4 + \tfrac{1}{2}K_6,\\ K_5' &= K_5 + K_6,\\ K_7' &= K_7 + K_{11} + K_{12},\\ K_8' &= K_8 - K_{11} - K_{12},\\ K_9' &= K_9 + K_{11} + \tfrac{1}{2}K_{12},\\ K_{10}' &= K_{10} + \tfrac{1}{2}K_{12}.\end{aligned} \tag{7.60}$$

Therefore under biaxial stress states the independent material functions reduce from twelve to nine.

Notice that when inserting (3.50) and (3.54) into (7.9) it has been assumed that the material functions are fully symmetric. Otherwise a generalized Hamilton-Cayley theorem by Rivlin [7.10] must be used instead. Also (7.59) is limited to a 2×2 matrix. Substituting a 3×3 matrix into it does not yield the correct value for ε_{33}.

Stress Relaxation. Using the same approach to the relaxation formulation (7.12) as used for creep the following expression can be obtained for a biaxial state of strain (for small strains) using a 2×2 matrix.

$$
\begin{aligned}
\boldsymbol{\sigma}(t) = & \int_0^t [\mathbf{I}\psi_1 \bar{\dot{\boldsymbol{\epsilon}}}(\xi_1) + \psi_2 \dot{\boldsymbol{\epsilon}}(\xi_1)]\, d\xi_1 \\
& + \int_0^t \int_0^t \{\mathbf{I}[\psi_3 \bar{\dot{\boldsymbol{\epsilon}}}(\xi_1)\, \bar{\dot{\boldsymbol{\epsilon}}}(\xi_2) + \psi_4' \overline{\dot{\boldsymbol{\epsilon}}(\xi_1)\, \dot{\boldsymbol{\epsilon}}(\xi_2)}] + \psi_5' \bar{\dot{\boldsymbol{\epsilon}}}(\xi_1)\, \dot{\boldsymbol{\epsilon}}(\xi_2)\}\, d\xi_1\, d\xi_2 \\
& + \int_0^t \int_0^t \int_0^t \{\mathbf{I}[\psi_7' \overline{\dot{\boldsymbol{\epsilon}}(\xi_1)\, \dot{\boldsymbol{\epsilon}}(\xi_2)\, \dot{\boldsymbol{\epsilon}}(\xi_3)} + \psi_8' \overline{\dot{\boldsymbol{\epsilon}}(\xi_1)\, \dot{\boldsymbol{\epsilon}}(\xi_2)}\, \bar{\dot{\boldsymbol{\epsilon}}}(\xi_3)] \\
& + \psi_9' \bar{\dot{\boldsymbol{\epsilon}}}(\xi_1)\, \bar{\dot{\boldsymbol{\epsilon}}}(\xi_2)\, \dot{\boldsymbol{\epsilon}}(\xi_3) + \psi_{10}' \overline{\dot{\boldsymbol{\epsilon}}(\xi_1)\, \dot{\boldsymbol{\epsilon}}(\xi_2)}\, \dot{\boldsymbol{\epsilon}}(\xi_3)\}\, d\xi_1\, d\xi_2\, d\xi_3,
\end{aligned}
\tag{7.61}
$$

where

$$
\begin{aligned}
\psi_3' &= \psi_3 - \tfrac{1}{2}\psi_6, \\
\psi_4' &= \psi_4 + \tfrac{1}{2}\psi_6, \\
\psi_5' &= \psi_5 + \psi_6, \\
\psi_7' &= \psi_7 + \psi_{11} + \psi_{12}, \\
\psi_8' &= \psi_8 - \psi_{11} - \psi_{12}, \\
\psi_9' &= \psi_9 + \psi_{11} + \tfrac{1}{2}\psi_{12}, \\
\psi_{10}' &= \psi_{10} + \tfrac{1}{2}\psi_{12}.
\end{aligned}
\tag{7.62}
$$

Thus under biaxial strain states the independent material functions reduce from twelve to nine.

In deriving (7.61) the strain equivalent of (3.50), (3.54) and (3.35) were employed. Since these forms result from taking one of the principal values (strain in this case) to be zero and limiting consideration to a 2×2 matrix it follows that the definition of biaxial strain to which (7.61) is applicable is a strain state in which one of the principal strains is zero. Equation (7.61) is not applicable to another possible definition of biaxial strain; i.e., a state of strain in which one principal strain is not prescribed while all other strains are prescribed. In this instance the unprescribed principal strain is generally not zero. Because of the limitation to a 2×2 matrix (7.61) will not yield the correct value of stress σ_{33} required to maintain $\varepsilon_{33} = 0$.

7.11 Constitutive Relations under Uniaxial Stress and Strain

Creep. The uniaxial constitutive equation for creep is obtained from (7.9) by substituting the 3×3 stress matrix

$$\boldsymbol{\sigma} = \begin{vmatrix} \sigma & 0 & 0 \\ 0 & 0 & 0 \\ 0 & 0 & 0 \end{vmatrix}$$

together with its products and corresponding traces from column 1 of Table 3.1 as follows:

$$\begin{aligned}
\boldsymbol{\epsilon}(t) = & \int_0^t [\mathbf{I} K_1 \dot{\sigma}(\xi_1) + K_2 \dot{\boldsymbol{\sigma}}(\xi_1)]\, d\xi_1 \\
& + \int_0^t \int_0^t [\mathbf{I}(K_3+K_4)\, \dot{\sigma}(\xi_1)\dot{\sigma}(\xi_2) + K_5\, \dot{\sigma}(\xi_1)\, \dot{\boldsymbol{\sigma}}(\xi_2) \\
& + K_6\, \dot{\boldsymbol{\sigma}}(\xi_1)\, \dot{\boldsymbol{\sigma}}(\xi_2)]\, d\xi_1\, d\xi_2 + \int_0^t \int_0^t \int_0^t [\mathbf{I}(K_7+K_8)\, \dot{\sigma}(\xi_1)\, \dot{\sigma}(\xi_2)\, \dot{\sigma}(\xi_3) \\
& + (K_9+K_{10})\, \dot{\sigma}(\xi_1)\, \dot{\sigma}(\xi_2)\, \dot{\boldsymbol{\sigma}}(\xi_3) + K_{11}\, \dot{\sigma}(\xi_1)\, \dot{\boldsymbol{\sigma}}(\xi_2)\, \dot{\boldsymbol{\sigma}}(\xi_3) \\
& + K_{12}\, \dot{\boldsymbol{\sigma}}(\xi_1)\, \dot{\boldsymbol{\sigma}}(\xi_2)\, \dot{\boldsymbol{\sigma}}(\xi_3)]\, d\xi_1\, d\xi_2\, d\xi_3.
\end{aligned} \tag{7.63}$$

There are nine independent kernel functions in (7.63). However, when (7.63) is expressed in terms of strain components the kernel functions can be combined into six independent functions. Note, if (7.59) for the 2×2 matrix were used to derive the uniaxial equation a different expression would result containing six independent kernel functions but the expression would be limited to a 2×2 matrix. Both expressions yield the same strain components, however.

Stress Relaxation. Introducing the 3×3 strain matrix for uniaxial strain

$$\boldsymbol{\epsilon} = \begin{vmatrix} \varepsilon & 0 & 0 \\ 0 & 0 & 0 \\ 0 & 0 & 0 \end{vmatrix}$$

and required products and traces into (7.12) yields a relation exactly like (7.63) except interchange ε and σ and substitute ψ for K.

7.12 Strain Components for Biaxial and Uniaxial Stress States, Compressible Material

In this section strain components will be derived for biaxial and uniaxial stress states only. The same principles may be applied to the general stress state.

Since the biaxial form given by (7.59) is limited to a 2×2 matrix the components described below will be obtained from the general equations for the 3×3 matrix. For biaxial stress

$$\boldsymbol{\sigma} = \begin{vmatrix} \sigma_{11} & \sigma_{12} & 0 \\ \sigma_{12} & \sigma_{22} & 0 \\ 0 & 0 & 0 \end{vmatrix} = \begin{vmatrix} \sigma_1 & \tau & 0 \\ \tau & \sigma_2 & 0 \\ 0 & 0 & 0 \end{vmatrix} \tag{7.64}$$

and the various products of $\boldsymbol{\sigma}$ and corresponding traces are given in column 4 of Table 3.1.

General Biaxial Equations. The strain components given by (7.9) for the biaxial stress state (7.64) are found by substituting appropriate values from column 4 of Table 3.1 for each component of stress into (7.9). For example, the required values from Table 3.1 for use in computing ε_{11} are as follows from column 4 of Table 3.1

$$\begin{aligned} \mathbf{I} = 1, \quad \boldsymbol{\sigma} = \sigma_1, \quad \boldsymbol{\sigma} &= \sigma_1 + \sigma_2, \quad \overline{\boldsymbol{\sigma\sigma}} = \sigma_1^2 + \sigma_2^2 + 2\tau^2, \\ \boldsymbol{\sigma\sigma} = \sigma_1^2 + \tau^2, &\quad \overline{\boldsymbol{\sigma\sigma\sigma}} = \sigma_1^3 + \sigma_2^3 + 3(\sigma_1 + \sigma_2)\tau^2, \\ \boldsymbol{\sigma\sigma\sigma} &= \sigma_1^3 + 2\sigma_1\tau^2 + \sigma_2\tau^2. \end{aligned} \tag{7.65}$$

Introducing (7.65) into (7.9) yields the following expression for the strain component $\varepsilon_{11}(t)$,

$$\begin{aligned} \varepsilon_{11}(t) = &\int_0^t [K_1(\dot\sigma_1 + \dot\sigma_2) + K_2\dot\sigma_1]\, d\xi_1 + \int_0^t\int_0^t [K_3(\dot\sigma_1 + \dot\sigma_2)^2 + K_4(\dot\sigma_1^2 + \dot\sigma_2^2 + 2\dot\tau^2) \\ &+ K_5(\dot\sigma_1 + \dot\sigma_2)\dot\sigma_1 + K_6(\dot\sigma_1^2 + \dot\tau^2)]\, d\xi_1\, d\xi_2 + \int_0^t\int_0^t\int_0^t \{K_7[\dot\sigma_1^3 + \dot\sigma_2^3 \\ &+ 3(\dot\sigma_1 + \dot\sigma_2)\dot\tau^2] + K_8[(\dot\sigma_1^2 + \dot\sigma_2^2 + 2\dot\tau^2)(\dot\sigma_1 + \dot\sigma_2)] + K_9(\dot\sigma_1 + \dot\sigma_2)^2\dot\sigma_1 \\ &+ K_{10}(\dot\sigma_1^2 + \dot\sigma_2^2 + 2\dot\tau^2)\dot\sigma_1 + K_{11}(\dot\sigma_1 + \dot\sigma_2)(\dot\sigma_1^2 + \dot\tau^2) \\ &+ K_{12}(\dot\sigma_1^3 + 2\dot\sigma_1\dot\tau^2 + \dot\sigma_2\dot\tau^2)\}\, d\xi_1\, d\xi_2\, d\xi_3. \end{aligned} \tag{7.66}$$

The other five components may be computed in a similar manner. After simplifying and rearranging all six components may be expressed as follows:

$$\varepsilon_{11}(t) = \int_0^t [F_1\dot{\sigma}_1+(F_1-G_1)\dot{\sigma}_2]\,d\xi_1+\int_0^t\int_0^t [F_2\dot{\sigma}_1^2+(F_2-G_3)\dot{\sigma}_2^2$$
$$+(2F_2-F_5-G_3)\dot{\sigma}_1\dot{\sigma}_2+F_5\dot{\tau}^2]\,d\xi_1\,d\xi_2+\int_0^t\int_0^t\int_0^t [F_3\dot{\sigma}_1^3+(F_3-G_4)\dot{\sigma}_2^3$$
$$+(3F_3-F_4+G_2-2G_4)\dot{\sigma}_2^2\dot{\sigma}_1+(3F_3-F_4-G_4)\dot{\sigma}_1^2\dot{\sigma}_2+F_4\dot{\sigma}_1\dot{\tau}^2$$
$$+(F_4-G_2)\dot{\sigma}_2\dot{\tau}^2]\,d\xi_1\,d\xi_2\,d\xi_3, \tag{7.67}$$

$$\varepsilon_{22}(t) = \int_0^t [F_1\dot{\sigma}_2+(F_1-G_1)\dot{\sigma}_1]\,d\xi_1+\int_0^t\int_0^t [F_2\dot{\sigma}_2^2+(F_2-G_3)\dot{\sigma}_1^2$$
$$+(2F_2-F_5-G_3)\dot{\sigma}_1\dot{\sigma}_2+F_5\dot{\tau}^2]\,d\xi_1\,d\xi_2$$
$$+\int_0^t\int_0^t\int_0^t [F_3\dot{\sigma}_2^3+(F_3-G_4)\dot{\sigma}_1^3+(3F_3-F_4+G_2-2G_4)\dot{\sigma}_1^2\dot{\sigma}_2$$
$$+(3F_3-F_4-G_4)\dot{\sigma}_2^2\dot{\sigma}_1+F_4\dot{\sigma}_2\dot{\tau}^2+(F_4-G_2)\dot{\sigma}_1\dot{\tau}^2]\,d\xi_1\,d\xi_2\,d\xi_3, \tag{7.68}$$

$$\varepsilon_{33}(t) = \int_0^t (F_1-G_1)(\dot{\sigma}_1+\dot{\sigma}_2)\,d\xi_1+\int_0^t\int_0^t [(F_2-G_3)(\dot{\sigma}_1^2+\dot{\sigma}_2^2)$$
$$+(2F_2-F_6-2G_3)\dot{\sigma}_1\dot{\sigma}_2+F_6\dot{\tau}^2]\,d\xi_1\,d\xi_2$$
$$+\int_0^t\int_0^t\int_0^t [(F_3-G_4)(\dot{\sigma}_1^3+\dot{\sigma}_2^3)+(3F_3-F_7-3G_4)(\dot{\sigma}_1\dot{\sigma}_2^2+\dot{\sigma}_1^2\dot{\sigma}_2)$$
$$+F_7(\dot{\sigma}_1+\dot{\sigma}_2)\,\dot{\tau}^2]\,d\xi_1\,d\xi_2\,d\xi_3, \tag{7.69}$$

$$\varepsilon_{12}(t) = \int_0^t G_1\dot{\tau}\,d\xi_1+\int_0^t\int_0^t G_3(\dot{\sigma}_1+\dot{\sigma}_2)\,\dot{\tau}\,d\xi_1\,d\xi_2+\int_0^t\int_0^t\int_0^t [G_2\dot{\tau}^3+G_4(\dot{\sigma}_1^2+\dot{\sigma}_2^2)\,\dot{\tau}$$
$$+(2G_4-G_2)\dot{\sigma}_1\dot{\sigma}_2\dot{\tau}]\,d\xi_1\,d\xi_2\,d\xi_3, \tag{7.70}$$

$$\varepsilon_{13} = \varepsilon_{23} = 0, \tag{7.71}$$

where

$$
\begin{aligned}
F_1 &= K_1+K_2,\\
F_2 &= K_3+K_4+K_5+K_6 = K_3'+K_4'+K_5',\\
F_3 &= K_7+K_8+K_9+K_{10}+K_{11}+K_{12} = K_7'+K_8'+K_9'+K_{10}',\\
F_4 &= 3K_7+2K_8+2K_{10}+K_{11}+2K_{12} = 3K_7'+2K_8'+2K_{10}',\\
F_5 &= 2K_4+K_6 = 2K_4',\\
F_6 &= 2K_4,\\
F_7 &= 3K_7+2K_8,\\
G_1 &= K_2,\\
G_2 &= 2K_{10}+K_{12} = 2K_{10}',\\
G_3 &= K_5+K_6 = K_5',\\
G_4 &= K_9+K_{10}+K_{11}+K_{12} = K_9'+K_{10}',
\end{aligned} \tag{7.72}
$$

and where σ_1, σ_2, τ and F_i, G_i are functions of time parameters ξ_i.

Note that the strain components corresponding to the 2×2 matrix, i.e., ε_{11}, ε_{22}, ε_{12} may be determined for the biaxial stress case from only nine independent kernel functions, F_1, F_2, F_3, F_4, F_5, G_1, G_2, G_3, G_4. However, the strain component ε_{33} requires two additional kernel functions F_6 and F_7.

Axial Plus Torsion. The components of strain corresponding to axial stress plus torsion,

$$\boldsymbol{\sigma} = \begin{vmatrix} \sigma & \tau & 0 \\ \tau & 0 & 0 \\ 0 & 0 & 0 \end{vmatrix},$$

may be obtained by setting $\sigma_1 = \sigma, \sigma_2 = 0$ in (7.67)–(7.71) with the following result:

$$
\begin{aligned}
\varepsilon_{11}(t) = &\int_0^t F_1\dot{\sigma}\,\mathrm{d}\xi_1 + \int_0^t\int_0^t (F_2\dot{\sigma}^2+F_5\dot{\tau}^2)\,\mathrm{d}\xi_1\,\mathrm{d}\xi_2\\
&+\int_0^t\int_0^t\int_0^t (F_3\dot{\sigma}^3+F_4\dot{\sigma}\dot{\tau}^2)\,\mathrm{d}\xi_1\,\mathrm{d}\xi_2\,\mathrm{d}\xi_3,
\end{aligned} \tag{7.73}
$$

$$
\begin{aligned}
\varepsilon_{22}(t) = &\int_0^t (F_1-G_1)\dot{\sigma}\,\mathrm{d}\xi_1 + \int_0^t\int_0^t [(F_2-G_3)\dot{\sigma}^2+F_5\dot{\tau}^2]\,\mathrm{d}\xi_1\,\mathrm{d}\xi_2\\
&+\int_0^t\int_0^t\int_0^t [(F_3-G_4)\dot{\sigma}^3+(F_4-G_2)\dot{\sigma}\dot{\tau}^2]\,\mathrm{d}\xi_1\,\mathrm{d}\xi_2\,\mathrm{d}\xi_3,
\end{aligned} \tag{7.74}
$$

$$\varepsilon_{33}(t) = \int_0^t (F_1-G_1)\dot{\sigma}\, d\xi_1 + \int_0^t\int_0^t [(F_2-G_3)\dot{\sigma}^2 + F_6\dot{\tau}^2]\, d\xi_1\, d\xi_2$$

$$+ \int_0^t\int_0^t\int_0^t [(F_3-G_4)\dot{\sigma}^3 + F_7\dot{\sigma}\dot{\tau}^2]\, d\xi_1\, d\xi_2\, d\xi_3, \tag{7.75}$$

$$\varepsilon_{12}(t) = \tfrac{1}{2}\gamma_1 = \int_0^t G_1\dot{\tau}\, d\xi_1 + \int_0^t\int_0^t G_3\dot{\sigma}\dot{\tau}\, d\xi_1\, d\xi_2$$

$$+ \int_0^t\int_0^t\int_0^t (G_2\dot{\tau}^3 + G_4\dot{\sigma}^2\dot{\tau})\, d\xi_1\, d\xi_2\, d\xi_3, \tag{7.76}$$

$$\varepsilon_{13} = \varepsilon_{23} = 0. \tag{7.77}$$

Pure Torsion. Components of strain corresponding to pure torsion

$$\boldsymbol{\sigma} = \begin{vmatrix} 0 & \tau & 0 \\ \tau & 0 & 0 \\ 0 & 0 & 0 \end{vmatrix},$$

are found by setting $\sigma_1 = \sigma_2 = 0$ in (7.67)–(7.71) or (7.73) to (7.77) as follows

$$\varepsilon_{11}(t) = \varepsilon_{22}(t) = \int_0^t\int_0^t F_5\dot{\tau}^2\, d\xi_1\, d\xi_2, \tag{7.78}$$

$$\varepsilon_{33}(t) = \int_0^t\int_0^t F_6\dot{\tau}^2\, d\xi_1\, d\xi_2, \tag{7.79}$$

$$\varepsilon_{12}(t) = \int_0^t G_1\dot{\tau}\, d\xi_1 + \int_0^t\int_0^t\int_0^t G_2\dot{\tau}^3\, d\xi_1\, d\xi_2\, d\xi_3, \tag{7.80}$$

$$\varepsilon_{13} = \varepsilon_{23} = 0. \tag{7.81}$$

Pure Axial Stress. Components of strain for pure uniaxial stress,

$$\boldsymbol{\sigma} = \begin{vmatrix} \sigma & 0 & 0 \\ 0 & 0 & 0 \\ 0 & 0 & 0 \end{vmatrix},$$

are found by setting $\sigma_1 = \sigma, \sigma_2 = \tau = 0$ in (7.67) to (7.71) or (7.73) to (7.77) as follows:

$$\varepsilon_{11}(t) = \int_0^t F_1\dot{\sigma}\, d\xi_1 + \int_0^t\int_0^t F_2\dot{\sigma}^2\, d\xi_1\, d\xi_2 + \int_0^t\int_0^t\int_0^t F_3\dot{\sigma}^3\, d\xi_1\, d\xi_2\, d\xi_3, \tag{7.82}$$

$$\varepsilon_{22}(t) = \varepsilon_{33}(t) = \int_0^t (F_1 - G_1)\,\dot{\sigma}\, d\xi_1 + \int_0^t\int_0^t (F_2 - G_3)\,\dot{\sigma}^2\, d\xi_1\, d\xi_2$$
$$+ \int_0^t\int_0^t\int_0^t (F_3 - G_4)\dot{\sigma}^3\, d\xi_1\, d\xi_2\, d\xi_3, \tag{7.83}$$

$$\varepsilon_{12} = \varepsilon_{13} = \varepsilon_{23} = 0. \tag{7.84}$$

7.13 Strain Components for Biaxial and Uniaxial Stress States, Linearly Compressible Material

Using the 3×3 matrix for the biaxial stress state

$$\boldsymbol{\sigma} = \begin{vmatrix} \sigma_1 & \tau & 0 \\ \tau & \sigma_2 & 0 \\ 0 & 0 & 0 \end{vmatrix}$$

and substituting $\boldsymbol{\sigma}$ and the several products of $\boldsymbol{\sigma}$ and corresponding traces from column 4 of Table 3.1 into (7.56) yields the following strain components.

General Biaxial Equations:

$$\varepsilon_{11}(t) = \int_0^t [K_1(\dot{\sigma}_1 + \dot{\sigma}_2) + K_2(\tfrac{2}{3}\dot{\sigma}_1 - \tfrac{1}{3}\dot{\sigma}_2)]\, d\xi_1$$
$$+ \int_0^t\int_0^t K_6(\tfrac{2}{9}\dot{\sigma}_1^2 - \tfrac{1}{9}\dot{\sigma}_2^2 - \tfrac{2}{9}\dot{\sigma}_1\dot{\sigma}_2 + \tfrac{1}{3}\dot{\tau}^2)\, d\xi_1\, d\xi_2$$
$$+ \int_0^t\int_0^t\int_0^t K_{12}(\tfrac{2}{3}\dot{\sigma}_1^3 - \tfrac{1}{3}\dot{\sigma}_2^3 + \sigma_1\dot{\sigma}_2^2 - \dot{\sigma}_1^2\dot{\sigma}_2 + 2\dot{\sigma}_1\dot{\tau}^2 - \dot{\sigma}_2\dot{\tau}^2)\, d\xi_1\, d\xi_2\, d\xi_3, \tag{7.85}$$

$$\begin{aligned}\varepsilon_{22}(t) = &\int_0^t [K_1(\dot\sigma_1+\dot\sigma_2)+K_2(\tfrac{2}{3}\dot\sigma_2-\tfrac{1}{3}\dot\sigma_1)]\,d\xi_1\\ &+\int_0^t\int_0^t K_6(\tfrac{2}{9}\dot\sigma_2^2-\tfrac{1}{9}\dot\sigma_1^2-\tfrac{2}{9}\dot\sigma_1\dot\sigma_2+\tfrac{1}{3}\dot\tau^2)\,d\xi_1\,d\xi_2\\ &+\int_0^t\int_0^t\int_0^t K_{12}(\tfrac{2}{3}\dot\sigma_2^3-\tfrac{1}{3}\dot\sigma_1^3+\dot\sigma_1^2\dot\sigma_2-\dot\sigma_2^2\dot\sigma_1+2\dot\sigma_2\dot\tau^2-\dot\sigma_1\dot\tau^2)\,d\xi_1\,d\xi_2\,d\xi_3,\end{aligned} \tag{7.86}$$

$$\begin{aligned}\varepsilon_{33}(t) = &\int_0^t (K_1-\tfrac{1}{3}K_2)(\dot\sigma_1+\dot\sigma_2)\,d\xi_1\\ &+\int_0^t\int_0^t K_6(-\tfrac{1}{9}\dot\sigma_1^2-\tfrac{1}{9}\dot\sigma_2^2+\tfrac{4}{9}\dot\sigma_1\dot\sigma_2-\tfrac{2}{3}\dot\tau^2)\,d\xi_1\,d\xi_2\\ &+\int_0^t\int_0^t\int_0^t K_{12}(-\tfrac{1}{3}\dot\sigma_1^3-\tfrac{1}{3}\dot\sigma_2^3-\dot\sigma_1\dot\tau^2-\dot\sigma_2\dot\tau^2)\,d\xi_1\,d\xi_2\,d\xi_3,\end{aligned} \tag{7.87}$$

$$\begin{aligned}\varepsilon_{12}(t) = &\int_0^t K_2\dot\tau\,d\xi_1+\int_0^t\int_0^t \tfrac{1}{3}K_6(\dot\sigma_1+\dot\sigma_2)\dot\tau\,d\xi_1\,d\xi_2\\ &+\int_0^t\int_0^t\int_0^t K_{12}(3\dot\tau^3+\dot\sigma_1^2\dot\tau+\dot\sigma_2^2\dot\tau-\dot\sigma_1\dot\sigma_2\dot\tau)\,d\xi_1\,d\xi_2\,d\xi_3,\end{aligned} \tag{7.88}$$

$$\varepsilon_{13} = \varepsilon_{23} = 0. \tag{7.89}$$

The trace $\bar{\epsilon}$ from (7.85), (7.86) and (7.87) is

$$\varepsilon_{11}+\varepsilon_{22}+\varepsilon_{33} = \int_0^t 3K_1(\dot\sigma_1+\dot\sigma_2)\,d\xi_1 \tag{7.90}$$

as required for the linearly compressible material. A comparison of (7.85) to (7.89) with the corresponding components of the general biaxial equation

(7.67) to (7.71) shows that all types of stress terms are represented in (7.85) to (7.89) except in ε_{33} (7.87) where terms involving $\dot{\sigma}_1\dot{\sigma}_2^2$ and $\dot{\sigma}_1^2\dot{\sigma}_2$ are missing.

Axial Stress Plus Torsion, Pure Torsion and Pure Axial Stress. The strain components corresponding to these states of stress are readily obtained from (7.85) to (7.88) by setting $\sigma_1 = \sigma$, $\sigma_2 = 0$ for axial stress plus torsion, $\sigma_1 = \sigma_2 = 0$ for pure torsion and $\sigma_2 = \tau = 0$ for pure axial stress. The resulting stress components show that for these states of stress all types of stress terms found in the corresponding component equations for the general compressible form (7.73) through (7.84) are found in the component equations for linearly compressible material under axial stress plus torsion, pure torsion and pure axial stress.

7.14 Stress Components for Biaxial and Uniaxial Strain States

The stress components resulting from given strain states may be derived in the same manner as the strain components were derived in Sections 7.12 and 7.13. Since all equations for the stress relaxation situation are exactly the same mathematically as for the creep situation the stress components for a given strain state may be obtained from the equations for the corresponding stress state by interchanging stress and strain type symbols. Hence these equations will not be repeated in this section except for the relaxation equivalent of the combinations of kernel functions given in (7.72). For stress relaxation the following notation will be used

$$
\begin{aligned}
\mathcal{F}_1 &= \psi_1+\psi_2\\
\mathcal{F}_2 &= \psi_3+\psi_4+\psi_5+\psi_6\\
\mathcal{F}_3 &= \psi_7+\psi_8+\psi_9+\psi_{10}+\psi_{11}+\psi_{12}\\
\mathcal{F}_4 &= 3\psi_7+2\psi_8+2\psi_{10}+\psi_{11}+2\psi_{12}\\
\mathcal{F}_5 &= 2\psi_4+\psi_6\\
\mathcal{F}_6 &= 2\psi_4\\
\mathcal{F}_7 &= 3\psi_7+2\psi_8\\
\mathcal{G}_1 &= \psi_2\\
\mathcal{G}_2 &= 2\psi_{10}+\psi_{12}\\
\mathcal{G}_3 &= \psi_5+\psi_6\\
\mathcal{G}_4 &= \psi_9+\psi_{10}+\psi_{11}+\psi_{12}
\end{aligned}
\tag{7.91}
$$

However, a physically different situation arises in seemingly simple strain states from that found in the corresponding simple stress states. Hence, the uniaxial stress relaxation situation will be discussed in detail as follows.

The components of stress corresponding to a pure uniaxial strain state,

$$\boldsymbol{\epsilon} = \begin{vmatrix} \varepsilon & 0 & 0 \\ 0 & 0 & 0 \\ 0 & 0 & 0 \end{vmatrix}, \tag{7.92}$$

are found by interchanging stress and strain type symbols in (7.82) to (7.84) as follows

$$\sigma_{11}(t) = \int_0^t \mathcal{F}_1 \dot{\varepsilon} \, d\xi_1 + \int_0^t \int_0^t \mathcal{F}_2 \dot{\varepsilon}^2 \, d\xi_1 \, d\xi_2 + \int_0^t \int_0^t \int_0^t \mathcal{F}_3 \dot{\varepsilon}^3 \, d\xi_1 \, d\xi_2 \, d\xi_3, \tag{7.93}$$

$$\sigma_{22}(t) = \sigma_{33}(t) = \int_0^t (\mathcal{F}_1 - \mathcal{G}_1)\dot{\varepsilon} \, d\xi_1 + \int_0^t \int_0^t (\mathcal{F}_2 - \mathcal{G}_3)\dot{\varepsilon}^2 \, d\xi_1 \, d\xi_2$$

$$+ \int_0^t \int_0^t \int_0^t (\mathcal{F}_3 - \mathcal{G}_4)\dot{\varepsilon}^3 \, d\xi_1 \, d\xi_2 \, d\xi_3, \tag{7.94}$$

$$\sigma_{12} = \sigma_{13} = \sigma_{23} = 0. \tag{7.95}$$

The above describes a uniaxial strain state; but it does not describe the strain state associated with the conventional stress relaxation test. In the conventional test an axial strain is prescribed and no other stresses or strains are imposed. This is an unconfined uniaxial stress-relaxation situation. It is not, however, a simple strain state, but a mixed boundary state consisting of prescribed axial strain and prescribed lateral stresses (zero stress). The resulting strain state is triaxial but varies with time. Even for constant axial strain the lateral strains vary with time.

Note for comparison that there is only one non-zero strain component ε in (7.92) and the lateral stress components σ_{22} and σ_{33} are functions of ε and time and hence not zero. To achieve a uniaxial strain state in a nonlinear viscoelastic material requires imposing lateral stresses which vary with strain and time as given by (7.94) as well as imposing the desired axial strain ε

In the conventional (or unconfined) state of "uniaxial" stress relaxation the strain tensor may be written as follows

$$\boldsymbol{\epsilon}(t) = \begin{vmatrix} \varepsilon(t) & 0 & 0 \\ 0 & \varkappa(t) & 0 \\ 0 & 0 & \varkappa(t) \end{vmatrix}, \tag{7.96}$$

where $\varepsilon(t)$ and $\varkappa(t)$ are prescribed axial and undetermined transverse strains respectively. Determining the products and traces of (7.96) from column 6 of Table 3.1 using appropriate strain symbols and inserting these into (7.12) the following components of the stress tensor are found,

$$\sigma_{11}(t) = \int_0^t [(\psi_1+\psi_2)\dot{\varepsilon}+2\psi_1\dot{\varkappa}]\, d\xi_1 + \int_0^t\int_0^t [(\psi_3+\psi_4+\psi_5+\psi_6)\,\dot{\varepsilon}^2+(4\psi_3+2\psi_5)\,\dot{\varepsilon}\dot{\varkappa}$$
$$+(4\psi_3+2\psi_4)\dot{\varkappa}^2]\, d\xi_1\, d\xi_2$$
$$+\int_0^t\int_0^t\int_0^t [(\psi_7+\psi_8+\psi_9+\psi_{10}+\psi_{11}+\psi_{12})\,\dot{\varepsilon}^3+(2\psi_8+4\psi_9+2\psi_{11})\dot{\varepsilon}^2\dot{\varkappa}$$
$$+(2\psi_8+4\psi_9+2\psi_{10})\,\dot{\varepsilon}\dot{\varkappa}^2+(2\psi_7+4\psi_8)\dot{\varkappa}^3]\, d\xi_1\, d\xi_2\, d\xi_3, \tag{7.97}$$

$$\sigma_{22}(t) = \sigma_{33}(t) = \int_0^t [(2\psi_1+\psi_2)\dot{\varkappa}+\psi_1\dot{\varepsilon}]\;\; d\xi_1 + \int_0^t\int_0^t [(\psi_3+\psi_4)\dot{\varepsilon}^2$$
$$+(4\psi_3+\psi_5)\,\dot{\varepsilon}\dot{\varkappa}+(4\psi_3+2\psi_4+2\psi_5+\psi_6)\dot{\varkappa}^2]\, d\xi_1\, d\xi_2$$
$$+\int_0^t\int_0^t\int_0^t [(\psi_7+\psi_8)\,\dot{\varepsilon}^3+(2\psi_8+\psi_9+\psi_{10})\,\dot{\varepsilon}^2\dot{\varkappa}+(2\psi_8+4\psi_9+\psi_{11})\dot{\varepsilon}\dot{\varkappa}^2$$
$$+(2\psi_7+4\psi_8+4\psi_9+2\psi_{10}+2\psi_{11}+\psi_{12})\,\dot{\varkappa}^3]\, d\xi_1\, d\xi_2\, d\xi_3, \tag{7.98}$$

$$\sigma_{12} = \sigma_{13} = \sigma_{23} = 0. \tag{7.99}$$

Since in this conventional (or unconfined) state of "uniaxial" stress relaxation the transverse stresses σ_{22}, σ_{33} are zero, $\varkappa(t)$ can be obtained as a function of the prescribed axial strain $\varepsilon(t)$ by setting (7.98) equal to zero. This relation for $\varkappa(t)$ can be obtained in principle but may be difficult in practice.

7.15 Approximating Nonlinear Constitutive Equations under Short Time Loading

In some engineering problems, the stress and strain under short duration loading are of interest. In this case, the nonlinear constitutive equations derived in the previous sections can be reduced to a somewhat simpler form. In the following, the approximation resulting from short time loading will be demonstrated using the nonlinear constitutive relation under uniaxial creep stress. Under this stress state, the creep strain in the stress direction is given by (7.82), which can be rewritten into the following form for convenience,

$$\varepsilon_{11}(t) = \int_0^t F_1(t-\xi_1)\dot{\sigma}(\xi_1)\,d\xi_1 + \int_0^t\int_0^t F_2(t-\xi_1, t-\xi_2)\dot{\sigma}(\xi_1)\dot{\sigma}(\xi_2)\,d\xi_1\,d\xi_2$$
$$+ \int_0^t\int_0^t\int_0^t F_3(t-\xi_1, t-\xi_2, t-\xi_3)\dot{\sigma}(\xi_1)\dot{\sigma}(\xi_2)\dot{\sigma}(\xi_3)\,d\xi_1\,d\xi_2\,d\xi_3. \tag{7.100}$$

Integration by parts of the first order term Δ_1 of (7.100) yields

$$\Delta_1 = F_1(0)\,\sigma(t) - \int_0^t \left[\frac{\partial}{\partial \xi_1} F_1(t-\xi_1)\right] \sigma(\xi_1)\,d\xi_1. \tag{7.101}$$

Integration by parts of the second order term of (7.100) yields

$$\Delta_2 = \int_0^t \left[F_2(t-\xi_1, 0)\,\sigma(t) - \int_0^t \frac{\partial}{\partial \xi_2} F_2(t-\xi_1, t-\xi_2)\,\sigma(\xi_2)\,d\xi_2\right] \dot{\sigma}(\xi_1)\,d\xi_1. \tag{7.102}$$

Integration by parts again of equation (7.102) yields

$$\Delta_2 = F_2(0, 0)\,\sigma^2(t) - \int_0^t \left[\frac{\partial}{\partial \xi_1} F_2(t-\xi_1, 0)\right] \sigma(t)\,\sigma(\xi_1)\,d\xi_1$$
$$- \int_0^t \left[\frac{\partial}{\partial \xi_2} F_2(0, t-\xi_2)\right] \sigma(\xi_2)\,\sigma(t)\,d\xi_2$$
$$+ \int_0^t\int_0^t \left[\frac{\partial^2}{\partial \xi_1\,\partial \xi_2} F_2(t-\xi_1, t-\xi_2)\right] \sigma(\xi_1)\,\sigma(\xi_2)\,d\xi_1\,d\xi_2. \tag{7.103}$$

In (7.103) the second and third terms can be combined into a single term due to the fact that F_2 is symmetric with respect to the time arguments and ξ_1 and ξ_2 are merely dummy variables. Furthermore if the last term in (7.103) can be neglected, which will be discussed later, then (7.103) can be reduced to the following form

$$\Delta_2 = F_2(0, 0)\sigma^2(t) - \int_0^t \left[\frac{\partial}{\partial \xi} F_2(t-\xi, 0)\right] [\sigma(t)\,\sigma(\xi) + \sigma(\xi)\,\sigma(t)]\, \mathrm{d}\xi + R_2(t, t), \tag{7.104}$$

where $R_2(t, t)$ is the remainder, which is equal to the last term of (7.103).

A similar procedure of integration by parts can be used for the third order term Δ_3 of (7.100). This results in

$$\Delta_3 = F_3(0, 0, 0)\,\sigma^3(t) - \int_0^t \left[\frac{\partial}{\partial \xi} F_3(t-\xi, 0, 0)\right] [\sigma(\xi)\,\sigma^2(t) + \sigma^2(t)\,\sigma(\xi) + \sigma(t)\,\sigma(\xi)\,\sigma(t)]\, \mathrm{d}\xi + R_3(t, t, t), \tag{7.105}$$

where $R_3(t, t, t)$ is the remainder. If the remainders R_2 and R_3 can be neglected, then the multiple integral representation of (7.100) can be reduced to a single integral representation as follows from (7.103), (7.104) and (7.105)

$$\begin{aligned} \varepsilon_{11}(t) = {} & F_1(0)\,\sigma(t) + F_2(0, 0)\,\sigma^2(t) + F_3(0, 0, 0)\,\sigma^3(t) \\ & - \int_0^t \{\dot{F}_1(t-\xi)\,\sigma(\xi) + \dot{F}_2(t-\xi, 0)\,[\sigma(t)\,\sigma(\xi) + \sigma(\xi)\,\sigma(t)] \\ & + \dot{F}_3(t-\xi, 0, 0)\,[\sigma^2(t)\,\sigma(\xi) + \sigma(\xi)\,\sigma^2(t) + \sigma(t)\,\sigma(\xi)\,\sigma(t)]\}\, \mathrm{d}\xi, \end{aligned} \tag{7.106}$$

where $\dot{F}_1(t-\xi) = (\partial/\partial\xi)\,F_1(t-\xi)$, etc. In (7.106) the first three terms represent the nonlinear "instantaneous response" or elastic response. The error resulting from neglecting the R_2 and R_3 terms depends on the material functions F_2 and F_3, on the stress intensity and on the length of time. This has been discussed in detail by Huang and Lee [7.11]. For example, consider R_2. Equating (7.103) and (7.104) yields

$$R_2(t, t) = \int_0^t \int_0^t \left[\frac{\partial^2}{\partial \xi_1\, \partial \xi_2} F_2(t-\xi_1, t-\xi_2)\right] \sigma(\xi_1)\,\sigma(\xi_2)\, \mathrm{d}\xi_1\, \mathrm{d}\xi_2. \tag{7.107}$$

If the maximum absolute value of the stress $\sigma(\xi_1)$, $\sigma(\xi_2)$ in the given time interval is less than say M then (7.107) becomes

$$R_2(t, t) \leqslant M^2 \int_0^t \int_0^t \left[\frac{\partial^2}{\partial \xi_1 \, \partial \xi_2} F_2(t-\xi_1, t-\xi_2) \right] d\xi_1 \, d\xi_2. \tag{7.108}$$

Integrating (7.108) twice yields

$$R_2(t, t) \leqslant M^2[F_2(0, 0) - 2F_2(0, t) + F_2(t, t)]. \tag{7.109}$$

If the material function F_2 can be approximated by

$$F_2(t_1, t_2) = b_0 + b_1(t_1 + t_2) + b_2(t_1^2 + t_2^2) + b_{12} t_1 t_2, \tag{7.110}$$

then substituting (7.110) into the right-hand side of (7.109) yields $R_2 \leqslant M^2|b_{12}|t^2$.

This provides an upper bound for R_2. It is to be noted that if the material function F_2 does not contain any cross product terms such as $b_{12}t_1t_2$, R_2 is zero.

The short time approximation was developed by Huang and Lee [7.11] and has been extended by Lubliner [7.12] to include large deformations.

7.16 **Superposed Small Loading on a Large Constant Loading**

Another possible simplification of the multiple integral representation which may be employed in certain circumstances, will be discussed in this section. In some engineering problems, the loading consists of a large constant load, such as dead weight, and a superimposed time-dependent small load. Under this condition, the multiple integral representation for nonlinear viscoelasticity can be reduced to a single integral representation. Again, creep under uniaxial stress will be discussed for illustration. Consider a given time-dependent stress which can be approximated by

$$\sigma(t) = \sigma_0 H(t) + \sigma_t(t), \tag{7.111}$$

where σ_0 is a constant and assumed to be much larger than $\sigma_t(t)$, which is time-dependent.

Inserting (7.111) into (7.100) yields

$$\varepsilon_{11}(t) = F_1(t)\,\sigma_0+F_2(t,\,t)\sigma_0^2+F_3(t,\,t,\,t)\sigma_0^3$$
$$+\int_0^t F_1(t-\xi)\dot{\sigma}_t(\xi)\,\mathrm{d}\xi+2\sigma_0\int_0^t F_2(t-\xi,\,0)\,\dot{\sigma}_t(\xi)\,\mathrm{d}\xi$$
$$+3\sigma_0^2\int_0^t F_3(t-\xi,\,0,\,0)\dot{\sigma}_t(\xi)\,\mathrm{d}\xi+R_4(t), \tag{7.112}$$

where the remainder R_4 in (7.112) contains double and triple integral terms which are dependent nonlinearly on $\sigma_t(t)$. If the remainder R_4 can be neglected then (7.112) becomes a single integral form. This kind of approximation was first proposed by Pipkin and Rivlin [7.13] and later extended by Appleby and Lee [7.14] and Goldberg and Lianis [7.15].

7.17 Other Representations

In addition to the multiple integral representation and its various approximate forms discussed in this chapter and in subsequent chapters, several other representations have been developed for describing the behavior of nonlinear viscoelastic materials.

Some of these are simple in form and may be readily adapted for laboratory characterization of nonlinear viscoelastic materials with relatively less effort than the multiple integral method. The ability of these theories to predict complex inputs is less satisfactory, however. On the other hand, some of these representations are able to describe the response to complex inputs because of their more complex formulation, though the effort required for a laboratory characterization is substantial.

In the following three representations are reviewed which permit fairly straightforward material characterization and have demonstrated the capability of predicting nonlinear viscoelastic behavior of at least some materials with fairly satisfactory accuracy. For derivations of these representations the reader is referred to the original papers cited.

7.18 Finite Linear Viscoelasticity

A finite linear viscoelasticity theory described by Lianis [7.16] and others [7.17–7.21] is an approximate form of the theory of finite viscoelasticity originally developed by Coleman and Noll [7.22]. The constitutive equation as used by Lianis is a stress relaxation formulation which contains three material constants plus four relaxation functions. For uniaxial stressing (unconfined conventional relaxation) this constitutive equation has the following form for an incompressible isotropic material

$$\sigma(t) = \left[\Omega^2 - \frac{1}{\Omega}\right]\left[a + b\left(\Omega^2 + \frac{2}{\Omega} - 3\right) + \frac{c}{\Omega}\right]$$
$$+2\int_0^t \psi_0(t-\xi)\frac{d}{d\xi}\left[\frac{\Omega^2(\xi)}{\Omega^2} - \frac{\Omega}{\Omega(\xi)}\right] d\xi$$
$$+2\int_0^t \psi_1(t-\xi)\frac{d}{d\xi}\left[\Omega^2(\xi) - \frac{1}{\Omega(\xi)}\right] d\xi$$
$$+2\int_0^t \psi_2(t-\xi)\frac{d}{d\xi}\left[\Omega^2(\xi) - \frac{1}{\Omega\Omega(\xi)}\right] d\xi$$
$$+\left(\Omega^2 - \frac{1}{\Omega}\right)\int_0^t \psi_3(t-\xi)\frac{d}{d\xi}\left[\Omega^2(\xi) + \frac{2}{\Omega(\xi)}\right] d\xi, \tag{7.113}$$

where $\sigma(t)$ is the uniaxial stress; $\Omega = \Omega(t)$ is the extension ratio at time t; $\Omega(\xi)$ is the extension ratio at any prior (generic) time ξ; a, b, c are time-independent material constants; and $\psi_0, \psi_1, \psi_2, \psi_3$ are time-dependent relaxation functions. The extension ratio $\Omega(t) = \varepsilon(t)+1$ is the ratio of sample length at time t to the initial length. Constants a, b, c are equilibrium values which remain after time effects have been dissipated under constant extension. These three constants and the four relaxation functions may be determined from a set of single step stress relaxation tests plus one two-step relaxation test [7.21] as follows.

Inserting into (7.113) a single step history of extension ratio defined as follows

$$\Omega(\xi) = 1, \quad [\varepsilon(\xi) = 0], \quad \xi < 0$$
$$\Omega(\xi) = \Omega_0 = \text{a constant}, \quad \xi > 0, \tag{7.114}$$

yields

$$\sigma(t) = \left\{\Omega_0^2 - \frac{1}{\Omega_0}\right\}\left\{\beta_1(t) + \frac{1}{\Omega_0}\beta_2(t) + (\Omega_0^2 - 1)\,\beta_3(t)\right\}, \tag{7.115}$$

where

$$\beta_1(t) = a-2b+2\psi_1(t)-2\psi_3(t)$$
$$\beta_2(t) = 2b+c+2\psi_0(t)+2\psi_2(t)+2\psi_3(t)$$
$$\beta_3(t) = b+2\psi_2(t)+\psi_3(t).$$

$\beta_1(t)$, $\beta_2(t)$ and $\beta_3(t)$ can be determined from a set of three relaxation tests at three different values of Ω_0. If these tests are carried to equilibrium, i.e., the stress no longer changes with time, then the values of a, b, c may be determined from equilibrium values of $\beta_1(\infty)$, $\beta_2(\infty)$, $\beta_3(\infty)$ by taking $\psi_0(\infty) = \psi_1(\infty) = \psi_2(\infty) = \psi_3(\infty) = 0$. Since there are four relaxation functions these can be determined from the three equations for β_1, β_2, β_3 only by obtaining a fourth relation. This may be found by using one two-step relaxation test [7.21].

The theory described above has been applied to several polymers under different extension histories with reasonable results [7.19–7.21]. No equivalent creep formulation has been reported.

7.19 Elastic Fluid Theory

A constitutive equation for an elastic fluid was developed by Bernstein, Kearsley and Zapas (BKZ) [7.23] and its applicability to elastomers was demonstrated by them. Under uniaxial stressing (unconfined extension or compression) the BKZ constitutive equation for an isotropic incompressible material has the form

$$\sigma(t) = -\int_0^t \dot{G}^*\left[\frac{\Omega(t)}{\Omega(\xi)},\ t-\xi\right]d\xi. \tag{7.116}$$

A special form of $\dot{G}^*$ as given in [7.24] is as follows

$$\dot{G}^*[\ ] = \left[\frac{\Omega^2(t)}{\Omega^2(\xi)}-\frac{\Omega(\xi)}{\Omega(t)}\right]F\left(\frac{\Omega(t)}{\Omega(\xi)}, t-\xi\right), \tag{7.117}$$

where $F(\ )$ is another function of $\Omega(t)/\Omega(\xi)$ and $t-\xi$, and Ω is the extension ratio. Under a single step of constant extension (or compression) as defined by (7.114) the relaxational stress given by (7.116) becomes

$$\sigma(t) = G^*(\Omega_0, t), \tag{7.118}$$

where

$$\frac{\partial G^*(\Omega_0, t)}{\partial t} = \dot{G}^*(\Omega_0, t).$$

In [7.23] the following form was obtained for $G^*(\Omega_0, t)$

$$G^*(\Omega_0, t) = \left(\Omega_0^2 - \frac{1}{\Omega_0}\right)\left[\beta_1(t) + \frac{1}{\Omega_0}\beta_2(t) + (\Omega_0^2 - 1)\beta_3(t)\right]. \qquad (7.119)$$

Comparison of (7.119) with (7.115) indicates that these two equations are identical for single step extension. The time functions $\beta_1(t)$, $\beta_2(t)$ and $\beta_3(t)$ may be determined from a set of three relaxation tests at three different values of extension ratio Ω_0.

The ability of this theory to describe behavior of nonlinear viscoelastic elastomers has been demonstrated in [7.18, 7.24, 7.25]. However, Smart and Williams [7.26] found this representation to be of little interest in describing behavior of polypropylene and poly (vinyl chloride) for moderate strains. No creep formulation corresponding to the above has been proposed.

7.20 Thermodynamic Constitutive Theory

The nonlinear constitutive theory presented by Schapery [7.27–7.30] was derived by the use of principles of thermodynamics of irreversible processes. It employed generalized coordinates (divided into hidden and observed) and generalized forces and used forms for free energy and entropy production suggested by an observation that nonlinear stress relaxation of various materials can be described in terms of the same time-dependent properties found in the linear range. Other work on constitutive equations based on thermodynamic principles includes that by Biot [7.31], Coleman [7.32] and Coleman and Gurtin [7.33].

A creep formulation employed by Schapery [7.28, 7.29] has the following form under constant temperature and uniaxial stress σ,

$$\varepsilon(t) = g_0 J_0\, \sigma(t) + g_1 \int_0^t \varphi(\zeta_\sigma - \zeta_\sigma')\, \frac{dg_2\, \sigma(\xi)}{d\xi}\, d\xi, \qquad (7.120)$$

where

$$\zeta_\sigma = \zeta_\sigma(t) = \int_0^t \frac{ds}{a_\sigma[\sigma(s)]}, \qquad (7.121a)$$

$$\zeta_\sigma' = \zeta_\sigma(\xi) = \int_0^\xi \frac{ds}{a_\sigma[\sigma(s)]}, \qquad (7.121b)$$

are reduced times and a_σ is a shift factor. At constant temperature g_0, g_1, g_2 and a_σ are functions of stress, which in turn is a function of time, that is all have the form $a_\sigma[\sigma(t)]$ as spelled out for a_σ in (7.121). J_0 is the time-independent compliance, $\varphi(t)$ is the creep compliance and ξ and s are generic times.

The corresponding unconfined stress relaxation formulation [7.27, 7.29] may be expressed as follows under constant temperature and uniaxial stress,

$$\sigma(t) = h_\infty E_\infty \varepsilon(t) + h_1 \int_0^t E_t(\zeta_\varepsilon - \zeta_\varepsilon') \frac{\mathrm{d}h_2\,\varepsilon(\xi)}{\mathrm{d}\xi}\,\mathrm{d}\xi, \tag{7.122a}$$

or alternatively

$$\sigma(t) = h_0 E_0 \varepsilon(t) - h_3 \int_0^t \psi(\zeta_\varepsilon - \zeta_\varepsilon') \frac{\mathrm{d}h_4\,\varepsilon(\xi)}{\mathrm{d}\xi}\,\mathrm{d}\xi, \tag{7.122b}$$

where

$$\zeta_\varepsilon = \zeta_\varepsilon(t) = \int_0^t \frac{\mathrm{d}s}{a_\varepsilon[\varepsilon(s)]}, \tag{7.123a}$$

$$\zeta_\varepsilon' = \zeta_\varepsilon'(\xi) = \int_0^\xi \frac{\mathrm{d}s}{a_\varepsilon[\varepsilon(s)]}, \tag{7.123b}$$

are reduced times and a_ε is a shift factor. At constant temperature h_0, h_1, h_2, h_3, h_4, h_∞, a_ε are functions of strain. E_∞ is the equilibrium (final value) of the modulus at constant stress, E_0 is the initial time-independent modulus, $E_t(t)$ is the transient modulus and $\psi(t)$ the relaxation modulus. These moduli are related as follows

$$E_0 = \psi(t) + E_t(t) + E_\infty. \tag{7.124}$$

None of the parameters in (7.120) and (7.122) are fundamental physical or thermodynamic constants. All must be determined from creep or relaxation or other mechanical tests of the material being considered.

Equations (7.120) and (7.122) are quite adjustable since in each there are five functions and a constant. In (7.120) these are stress functions $g_0(\sigma)$, $g_1(\sigma)$, $g_2(\sigma)$, $a_\sigma(\sigma)$, time function $\varphi(t)$ and constant J_0. These may be adjusted to describe various types of nonlinear behavior including linear elastic behavior. Defining $g_0 = g_1 = g_2 = a_\sigma = 1$ reduces (7.120) to the linear integral relaxation formulation (5.104). Similarly defining $h_0 = h_3 = h_4 = a_\varepsilon = 1$ reduces (7.122b) to the linear integral relaxation formulation (5.106).

Under constant stress, $\sigma(t) = \sigma_0 H(t)$, (7.120) and (7.121) reduce to

$$\varepsilon(t) = g_0(\sigma_0)\, J_0\sigma_0 + g_1(\sigma_0)\, g_2(\sigma_0)\varphi\left(\frac{t}{a_\sigma(\sigma_0)}\right)\sigma_0. \tag{7.125}$$

As shown in Chapter 8 the creep behavior of many materials under constant stress can be represented for modest strains by (8.121),

$$\varepsilon(t) = \varepsilon^0(\sigma) + \varepsilon^+(\sigma)t^n, \tag{7.126}$$

where $\varepsilon^0(\sigma)$, $\varepsilon^+(\sigma)$ are nonlinear functions of stress at constant temperature and n is a constant. For example, according to (8.54) $\varepsilon^0(\sigma)$, $\varepsilon^+(\sigma)$ may be expressed as follows:

$$\varepsilon^0(\sigma) = F_1^0\sigma + F_2^0\sigma^2 + F_3^0\sigma^3, \tag{7.126a}$$

$$\varepsilon^+(\sigma) = F_1^+\sigma + F_2^+\sigma^2 + F_3^+\sigma^3. \tag{7.126b}$$

The functions $g_0 J_0$, $g_1 g_2$ and $\varphi\,(t/a_\sigma)$ in (7.125) corresponding to (7.126) may be expressed as follows

$$g_0 J_0 = \varepsilon^0/\sigma_0, \tag{7.127a}$$

$$g_1 g_2/a_\sigma^n = \varepsilon^+/\sigma_0, \tag{7.127b}$$

$$a_\sigma^n \varphi\left(\frac{t}{a_\sigma}\right) = t^n, \tag{7.127c}$$

It is evident that determination of g_1, g_2, and a_σ from ε^+/σ_0 in (7.127b) using data from constant stress creep tests is not unique.

By assuming $g_1 = a_\sigma = 1$ and restating (7.127a), (7.127b) and (7.127c) for variable stress as

$$g_0 J_0 = \varepsilon^0(\sigma)/\sigma,$$

$$g_2 = \varepsilon^+(\sigma)/\sigma,$$

$$\varphi(t) = t^n,$$

respectively, and using

$$\frac{d\varepsilon^+(\sigma)}{dt} = \frac{d\varepsilon^+(\sigma)}{d\sigma}\frac{d\sigma}{dt}$$

equation (7.120) may be expressed as follows:

$$\varepsilon(t) = \varepsilon^0(\sigma) + \int_0^t (t-\xi)^n \frac{d\varepsilon^+(\sigma)}{d\sigma} \dot{\sigma}(\xi) \, d\xi. \tag{7.128}$$

This expression is equivalent to the modified superposition representation discussed in Chapter 9 and Appendix A5.

A more detailed discussion of material characterization is found in [7.29]. Thermodynamic type relations have been employed to describe several polymeric materials by Schapery [7.28, 7.29], Stafford [7.25] and Smart and Williams [7.26].

CHAPTER 8

NONLINEAR CREEP AT CONSTANT STRESS AND RELAXATION AT CONSTANT STRAIN

8.1 Introduction

In this chapter the constitutive equations for constant stress or strain are presented for compressible material using both the 3×3 tensor matrix and the 2×2 matrix. Also the linearly compressible material is presented. These equations are followed in each case by expressions for components of strain or stress for creep or relaxation under several typical states of stress or strain. In several instances the expressions for the components are presented for both creep and relaxation as a convenience to the reader. However, the expressions for creep and corresponding relaxation are identical except for symbols used. Hence, not all are repeated. For example the components for the most general constitutive equation under the most general stress state for creep is not presented for relaxation because of the length of the expressions.

When a state of stress $\boldsymbol{\sigma}$ (or strain $\boldsymbol{\epsilon}$) is applied abruptly at time $t = 0$ and is held constant for a stated period, then within that period all time parameters in the kernel functions reduce to t instead of $(t-\xi_i)$ and the integrals, in (7.9) for example, may be evaluated by means of the Dirac delta function $\delta(t)$ for which

$$\dot{\boldsymbol{\sigma}}(\xi) = \boldsymbol{\sigma}\delta(\xi) \tag{8.1}$$

for creep, or

$$\dot{\boldsymbol{\epsilon}}(\xi) = \boldsymbol{\epsilon}\delta(\xi) \tag{8.2}$$

for relaxation, see Appendix A3 for details of use of $\delta(t)$.

8.2 Constitutive Equations for 3×3 Matrix

Creep. Introducing (8.1) with $\boldsymbol{\sigma}$ a constant into (7.9) yields after integration the following expression for creep at constant stress $\boldsymbol{\sigma}$

$$\begin{aligned}\boldsymbol{\epsilon}(t) = \mathbf{I}(&K_1\bar{\boldsymbol{\sigma}}+K_3\bar{\boldsymbol{\sigma}}\bar{\boldsymbol{\sigma}}+K_4\overline{\boldsymbol{\sigma}\boldsymbol{\sigma}}+K_7\overline{\boldsymbol{\sigma}\boldsymbol{\sigma}\boldsymbol{\sigma}}+K_8\overline{\boldsymbol{\sigma}\boldsymbol{\sigma}}\bar{\boldsymbol{\sigma}})\\ &+\boldsymbol{\sigma}(K_2+K_5\bar{\boldsymbol{\sigma}}+K_9\bar{\boldsymbol{\sigma}}\,\bar{\boldsymbol{\sigma}}+K_{10}\overline{\boldsymbol{\sigma}\boldsymbol{\sigma}})\\ &+\boldsymbol{\sigma}^2(K_6+K_{11}\bar{\boldsymbol{\sigma}})+\boldsymbol{\sigma}^3K_{12},\end{aligned} \tag{8.3}$$

where $K_\alpha = K_\alpha(t)$ are the functions of time t and material for creep.

Relaxation: Inserting (8.2) with $\boldsymbol{\epsilon}$ a constant into (7.12) yields the following expression for stress relaxation at constant strain $\boldsymbol{\epsilon}$ after integration

$$\begin{aligned}\boldsymbol{\sigma}(t) = \mathbf{I}(&\psi_1\bar{\boldsymbol{\epsilon}}+\psi_3\bar{\boldsymbol{\epsilon}}\,\bar{\boldsymbol{\epsilon}}+\psi_4\overline{\boldsymbol{\epsilon\epsilon}}+\psi_7\overline{\boldsymbol{\epsilon\epsilon\epsilon}}+\psi_8\overline{\boldsymbol{\epsilon\epsilon}}\,\bar{\boldsymbol{\epsilon}})\\ &+\boldsymbol{\epsilon}(\psi_2+\psi_5\bar{\boldsymbol{\epsilon}}+\psi_9\bar{\boldsymbol{\epsilon}}\,\bar{\boldsymbol{\epsilon}}+\psi_{10}\overline{\boldsymbol{\epsilon\epsilon}})\\ &+\boldsymbol{\epsilon}^2(\psi_6+\psi_{11}\bar{\boldsymbol{\epsilon}})+\boldsymbol{\epsilon}^3\psi_{12},\end{aligned} \tag{8.4}$$

where $\psi_\alpha = \psi_\alpha\,(t)$ are functions of time and material for relaxation.

8.3 Components of Strain for Creep at Constant Stress

Triaxial Stress State. The most general state of triaxial stress is given by the stress tensor

$$\boldsymbol{\sigma} = \begin{vmatrix}\sigma_{11} & \sigma_{12} & \sigma_{13}\\ \sigma_{21} & \sigma_{22} & \sigma_{23}\\ \sigma_{31} & \sigma_{32} & \sigma_{33}\end{vmatrix}. \tag{8.5}$$

Computations for this most general state of stress are seldom required; when required it is usually simpler to first transform the coordinates so as to express the state of stress in terms of principal stresses as detailed in the next item. Introducing the general state of stress (8.5) into (8.3) yields six equations for the six components of strain. As an illustration one component ε_{11} of the strain tensor $\boldsymbol{\epsilon}$ will be derived as follows. To determine the expression for ε_{11} it is necessary to formulate the values of the following traces and products of traces corresponding to the general state of stress (8.5), $\bar{\boldsymbol{\sigma}}, \overline{\boldsymbol{\sigma\sigma}}, \overline{\boldsymbol{\sigma\sigma\sigma}}$. It is also necessary to determine the value of the component in the first row–first column position for the tensor and tensor products, $\boldsymbol{\sigma}, \boldsymbol{\sigma\sigma}$, and $\boldsymbol{\sigma\sigma\sigma}$. Some required products of the above quantities are equal to the simple product of the quantities. The traces are as follows: $\bar{\boldsymbol{\sigma}}$ as given by (3.25) is

$$\bar{\boldsymbol{\sigma}} = \sigma_{ii} = \sigma_{11}+\sigma_{22}+\sigma_{33}, \tag{8.6}$$

$\overline{\boldsymbol{\sigma\sigma}}$ as given by (3.28) is

$$\overline{\boldsymbol{\sigma\sigma}} = \sigma_{ij}\sigma_{ji} = \sigma_{11}^2+\sigma_{22}^2+\sigma_{33}^2+2\sigma_{12}^2+2\sigma_{13}^2+2\sigma_{23}^2, \tag{8.7}$$

and $\overline{\boldsymbol{\sigma\sigma\sigma}}$ as given by (3.31) is

$$\begin{aligned}\overline{\boldsymbol{\sigma\sigma\sigma}} = \sigma_{ip}\sigma_{pq}\sigma_{qi} &= \sigma_{11}^3+\sigma_{22}^3+\sigma_{33}^3+3\sigma_{12}^2(\sigma_{11}+\sigma_{22}),\\ &+3\sigma_{13}^2(\sigma_{11}+\sigma_{33})+3\sigma_{23}^2(\sigma_{22}+\sigma_{33})+6\sigma_{12}\sigma_{13}\sigma_{23}.\end{aligned} \tag{8.8}$$

The components of the tensors and products corresponding to ε_{11} are the following:

$$\mathbf{I}: 1, \tag{8.9}$$

$$\boldsymbol{\sigma}: \sigma_{11}. \tag{8.10}$$

From (3.27) using the symmetry property of the stress tensor the term in the first row–first column of the product $\boldsymbol{\sigma\sigma}$ is given by

$$\boldsymbol{\sigma\sigma}: \sigma_{11}^2+\sigma_{12}^2+\sigma_{13}^2. \tag{8.11}$$

The component in the first row–first column of the triple product $\boldsymbol{\sigma\sigma\sigma}$ is given by Ω_{11} in (3.30a),

$$\begin{aligned}\boldsymbol{\sigma\sigma\sigma}: \Omega_{11} = {} & \sigma_{11}^3+2\sigma_{11}(\sigma_{12}^2+\sigma_{13}^2)+\sigma_{22}\sigma_{12}^2+\sigma_{33}\sigma_{13}^2 \\ & +2\sigma_{21}\sigma_{13}\sigma_{32}.\end{aligned} \tag{8.12}$$

Substituting (8.6) through (8.12) into (8.3) yields ε_{11},

$$\varepsilon_{11}(t) = [\alpha_1]+\sigma_{11}[\alpha_2]+(\sigma_{11}^2+\sigma_{12}^2+\sigma_{13}^2)[\alpha_3]+\Omega_{11}[\alpha_4], \tag{8.13}$$

where

$$\begin{aligned}[\alpha_1] &= K_1\bar{\boldsymbol{\sigma}}+K_3(\bar{\boldsymbol{\sigma}})^2+K_4\overline{\boldsymbol{\sigma\sigma}}+K_7\overline{\boldsymbol{\sigma\sigma\sigma}}+K_8\overline{\boldsymbol{\sigma\sigma}}\,\bar{\boldsymbol{\sigma}}, \\ [\alpha_2] &= K_2+K_5\bar{\boldsymbol{\sigma}}+K_9(\bar{\boldsymbol{\sigma}})^2+K_{10}\overline{\boldsymbol{\sigma\sigma}}, \\ [\alpha_3] &= K_6+K_{11}\bar{\boldsymbol{\sigma}}, \\ [\alpha_4] &= K_{12},\end{aligned} \tag{8.14}$$

and where the expressions for $\bar{\boldsymbol{\sigma}}$, $\overline{\boldsymbol{\sigma\sigma}}$, $\overline{\boldsymbol{\sigma\sigma\sigma}}$ and Ω_{11} are given in (8.6), (8.7), (8.8) and (8.12) respectively. The $[\alpha]$ terms are common to all stress components, but their coefficients differ.

The other strain components of $\boldsymbol{\epsilon}$ may be obtained in a similar manner as follows considering the symmetry of the stress tensor,

$$\begin{aligned}\varepsilon_{22}(t) &= [\alpha_1]+\sigma_{22}[\alpha_2]+(\sigma_{12}^2+\sigma_{22}^2+\sigma_{32}^2)\,[\alpha_3]+\Omega_{22}[\alpha_4], \\ \varepsilon_{33}(t) &= [\alpha_1]+\sigma_{33}[\alpha_2]+(\sigma_{13}^2+\sigma_{23}^2+\sigma_{33}^2)\,[\alpha_3]+\Omega_{33}[\alpha_4], \\ \varepsilon_{12}(t) &= \varepsilon_{21}(t) = \tfrac{1}{2}\gamma_1(t) = \sigma_{12}[\alpha_2]+[\sigma_{12}(\sigma_{11}+\sigma_{22})+\sigma_{13}\sigma_{23}]\,[\alpha_3]+\Omega_{12}[\alpha_4] \\ \varepsilon_{13}(t) &= \varepsilon_{31}(t) = \tfrac{1}{2}\gamma_2(t) = \sigma_{13}[\alpha_2]+[\sigma_{13}(\sigma_{11}+\sigma_{33})+\sigma_{12}\sigma_{23}]\,[\alpha_3]+\Omega_{13}[\alpha_4] \\ \varepsilon_{23}(t) &= \varepsilon_{32}(t) = \tfrac{1}{2}\gamma_3(t) = \sigma_{23}[\alpha_2]+[\sigma_{23}(\sigma_{22}+\sigma_{33})+\sigma_{12}\sigma_{13}]\,[\alpha_3]+\Omega_{23}[\alpha_4],\end{aligned} \tag{8.15}$$

where the $[\alpha]$ terms are as in (8.14), and the Ω_{ij} terms are as given in (3.30a), and γ_i are components of conventional shear strain.

Triaxial Principal Stresses. The components of strain corresponding to the stress state given by

$$\boldsymbol{\sigma} = \begin{vmatrix} \sigma_{\mathrm{I}} & 0 & 0 \\ 0 & \sigma_{\mathrm{II}} & 0 \\ 0 & 0 & \sigma_{\mathrm{III}} \end{vmatrix} \tag{8.16}$$

may be determined from (8.13) through (8.15) by setting $\sigma_{11} = \sigma_{\mathrm{I}}$, $\sigma_{22} = \sigma_{\mathrm{II}}$, $\sigma_{33} = \sigma_{\mathrm{III}}$, $\sigma_{12} = \sigma_{13} = \sigma_{23} = 0$ or by employing the products and traces of (8.16) from column 6 of Table 3.1 in (8.3) with the following result.

$$\varepsilon_{\mathrm{I}}(t) = \mathsf{B}_1 + \sigma_{\mathrm{I}}\mathsf{B}_2 + \sigma_{\mathrm{I}}^2\mathsf{B}_3 + \sigma_{\mathrm{I}}^3 K_{12}, \tag{8.17}$$

$$\varepsilon_{\mathrm{II}}(t) = \mathsf{B}_1 + \sigma_{\mathrm{II}}\mathsf{B}_2 + \sigma_{\mathrm{II}}^2\mathsf{B}_3 + \sigma_{\mathrm{II}}^3 K_{12}, \tag{8.18}$$

$$\varepsilon_{\mathrm{III}}(t) = \mathsf{B}_1 + \sigma_{\mathrm{III}}\mathsf{B}_2 + \sigma_{\mathrm{III}}^2\mathsf{B}_3 + \sigma_{\mathrm{III}}^3 K_{12}, \tag{8.19}$$

$$\varepsilon_{12} = \varepsilon_{13} = \varepsilon_{23} = 0, \tag{8.20}$$

where

$$\begin{aligned} \mathsf{B}_1 &= K_1(\sigma_{\mathrm{I}}+\sigma_{\mathrm{II}}+\sigma_{\mathrm{III}}) + K_3(\sigma_{\mathrm{I}}+\sigma_{\mathrm{II}}+\sigma_{\mathrm{III}})^2 + K_4(\sigma_{\mathrm{I}}^2+\sigma_{\mathrm{II}}^2+\sigma_{\mathrm{III}}^2) \\ &\quad + K_7(\sigma_{\mathrm{I}}^3+\sigma_{\mathrm{II}}^3+\sigma_{\mathrm{III}}^3) + K_8(\sigma_{\mathrm{I}}^2+\sigma_{\mathrm{II}}^2+\sigma_{\mathrm{III}}^2)(\sigma_{\mathrm{I}}+\sigma_{\mathrm{II}}+\sigma_{\mathrm{III}}), \\ \mathsf{B}_2 &= K_2 + K_5(\sigma_{\mathrm{I}}+\sigma_{\mathrm{II}}+\sigma_{\mathrm{III}}) + K_9(\sigma_{\mathrm{I}}+\sigma_{\mathrm{II}}+\sigma_{\mathrm{III}})^2 + K_{10}(\sigma_{\mathrm{I}}^2+\sigma_{\mathrm{II}}^2+\sigma_{\mathrm{III}}^2), \\ \mathsf{B}_3 &= K_6 + K_{11}(\sigma_{\mathrm{I}}+\sigma_{\mathrm{II}}+\sigma_{\mathrm{III}}). \end{aligned} \tag{8.21}$$

Hydrostatic Stress (Either Tensile or Compressive): In this state of stress the three principal stresses are equal. Thus $\sigma_{\mathrm{I}} = \sigma_{\mathrm{II}} = \sigma_{\mathrm{III}} = \sigma_v$ and the state of stress is given from (8.16)

$$\boldsymbol{\sigma} = \begin{vmatrix} \sigma_v & 0 & 0 \\ 0 & \sigma_v & 0 \\ 0 & 0 & \sigma_v \end{vmatrix}, \tag{8.22}$$

Setting $\sigma_{\mathrm{I}} = \sigma_{\mathrm{II}} = \sigma_{\mathrm{III}} = \sigma_v$ in (8.17) through (8.21) yields the following for the strain in any direction in an isotropic material under hydrostatic stress.

$$\begin{aligned} \varepsilon_v(t) = \varepsilon_{\mathrm{I}} = \varepsilon_{\mathrm{II}} = \varepsilon_{\mathrm{III}} &= (3K_1+K_2)\sigma_v + (9K_3+3K_4+3K_5+K_6)\sigma_v^2 \\ &\quad + (3K_7+9K_8+9K_9+3K_{10}+3K_{11}+K_{12})\sigma_v^3. \end{aligned} \tag{8.23}$$

Thus the three principal strains are the same; in fact the strain in all directions is the same in an isotropic material. In conventional notation for pressure $p = -\sigma_v$ the change in volume $\Delta v/v = \varepsilon_{kk}$ for small strains is three times (8.23) or

$$\begin{aligned} \Delta v/v = \varepsilon_{kk} &= -p(9K_1+3K_2) + p^2(27K_3+9K_4+9K_5+3K_6) - p^3(9K_7+27K_8 \\ &\quad + 27K_9+9K_{10}+9K_{11}+3K_{12}). \end{aligned} \tag{8.24}$$

Biaxial Stress State. Components of strain for the biaxial stress state given by the 3×3 matrix,

$$\boldsymbol{\sigma} = \begin{vmatrix} \sigma_1 & \tau & 0 \\ \tau & \sigma_2 & 0 \\ 0 & 0 & 0 \end{vmatrix}, \tag{8.25}$$

may be determined by inserting (8.1) into (7.67) through (7.71) and integrating or by employing the products and traces of (8.25) from column 4 of Table 3.1 in (8.3), or by setting $\sigma_{33} = \sigma_{31} = \sigma_{32} = 0$ in (8.13) and (8.14) as follows. The first procedure yields the following after rearrangement,

$$\varepsilon_{11}(t) = \{A\}+\sigma_1\{B\}, \tag{8.26}$$

$$\varepsilon_{22}(t) = \{A\}+\sigma_2\{B\}, \tag{8.27}$$

$$\varepsilon_{33}(t) = \{A\}+\{C\}, \tag{8.28}$$

$$\varepsilon_{12}(t) = (\tfrac{1}{2})\gamma_1(t) = \tau\{B\}, \tag{8.29}$$

$$\varepsilon_{13} = \varepsilon_{23} = 0, \tag{8.30}$$

where

$$\begin{aligned}\{A\} = {} & (F_1-G_1)(\sigma_1+\sigma_2)+(F_2-G_3)(\sigma_1{}^2+\sigma_2{}^2)+(F_3-G_4)(\sigma_1{}^3+\sigma_2{}^3) \\ & +(2F_2-F_5-2G_3)\sigma_1\sigma_2+(3F_3-F_4+G_2-3G_4)(\sigma_1\sigma_2^2+\sigma_1^2\sigma_2) \\ & +F_5\tau^2+(F_4-G_2)(\sigma_1+\sigma_2)\tau^2,\end{aligned} \tag{8.31}$$

$$\{B\} = G_1+G_3(\sigma_1+\sigma_2)+G_4(\sigma_1+\sigma_2)^2+G_2(\tau^2-\sigma_1\sigma_2), \tag{8.32}$$

$$\{C\} = (F_6-F_5)(\tau^2-\sigma_1\sigma_2)+(F_7-F_4+G_2)(\sigma_1+\sigma_2)(\tau^2-\sigma_1\sigma_2). \tag{8.33}$$

See (7.72) for definition of F_i and G_i in terms of K_i.

Note that components ε_{11}, ε_{22}, ε_{12} can be determined from only nine kernel functions F_1, F_2, F_3, F_4, F_5, G_1, G_2, G_3, G_4 but ε_{33} requires two more kernel functions F_6, and F_7. It may be observed that the normal strains ε_{11}, ε_{22}, ε_{33} contain a term involving τ^2, $F_5\tau^2$ for ε_{11} and ε_{22} and $F_6\tau^2$ for ε_{33}. This indicates that shear stress affects the normal strains even in the absence of normal stresses. The shear stress τ appears only in the form τ^2 in any term of the expressions for normal strains. Thus a shear stress (or torque) of either sense has the same effect on the normal strains.

Biaxial Principal Stress State. The components of strain for biaxial stress in terms of principal stresses

$$\boldsymbol{\sigma} = \begin{vmatrix} \sigma_{\mathrm{I}} & 0 & 0 \\ 0 & \sigma_{\mathrm{II}} & 0 \\ 0 & 0 & 0 \end{vmatrix} \tag{8.34}$$

may be obtained from (8.26) through (8.33) by setting $\sigma_1 = \sigma_{\mathrm{I}}$, $\sigma_2 = \sigma_{\mathrm{II}}$ and $\tau = 0$ as follows.

$$\varepsilon_{11}(t) = \{A'\}+\sigma_{\mathrm{I}}\{B'\}, \tag{8.35}$$

$$\varepsilon_{22}(t) = \{A'\}+\sigma_{\mathrm{II}}\{B'\}, \tag{8.36}$$

$$\varepsilon_{33}(t) = \{A'\}+\{C'\}, \tag{8.37}$$

$$\varepsilon_{12} = \varepsilon_{13} = \varepsilon_{23} = 0, \tag{8.38}$$

where

$$\begin{aligned}\{A'\} = {} & (F_1-G_1)(\sigma_{\mathrm{I}}+\sigma_{\mathrm{II}})+(F_2-G_3)(\sigma_{\mathrm{I}}^2+\sigma_{\mathrm{II}}^2)\\ & +(F_3-G_4)(\sigma_{\mathrm{I}}^3+\sigma_{\mathrm{II}}^3)+(2F_2-F_5-2G_3)\sigma_{\mathrm{I}}\sigma_{\mathrm{II}}\\ & +(3F_3-F_4+G_2-3G_4)(\sigma_{\mathrm{I}}\sigma_{\mathrm{II}}^2+\sigma_{\mathrm{I}}^2\sigma_{\mathrm{II}}),\end{aligned} \tag{8.39}$$

$$\{B'\} = G_1+G_3(\sigma_{\mathrm{I}}+\sigma_{\mathrm{II}})+G_4(\sigma_{\mathrm{I}}+\sigma_{\mathrm{II}})^2-G_2\sigma_{\mathrm{I}}\sigma_{\mathrm{II}}, \tag{8.40}$$

$$\{C'\} = (F_5-F_6)\sigma_{\mathrm{I}}\sigma_{\mathrm{II}}+(F_4-F_7-G_2)(\sigma_{\mathrm{I}}\sigma_{\mathrm{II}}^2+\sigma_{\mathrm{I}}^2\sigma_{\mathrm{II}}). \tag{8.41}$$

Axial Stress plus Torsion (Shear): In a similar manner the components of strain for the state of stress given by

$$\boldsymbol{\sigma} = \begin{vmatrix}\sigma & \tau & 0\\ \tau & 0 & 0\\ 0 & 0 & 0\end{vmatrix}, \tag{8.42}$$

may be obtained from (8.26) through (8.33) by setting $\sigma_1 = \sigma$ and $\sigma_2 = 0$ as follows

$$\varepsilon_{11}(t) = F_1\sigma+F_2\sigma^2+F_3\sigma^3+F_4\sigma\tau^2+F_5\tau^2, \tag{8.43}$$

$$\varepsilon_{22}(t) = (F_1-G_1)\sigma+(F_2-G_3)\sigma^2+(F_3-G_4)\sigma^3+(F_4-G_2)\sigma\tau^2+F_5\tau^2, \tag{8.44}$$

$$\varepsilon_{33}(t) = (F_1-G_1)\sigma+(F_2-G_3)\sigma^2+(F_3-G_4)\sigma^3+F_7\sigma\tau^2+F_6\tau^2, \tag{8.45}$$

$$\varepsilon_{12}(t) = (\tfrac{1}{2})\gamma_1(t) = G_1\tau+G_2\tau^3+G_3\sigma\tau+G_4\sigma^2\tau, \tag{8.46}$$

$$\varepsilon_{13} = \varepsilon_{23} = 0. \tag{8.47}$$

Pure Torsion (Shear). The state of stress for pure shear is given by

$$\boldsymbol{\sigma} = \begin{vmatrix}0 & \tau & 0\\ \tau & 0 & 0\\ 0 & 0 & 0\end{vmatrix} \tag{8.48}$$

Strain components for this state of stress may be obtained by setting $\sigma_1 =$

$\sigma_2 = 0$ in (8.26) through (8.33) as follows:

$$\varepsilon_{11}(t) = \varepsilon_{22}(t) = F_5\tau^2 \tag{8.49}$$

$$\varepsilon_{33}(t) = F_6\tau^2, \tag{8.50}$$

$$\varepsilon_{12}(t) = (\tfrac{1}{2})\gamma_1 = G_1\tau + G_2\tau^3, \tag{8.51}$$

$$\varepsilon_{13} = \varepsilon_{23} = 0. \tag{8.52}$$

It is interesting to note from (8.49) and (8.50) that under pure shear the normal strains ε_{11}, ε_{22}, and ε_{33}, and therefore the change in volume in general are not zero but proportional to τ^2 (unless $F_6 = -2F_5$). Experiments by Onaran and Findley [8.1], Lenoe, Heller and Freudenthal [8.2] and Findley and Stanley [8.3] have demonstrated this effect.

Pure Axial Stress (*Tension or Compression*). The components of strain for pure axial stress,

$$\boldsymbol{\sigma} = \begin{vmatrix} \sigma & 0 & 0 \\ 0 & 0 & 0 \\ 0 & 0 & 0 \end{vmatrix}, \tag{8.53}$$

are found from (8.26) through (8.33) by setting $\sigma_1 = \sigma$, $\sigma_2 = \tau = 0$ as follows

$$\varepsilon_{11}(t) = F_1\sigma + F_2\sigma^2 + F_3\sigma^3, \tag{8.54}$$

$$\varepsilon_{22}(t) = \varepsilon_{33}(t) = (F_1 - G_1)\sigma + (F_2 - G_3)\sigma^2 + (F_3 - G_4)\sigma^3, \tag{8.55}$$

$$\varepsilon_{12} = \varepsilon_{13} = \varepsilon_{23} = 0. \tag{8.56}$$

Equations (8.54) and (8.55) show that both even and odd powers of σ may appear in the expression for pure axial stress (and other stress states involving normal stress components). This means that the axial strain resulting from a tensile stress, σ, is not necessarily of the same magnitude as one resulting from a compressive stress, $(-\sigma)$, see for example [8.4, 8.5].

A popular form of expression for nonlinear stress dependence, the power function σ^m, can also describe nonlinear behavior in tension. However, this function cannot describe behavior of a material which has different creep in tension and compression. Also, negative stresses require special interpretation of σ^m.

8.4 Components of Stress for Relaxation at Constant Strain

Since the equations for relaxation are mathematically the same as those for creep the components of stress for relaxation under any given state of strain $\boldsymbol{\epsilon}$ can be obtained from the equations for components of strain for the corresponding state of stress $\boldsymbol{\sigma}$ given above. It is only necessary to interchange creep and relaxation symbols. In view of this fact only a few states of strains will be presented in detail in this section for convenience of the reader.

Triaxial Principal Strains. The components of stress corresponding to the strain state,

$$\boldsymbol{\epsilon} = \begin{vmatrix} \varepsilon_{\mathrm{I}} & 0 & 0 \\ 0 & \varepsilon_{\mathrm{II}} & 0 \\ 0 & 0 & \varepsilon_{\mathrm{III}} \end{vmatrix}, \tag{8.57}$$

may be determined by employing the products and traces of (8.57) corresponding to column 6 of Table 3.1 in (8.4) with the following result,

$$\sigma_{\mathrm{I}}(t) = \beta_1 + \varepsilon_{\mathrm{I}}\beta_2 + \varepsilon_{\mathrm{I}}^2\beta_3 + \varepsilon_{\mathrm{I}}^3\psi_{12}, \tag{8.58}$$

$$\sigma_{\mathrm{II}}(t) = \beta_1 + \varepsilon_{\mathrm{II}}\beta_2 + \varepsilon_{\mathrm{II}}^2\beta_3 + \varepsilon_{\mathrm{II}}^3\psi_{12}, \tag{8.59}$$

$$\sigma_{\mathrm{III}}(t) = \beta_1 + \varepsilon_{\mathrm{III}}\beta_2 + \varepsilon_{\mathrm{III}}^2\beta_3 + \varepsilon_{\mathrm{III}}^3\psi_{12}, \qquad (8.60$$

$$\sigma_{12} = \sigma_{13} = \sigma_{23} = 0, \tag{8.61}$$

where

$$\beta_1 = \psi_1(\varepsilon_{\mathrm{I}} + \varepsilon_{\mathrm{II}} + \varepsilon_{\mathrm{III}}) + \psi_3(\varepsilon_{\mathrm{I}} + \varepsilon_{\mathrm{II}} + \varepsilon_{\mathrm{III}})^2 + \psi_4(\varepsilon_{\mathrm{I}}^2 + \varepsilon_{\mathrm{II}}^2 + \varepsilon_{\mathrm{III}}^2)$$
$$+\psi_7(\varepsilon_{\mathrm{I}}^3 + \varepsilon_{\mathrm{II}}^3 + \varepsilon_{\mathrm{III}}^3) + \psi_8(\varepsilon_{\mathrm{I}}^2 + \varepsilon_{\mathrm{II}}^2 + \varepsilon_{\mathrm{III}}^2)(\varepsilon_{\mathrm{I}} + \varepsilon_{\mathrm{II}} + \varepsilon_{\mathrm{III}}), \tag{8.62}$$

$$\beta_2 = \psi_2 + \psi_5(\varepsilon_{\mathrm{I}} + \varepsilon_{\mathrm{II}} + \varepsilon_{\mathrm{III}}) + \psi_9(\varepsilon_{\mathrm{I}} + \varepsilon_{\mathrm{II}} + \varepsilon_{\mathrm{III}})^2 + \psi_{10}(\varepsilon_{\mathrm{I}}^2 + \varepsilon_{\mathrm{II}}^2 + \varepsilon_{\mathrm{III}}{}^2), \tag{8.63}$$

$$\beta_3 = \psi_6 + \psi_{11}(\varepsilon_{\mathrm{I}} + \varepsilon_{\mathrm{II}} + \varepsilon_{\mathrm{III}}). \tag{8.64}$$

Biaxial Strain State. Strain components for the biaxial strain state in terms of the 3×3 matrix,

$$\boldsymbol{\epsilon} = \begin{vmatrix} \varepsilon_{11} & \varepsilon_{12} & 0 \\ \varepsilon_{12} & \varepsilon_{22} & 0 \\ 0 & 0 & 0 \end{vmatrix}, \tag{8.65}$$

may be determined by inserting the products and traces of (8.65) corresponding to column 4 of Table 3.1 into (8.4) as follows

$$\sigma_{11}(t) = \{\mathsf{A}\} + \varepsilon_{11}\{\mathsf{B}\}, \tag{8.66}$$

$$\sigma_{22}(t) = \{\mathsf{A}\} + \varepsilon_{22}\{\mathsf{B}\}, \tag{8.67}$$

$$\sigma_{33}(t) = \{\mathsf{A}\} + \{\mathsf{C}\}, \tag{8.68}$$

$$\sigma_{12}(t) = \varepsilon_{12}\{\mathsf{B}\}, \tag{8.69}$$

$$\sigma_{13} = \sigma_{23} = 0, \tag{8.70}$$

where

$$\begin{aligned}\{A\} = {} & ((\mathcal{F}_1-\mathcal{G}_1)(\varepsilon_{11}+\varepsilon_{22})+((\mathcal{F}_2-\mathcal{G}_3)(\varepsilon_{11}^2+\varepsilon_{22}^2)+((\mathcal{F}_3-\mathcal{G}_4)\\ & (\varepsilon_{11}^3+\varepsilon_{22}^3)+(2\mathcal{F}_2-\mathcal{F}_5-2\mathcal{G}_3)\,\varepsilon_{11}\varepsilon_{22}+(3\mathcal{F}_3-\mathcal{F}_4+\mathcal{G}_2-3\mathcal{G}_4)\\ & (\varepsilon_{11}\varepsilon_{22}^2+\varepsilon_{11}^2\varepsilon_{22})+((\mathcal{F}_4-\mathcal{G}_2)(\varepsilon_{11}+\varepsilon_{22})\varepsilon_{12}^2+\mathcal{F}_5\varepsilon_{12}^2,\end{aligned} \tag{8.71}$$

$$\{B\} = \mathcal{G}_1+\mathcal{G}_3(\varepsilon_{11}+\varepsilon_{22})+\mathcal{G}_4(\varepsilon_{11}+\varepsilon_{22})^2+\mathcal{G}_2(\varepsilon_{12}^2-\varepsilon_{11}\varepsilon_{22}), \tag{8.72}$$

$$\{C\} = ((\mathcal{F}_6-\mathcal{F}_5)(\varepsilon_{12}^2-\varepsilon_{11}\varepsilon_{22})+((\mathcal{F}_7-\mathcal{F}_4+\mathcal{G}_2)(\varepsilon_{11}+\varepsilon_{22})(\varepsilon_{12}^2-\varepsilon_{11}\varepsilon_{22}). \tag{8.73}$$

See (7.91) for definition of $\mathcal{F}_i$ and $\mathcal{G}_i$ in terms of ψ_i.

It should be noted that $\sigma_{33}(t)$ is not zero and is a function of time. Thus biaxial strain states given by (8.65) do not arise naturally in engineering structures when strains ε_{11}, ε_{22}, and ε_{12} are imposed. It is also necessary to impose a time varying stress $\sigma_{33}(t)$ as described by (8.68).

Axial Strain plus Shear Strain. The stress components corresponding to the strain state

$$\boldsymbol{\epsilon} = \begin{vmatrix} \varepsilon_{11} & \varepsilon_{12} & 0 \\ \varepsilon_{12} & 0 & 0 \\ 0 & 0 & 0 \end{vmatrix} \tag{8.74}$$

may be obtained from (8.66) through (8.73) by setting $\varepsilon_{22} = 0$. Note that neither $\sigma_{22}(t)$ nor $\sigma_{33}(t)$ are zero. Thus these stresses must be imposed in addition to ε_{11} and ε_{12} if the state of strain (8.74) is to result.

Pure Shear Strain. The stress components for pure shear strain

$$\boldsymbol{\epsilon} = \begin{vmatrix} 0 & \varepsilon_{12} & 0 \\ \varepsilon_{12} & 0 & 0 \\ 0 & 0 & 0 \end{vmatrix} \tag{8.75}$$

may be obtained from (8.66) through (8.73) by setting $\varepsilon_{11} = \varepsilon_{22} = 0$ as follows

$$\sigma_{11}(t) = \sigma_{22}(t) = \mathcal{F}_5\varepsilon_{12}^2, \tag{8.76}$$

$$\sigma_{33}(t) = \mathcal{F}_6\varepsilon_{12}{}^2, \tag{8.77}$$

$$\sigma_{12}(t) = \mathcal{G}_1\varepsilon_{12}+\mathcal{G}_2\varepsilon_{12}^3, \tag{8.78}$$

$$\sigma_{13} = \sigma_{23} = 0. \tag{8.79}$$

Note again that time varying stresses σ_{11}, σ_{22} and σ_{33} as given by (8.76) and (8.77) must be imposed in addition to ε_{12} in order to achieve a state of pure shear strain.

Pure Axial Strain. The stress components resulting from pure axial strain

$$\boldsymbol{\epsilon} = \begin{vmatrix} \varepsilon & 0 & 0 \\ 0 & 0 & 0 \\ 0 & 0 & 0 \end{vmatrix} \tag{8.80}$$

are obtained from (8.66) through (8.73) by setting $\varepsilon_{22} = \varepsilon_{12} = 0$ with the result

$$\sigma_{11}(t) = \mathcal{F}_1\varepsilon + \mathcal{F}_2\varepsilon^2 + \mathcal{F}_3\varepsilon^3, \tag{8.81}$$

$$\sigma_{22}(t) = \sigma_{33}(t) = (\mathcal{F}_1 - \mathcal{G}_1)\varepsilon + (\mathcal{F}_2 - \mathcal{G}_3)\varepsilon^2 + (\mathcal{F}_3 - \mathcal{G}_4)\varepsilon^3, \tag{8.82}$$

$$\sigma_{12} = \sigma_{13} = \sigma_{23} = 0. \tag{8.83}$$

In this instance time varying stresses $\sigma_{22}(t) = \sigma_{33}(t)$ must be imposed as given by (8.82) in addition to strain ε in order that pure axial strain (8.80) result.

8.5 Biaxial Constitutive Equations for 2×2 Matrix

Both stress and strain tensors are limited to a 2×2 matrix in the following expressions:

Creep. The constitutive equation for creep under constant biaxial stress results from (7.59) by integrating with the use of (8.1),

$$\begin{aligned} \boldsymbol{\epsilon}(t) = \mathbf{I}(&K_1\bar{\boldsymbol{\sigma}} + K_3'\bar{\boldsymbol{\sigma}}\,\bar{\boldsymbol{\sigma}} + K_4'\overline{\boldsymbol{\sigma}\boldsymbol{\sigma}} + K_7'\overline{\boldsymbol{\sigma}\boldsymbol{\sigma}\boldsymbol{\sigma}} + K_8'\overline{\boldsymbol{\sigma}\boldsymbol{\sigma}}\bar{\boldsymbol{\sigma}}) \\ &+ \boldsymbol{\sigma}(K_2 + K_5'\bar{\boldsymbol{\sigma}} + K_9'\bar{\boldsymbol{\sigma}}\,\bar{\boldsymbol{\sigma}} + K_{10}'\overline{\boldsymbol{\sigma}\boldsymbol{\sigma}}), \end{aligned} \tag{8.84}$$

where K_3' etc. are as given in (7.60). A comparison with (8.3) shows that the first two lines of (8.3) are the same as (8.84) except for the primes, but the K_6, K_{11} and K_{12} terms are missing from (8.84).

Relaxation. Similarly employing (8.2) and (7.61) yields the constitutive equation for relaxation under constant biaxial strain.

$$\begin{aligned} \boldsymbol{\sigma}(t) = \mathbf{I}(&\psi_1\bar{\boldsymbol{\epsilon}} + \psi_3'\bar{\boldsymbol{\epsilon}}\,\bar{\boldsymbol{\epsilon}} + \psi_4'\overline{\boldsymbol{\epsilon}\boldsymbol{\epsilon}} + \psi_7'\overline{\boldsymbol{\epsilon}\boldsymbol{\epsilon}\boldsymbol{\epsilon}} + \psi_8'\overline{\boldsymbol{\epsilon}\boldsymbol{\epsilon}}\,\bar{\boldsymbol{\epsilon}}) \\ &+ \boldsymbol{\epsilon}(\psi_2 + \psi_5'\bar{\boldsymbol{\epsilon}} + \psi_9'\bar{\boldsymbol{\epsilon}}\,\bar{\boldsymbol{\epsilon}} + \psi_{10}'\boldsymbol{\epsilon}\boldsymbol{\epsilon}), \end{aligned} \tag{8.85}$$

where ψ_3' etc. are given in (7.62). Again comparison with (8.4) shows that the first two lines of (8.4) are the same as (8.85) except for primes, but the ψ_6, ψ_{11} and ψ_{12} terms are missing from (8.85).

8.6 Components of Strain (or Stress) for Biaxial States for 2×2 Matrix

The components of strain for creep under constant stress for any biaxial state may be obtained from (8.84) by using the appropriate stress matrix together with its products and traces as given in the appropriate column of Table 3.1. It must be noted that only the components in Table 3.1 corresponding to the 2×2 matrix may be employed. The results for ε_{11}, ε_{22}, and ε_{12} are identical to those determined from the 3×3 matrix in Section 8.3 and hence will not be repeated here except for one example.

Creep Under Biaxial Stress. For the stress state given by the 2×2 matrix

$$\boldsymbol{\sigma} = \begin{vmatrix} \sigma_1 & \tau \\ \tau & \sigma_2 \end{vmatrix} \tag{8.86}$$

The strain components as determined from (8.84) are as follows:

$$\varepsilon_{11}(t) = \{A\}+\sigma_1\{B\}, \tag{8.87}$$

$$\varepsilon_{22}(t) = \{A\}+\sigma_2\{B\}, \tag{8.88}$$

$$\varepsilon_{12}(t) = \tfrac{1}{2}\gamma(t) = \tau\{B\}, \tag{8.89}$$

where

$$\begin{aligned}\{A\} = {} & (F_1-G_1)(\sigma_1+\sigma_2)+(F_2-G_3)(\sigma_1^2+\sigma_2^2)+(F_3-G_4)(\sigma_1^3+\sigma_2^3)\\ & +(2F_2-F_5-2G_3)\sigma_1\sigma_2+(3F_3-F_4+G_2-3G_4)(\sigma_1\sigma_2^2+\sigma_1^2\sigma_2)\\ & +F_5\tau^2+(F_4-G_2)(\sigma_1+\sigma_2)\tau^2,\end{aligned} \tag{8.90}$$

$$\{B\} = G_1+G_3(\sigma_1+\sigma_2)+G_4(\sigma_1+\sigma_2)^2+G_2(\tau^2-\sigma_1\sigma_2), \tag{8.91}$$

and where F_i and G_i are as given in (7.72). If the 3×3 matrix were employed in (8.84) and $\varepsilon_{33}(t)$ computed, $\varepsilon_{33}(t) = \{A\}$ would result, which is incorrect as shown by comparison with (8.28).

Relaxation Under Biaxial Stress. Components of stress for relaxation under any biaxial strain state may be computed from (8.85) in a similar manner to that described above for creep. The results for σ_{11}, σ_{22}, and σ_{12} are identical to those determined from the 3×3 matrix as described in Section 8.4 and will not be repeated here. The correct value of σ_{33} cannot be determined from (8.85) since this equation is limited to the 2×2 matrix. The correct value is obtained from the 3×3 matrix equations as in Section 8.4.

8.7 Constitutive Equations for Linearly Compressible Material

Creep. For creep under constant stress $\boldsymbol{\sigma}$ the use of (8.1) in (7.56) yields after integration

$$\begin{aligned}\boldsymbol{\epsilon}(t) = \mathbf{I}K_1\bar{\boldsymbol{\sigma}}+K_2(\boldsymbol{\sigma}-\mathbf{I}\sigma_v)+K_6[\boldsymbol{\sigma\sigma}-2\boldsymbol{\sigma}\sigma_v-\tfrac{1}{3}\mathbf{I}(\overline{\boldsymbol{\sigma\sigma}}-6\sigma_v^2)]\\ +K_{12}(\boldsymbol{\sigma\sigma\sigma}+\overline{\boldsymbol{\sigma\sigma}}\boldsymbol{\sigma}-3\boldsymbol{\sigma\sigma}\sigma_v-\tfrac{1}{3}\mathbf{I}\overline{\boldsymbol{\sigma\sigma\sigma}}).\end{aligned} \tag{8.92}$$

Relaxation. Similarly for relaxation under constant strain $\boldsymbol{\epsilon}$ the use of (8.2) in (7.58) yields

$$\begin{aligned}\boldsymbol{\sigma}(t) = \mathbf{I}\psi_1\bar{\boldsymbol{\epsilon}}+\psi_2(\boldsymbol{\epsilon}-\mathbf{I}\varepsilon_v)+\psi_6[\boldsymbol{\epsilon\epsilon}-2\boldsymbol{\epsilon}\varepsilon_v-\tfrac{1}{3}\mathbf{I}(\overline{\boldsymbol{\epsilon\epsilon}}-6\varepsilon_v^2)]\\ +\psi_{12}(\boldsymbol{\epsilon\epsilon\epsilon}+\overline{\boldsymbol{\epsilon\epsilon}}\boldsymbol{\epsilon}-3\boldsymbol{\epsilon\epsilon}\varepsilon_v-\tfrac{1}{3}\mathbf{I}\overline{\boldsymbol{\epsilon\epsilon\epsilon}}).\end{aligned} \tag{8.93}$$

8.8 Components of Strain for Creep of Linearly Compressible Material

Triaxial Principal Stresses. The components of strain for the state of stress represented by

$$\boldsymbol{\sigma} = \begin{vmatrix} \sigma_{\mathrm{I}} & 0 & 0 \\ 0 & \sigma_{\mathrm{II}} & 0 \\ 0 & 0 & \sigma_{\mathrm{III}} \end{vmatrix} \tag{8.94}$$

are obtained from (8.92) by employing the products and traces of (8.94) in (8.92) as follows

$$\varepsilon_{11}(t) = B_1'+\sigma_{\mathrm{I}}B_2'+\sigma_{\mathrm{I}}^2B_3'+\sigma_{\mathrm{I}}^3K_{12}, \tag{8.95}$$

$$\varepsilon_{22}(t) = B_1'+\sigma_{\mathrm{II}}B_2'+\sigma_{\mathrm{II}}^2B_3'+\sigma_{\mathrm{II}}^3K_{12}, \tag{8.96}$$

$$\varepsilon_{33}(t) = B_1'+\sigma_{\mathrm{III}}B_2'+\sigma_{\mathrm{III}}^2B_3'+\sigma_{\mathrm{III}}^3K_{12}, \tag{8.97}$$

$$\varepsilon_{12} = \varepsilon_{13} = \varepsilon_{23} = 0, \tag{8.98}$$

where

$$\begin{aligned}B_1' = (K_1-\tfrac{1}{3}K_2)(\sigma_{\mathrm{I}}+\sigma_{\mathrm{II}}+\sigma_{\mathrm{III}})-\tfrac{1}{3}K_6(\sigma_{\mathrm{I}}^2+\sigma_{\mathrm{II}}^2+\sigma_{\mathrm{III}}^2)\\ +\tfrac{2}{9}K_6(\sigma_{\mathrm{I}}+\sigma_{\mathrm{II}}+\sigma_{\mathrm{III}})^2-\tfrac{1}{3}K_{12}(\sigma_{\mathrm{I}}^3+\sigma_{\mathrm{II}}^3+\sigma_{\mathrm{III}}^3),\end{aligned} \tag{8.99}$$

$$B_2' = K_2-\tfrac{2}{3}K_6(\sigma_{\mathrm{I}}+\sigma_{\mathrm{II}}+\sigma_{\mathrm{III}})+K_{12}(\sigma_{\mathrm{I}}^2+\sigma_{\mathrm{II}}^2+\sigma_{\mathrm{III}}^2), \tag{8.100}$$

$$B_3' = K_6-K_{12}(\sigma_{\mathrm{I}}+\sigma_{\mathrm{II}}+\sigma_{\mathrm{III}}). \tag{8.101}$$

Comparison of (8.95) through (8.101) with the same state of stress for compressible material shows similar relations except that there are four kernel functions in the linearly compressible compared to twelve for compressible and there are no product terms of the form $\sigma_k^2\sigma_l$ in B_1' and no products of the form $\sigma_k\sigma_l$ in B_2'.

Biaxial Stress State. Components of strain for the state of stress

$$\boldsymbol{\sigma} = \begin{vmatrix} \sigma_1 & \tau & 0 \\ \tau & \sigma_2 & 0 \\ 0 & 0 & 0 \end{vmatrix} \tag{8.102}$$

may be obtained from (7.85) through (7.89) by employing (8.1) and integrating as follows after rearrangement

$$\varepsilon_{11}(t) = \{\mathsf{A}'\}+\sigma_1\{\mathsf{B}'\}, \tag{8.103}$$
$$\varepsilon_{22}(t) = \{\mathsf{A}'\}+\sigma_2\{\mathsf{B}'\}, \tag{8.104}$$
$$\varepsilon_{33}(t) = \{\mathsf{A}'\}+\{\mathsf{C}'\}, \tag{8.105}$$
$$\varepsilon_{12}(t) = \tfrac{1}{2}\gamma_1(t) = \tau\{\mathsf{B}'\}, \tag{8.106}$$
$$\varepsilon_{13} = \varepsilon_{12} = 0, \tag{8.107}$$

where

$$\begin{aligned}\{\mathsf{A}'\} = {} & (K_1-\tfrac{1}{3}K_2)(\sigma_1+\sigma_2)-\tfrac{1}{9}K_6[(\sigma_1^2+\sigma_2^2)+5\sigma_1\sigma_2-3\tau^2] \\ & -\tfrac{1}{3}K_{12}[(\sigma_1^3+\sigma_2^3)+3(\sigma_1+\sigma_2)\tau^2],\end{aligned} \tag{8.108}$$
$$\{\mathsf{B}'\} = K_2+\tfrac{1}{3}K_6(\sigma_1+\sigma_2)+K_{12}(\sigma_1^2+\sigma_2^2-\sigma_1\sigma_2+3\tau^2), \tag{8.109}$$
$$\{\mathsf{C}'\} = K_6(\sigma_1\sigma_2-\tau^2). \tag{8.110}$$

In this instance no additional kernel functions are required for ε_{33} beyond the four required for ε_{11}, ε_{22} and ε_{12}. This was not true for the compressible form (8.26)–(8.33).

Axial Stress plus Torsion (Shear). Strain components for the stress state

$$\boldsymbol{\sigma} = \begin{vmatrix} \sigma & \tau & 0 \\ \tau & 0 & 0 \\ 0 & 0 & 0 \end{vmatrix} \tag{8.111}$$

are found from (8.102) through (8.110) by setting $\sigma_1 = \sigma$, $\sigma_2 = 0$ and rearranging as follows:

$$\varepsilon_{11}(t) = (K_1+\tfrac{2}{3}K_2)\sigma+\tfrac{2}{9}K_6\sigma^2+\tfrac{2}{3}K_{12}\sigma^3+2K_{12}\sigma\tau^2+\tfrac{1}{3}K_6\tau^2, \tag{8.112}$$
$$\varepsilon_{22}(t) = (K_1-\tfrac{1}{3}K_2)\sigma-\tfrac{1}{9}K_6\sigma^2-\tfrac{1}{3}K_{12}\sigma^3-K_{12}\sigma\tau^2+\tfrac{1}{3}K_6\tau^2, \tag{8.113}$$
$$\varepsilon_{33}(t) = (K_1-\tfrac{1}{3}K_2)\sigma-\tfrac{1}{9}K_6\sigma^2-\tfrac{1}{3}K_{12}\sigma^3-K_{12}\sigma\tau^2-\tfrac{2}{3}K_6\tau^2, \tag{8.114}$$
$$\varepsilon_{12}(t) = \tfrac{1}{2}\gamma_1(t) = K_2\tau+3K_{12}\tau^3+\tfrac{1}{3}K_6\sigma\tau+K_{12}\sigma^2\tau, \tag{8.115}$$
$$\varepsilon_{13} = \varepsilon_{23} = 0. \tag{8.116}$$

Comrparing (8.112) through (8.116) with the corresponding equations for the compressible material (8.43) through (8.47) shows that the same forms of stress terms are present in both.

Pure Shear and Pure Axial Stress. As before, the components of strain for these states of stress may be found from (8.112) through (8.116) by setting $\sigma = 0$ or $\tau = 0$ for pure shear or pure axial stress respectively.

8.9 Components of Stress for Relaxation of Linearly Compressible Material

Since these components can be obtained from those for creep as given for linearly compressible material in Section 8.8 by changing from creep to relaxation symbols they will not be repeated here.

8.10 Poisson's Ratio

Poisson's ratio ν is a material constant for linear elastic materials and is defined as the negative ratio of the transverse strain, ε_{22}, to the longitudinal strain, ε_{11}, under uniaxial stressing in the longitudinal direction. For visco-elastic material ν is in general a time-dependent quantity as defined above. Forming the ratio of (8.55) (5.54) yields

$$\frac{\varepsilon_{22}}{\varepsilon_{11}}(t) = \frac{(F_1-G_1)\sigma+(F_2-G_3)\sigma^2+(F_3-G_4)\sigma^3}{F_1\sigma+F_2\sigma^2+F_3\sigma^3}, \tag{8.117}$$

which is both stress and time dependent since F_α, G_α are time dependent. When time-independent terms F^0, G^0 may be separated out from F_α and G_α then a time-independent but stress-dependent ratio results

$$\frac{\varepsilon_{22}^0}{\varepsilon_{11}^0} = \frac{(F_1^0-G_1^0)\sigma+(F_2^0-G_3^0)\sigma^2+(F_3^0-G_4^0)\sigma^3}{F_1^0\sigma+F_2^0\sigma^2+F_3^0\sigma^3}. \tag{8.118}$$

For a linear material the ratio becomes

$$\frac{\varepsilon_{22}}{\varepsilon_{11}}(t) = 1-\frac{G_1}{F_1}, \tag{8.119}$$

which is stress independent but time dependent. If time-independent terms are separated out of (8.119) then for a linear material there is an equivalent to Poisson's ratio which is a material constant.

$$\frac{\varepsilon_{22}}{\varepsilon_{11}} = 1-\frac{G_1^0}{F_1^0} \tag{8.120}$$

The time dependence of Poisson's ratio has been demonstrated by Freudenthal [8.6].

8.11 Time Functions

As indicated in Chapter 5 viscoelastic behavior has been described by various mechanical models composed of linear springs and dashpots (in the linear theory). For models consisting of a small number of such elements the creep-time curve starts at time $t = 0+$ at a finite rate of creep and after a period approaches asymptotically either a constant strain or a constant rate of creep. Neither of these conditions accurately describes the actual behavior of many viscoelastic materials. Many plastics and also concrete, wood and some metals show a very rapid (essentially infinite) rate of creep at the start, within the sensitivity of experimental observation. Also within the small strain range the creep rate decreases continuously as shown in Figure 8.1, for example. Thus simple mechanical models do not yield a very

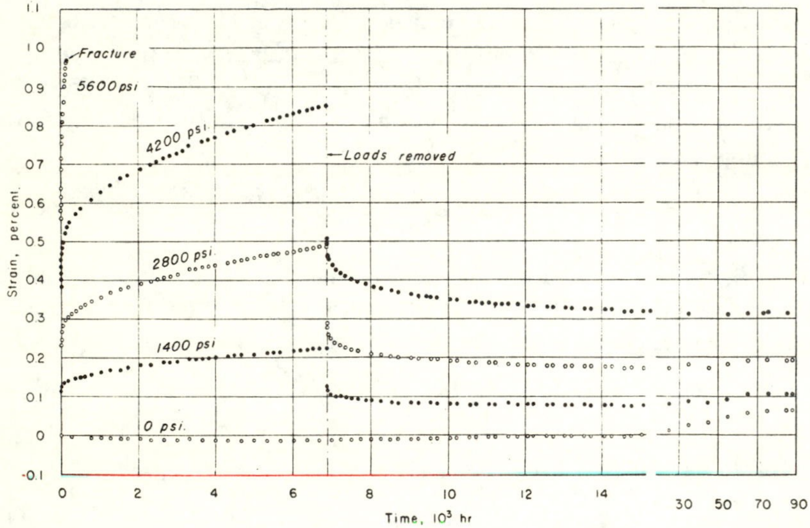

Fig. 8.1. Creep and Recovery of Asbestos Laminate for Constant Stress at 77°F and 50 per cent. R.H. From [8.8], courtesy ASTM.

accurate description–nor are they very satisfactory for predicting creep behavior beyond the time scale of the observations. Of course, a mechanical model composed of a large number of mechanical elements having a suitable distribution of retardation times can be proportioned so as to describe a given creep curve over any given range of time. For example (5.64b) shows the retardation spectrum equivalent to the power function of time (8.121) given below.

Some mathematical models (especially a power function of time) are capable of describing the main features of creep behavior of viscoelastic materials with good accuracy both in the early stage and over a wide time span. For materials whose creep response may be described by a separable time-independent and time-dependent strain the following expression has often been found to yield a good description of creep of viscoelastic materials at constant stress,

$$\varepsilon = \varepsilon^0 + \varepsilon^+ t^n, \tag{8.121}$$

where ε is the strain, t the time, n a constant independent of stress, ε^0 is the time-independent strain, and ε^+ is the coefficient of the time-dependent term. ε^0 and ε^+ are functions of stress. The symbol t may be taken to represent a dimensionless time ratio t/t_0, where t_0 is unit time. In this case ε^+ has the dimensions of strain, i.e., dimensionless. When t is taken to have the dimensions of time then ε^+ has the dimensions of $(\text{time})^{-n}$.

It has been found from creep of several plastics in tension, compression, torsion and combined tension and torsion that the creep-time curve may be represented with good accuracy by (8.121) with a constant value of n for a given material independent of stress or state of combined stress. This is shown in Fig. 8.2 for two plastics in creep under combined tension and torsion with different stresses. Rearranging (8.121) and taking logarithms yields

$$\log (\varepsilon - \varepsilon^0) = \log \varepsilon^+ + n \log t. \tag{8.121a}$$

Thus, if $\log (\varepsilon - \varepsilon^0)$ is plotted versus $\log t$ as in Fig. 8.2 (8.121a) predicts a straight line of slope n and intercept at unit time of ε^+. Since all curves in Fig. 8.2(a) and all curves in Fig. 8.2(b) are parallel, n is independent of stress and state of stress.

That (8.121) describes creep behavior of plastics reasonably well is shown in Fig. 8.2 by the good agreement between the test data and the solid line representing (8.121). This equation has been employed to describe the creep behavior under constant stress of many different types of plastics and some other materials including wood, concrete and some metals. Within the range of stress and strain of interest in structural applications the description of creep of many plastics by (8.121) is quite satisfactory. The values of constants in (8.121) for a number of different plastics, laminated plastics, wood, concrete and some metals are shown in Table 8.1 for particular values of stress.

When creep must be considered in design it is usual that the item must remain in service for an extended period of time, usually longer than it is

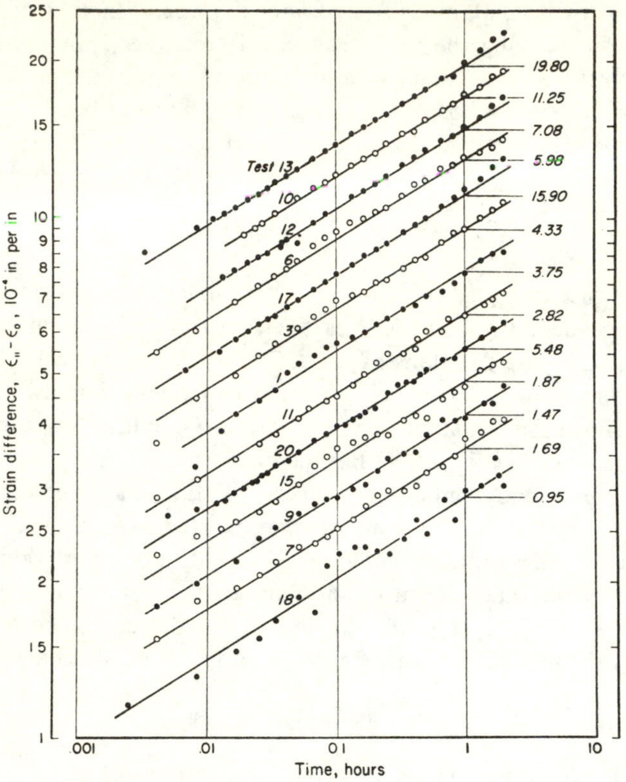

Fig. 8.2. Log Strain Difference $\varepsilon - \varepsilon^0$ Versus Log Time t for Two Plastics at 75°F and 50 per cent R.H.

(a) Tensile strain component for creep of PVC under combined tension and torsion. From [8.1], courtesy Soc. of Rheology. (Note: numbers on right indicate strain at one hour. To determine strain, shift vertical scale to match this number.)

practical to run creep experiments on the material to be employed. Thus, it is necessary to extrapolate the information obtained from relatively short-time laboratory creep tests in order to predict the probable behavior in service. Hence, the accuracy with which a creep equation describes the time dependence is an important consideration.

Long-time creep experiments have shown (8.121) to be especially suitable for this purpose. For example, creep data from several fabric and paper laminated plastics were found to be describable by (8.121). Data for these experiments and constants corresponding to (8.121) were determined from 2000 hr. of creep and published [8.7]. The tests, however, were continued

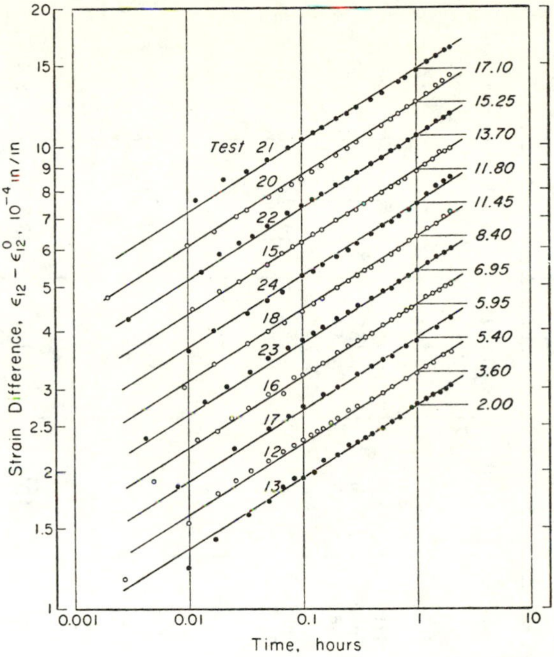

(b) Shearing strain component of creep of polyurethane under combined tension and torsion. From [8.5], courtesy ASME. (Note: numbers on right indicate strain at one hour. To determine strain, shift vertical scale to match this number.)

uninterrupted for almost 90,000 hr. Subsequently, these data were compared [8.8] with predictions from (8.121) based on the constants previously determined in [8.7]. Some of the results are shown in Fig. 8.3. The prediction of (8.121) shown in Fig. 8.3 is very satisfactory. The corresponding prediction using representations of the first 2000 hr. by several simple mechanical models was not as satisfactory, as shown in Fig. 8.4 for one test. The failure of these mechanical models to yield a satisfactory prediction resulted from the fact that their predicted creep was asymptotic either to a constant strain or a constant rate of creep, neither of which was representative of the actual behavior.

A recent report [8.9] also showed that (8.121) was suitable for predicting long time creep behavior of PVC and polyethylene. In [8.9] the first 2000 hr. of tensile creep data were fitted by (8.121) then the resulting equation was employed to predict the creep behavior up to 132,000 hr. (about 16 yr.) as shown for PVC in Fig. 8.5. Some of the small difference between (8.121) and

Table 8.1

Constants for Creep Equation $\varepsilon = \varepsilon^0 + \varepsilon^+ t^n$

Material	Temp., °F	R.H., %	Stress,* ksi	ε^0, 10^{-6} in./in.	ε^+, 10^{-6} in./in.	n	References	Figure numbers
Polyethylene	77	50	0.225T	8000	6100	0.089	[8.25]	8.7a
Poly [vinyl chloride]	77	50	1.T	2177	188	0.155	[8.1]	8.7b
Polyurethane (full density)	75	50	2.T	3810	385	0.155	[8.5]	8.7d
	75	50	2.C	−3645	−303	0.155	[8.5]	
	75	50	1.5TOR	3960	360	0.155	[8.5]	8.7e
Polycarbonate	75	50	2.T	5667	327	0.15	[8.26]	
	75	50	1.5TOR	6116	410	0.15	[8.26]	5.6
Polychlorotrifluoroethylene (crystalline)	77	50	2.7T	1000	3000	0.087	[8.25]	
Grade C Canvas Laminate (cresol and formaldehyde resin)	77	50	3.8T	3100	1240	0.151	[8.8]	8.4
Glass Laminate (181 fabric, polyester resin)	73	50	10T**	2438**	346**	0.09	[8.27]	
	73	wet†	10T**	4130**	144**	0.21	[8.27]	

Glass Laminate	73	50	10T**	2340**	101**	0.16	[8.27]	
Red Oak	80	12††	0.78T***	6010	3950	0.22	[8.21]	8.8a
Copper, oxygen free‡‡	329	–	10T	$\sigma/E+70$	38.5	0.453	[8.22]	8.8b
Aluminium 2618-T61	392 392	– –	20T 11.5TOR	2082 1667	236 258	0.200 0.338	[8.23] [8.23]	8.8c
Nickel, high purity	1292	–	5.2T		18,570	0.28	[8.28]	
Stainless Steel 304‡‡‡	1105	–	10.628T‡ 6.111TOR‡	2150 1655	1422 1294	0.195 0.195	[8.29] [8.29]	2.1
Asphalt concrete	75		0.0318C	0	−9060	0.182	[8.24]	8.8d

* Type of stressing is indicated by T for tension, C for compression and TOR for torsion.
** Selected from relation given in [8.27].
*** Tension perpendicular to the grain.
† Immersed in water.
†† Moisture content in percent.
‡ Combined tension and torsion.
‡‡ Prestrained 8 percent at room temperature.
‡‡‡ Prestrained at room temperature.

Note: For constants given time is in hours.

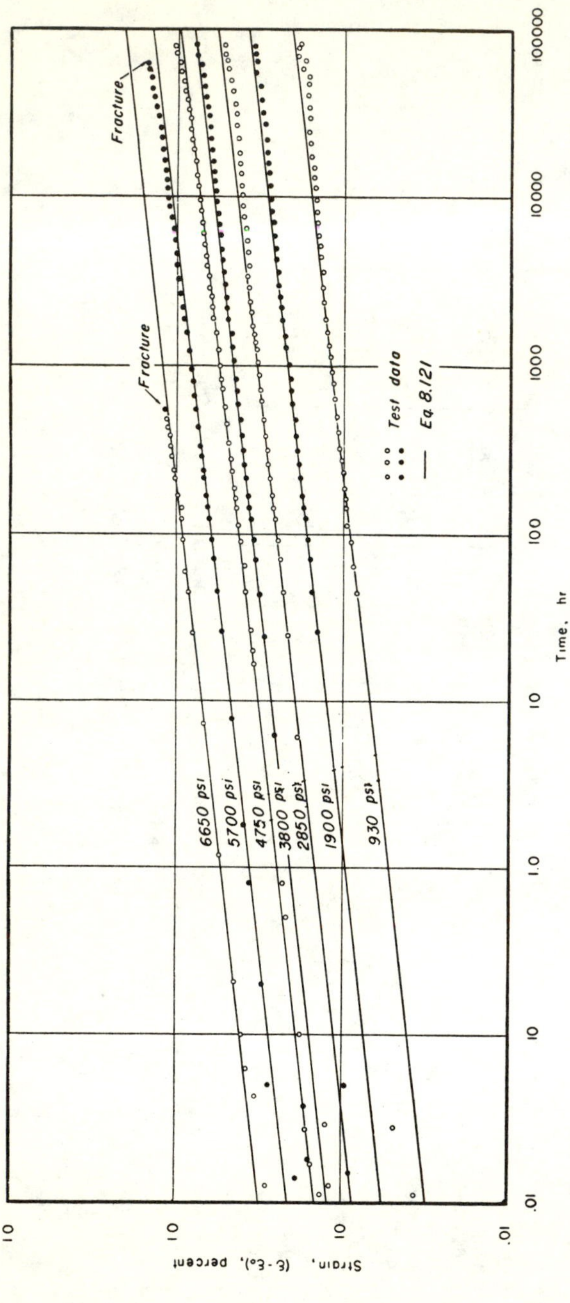

Fig. 8.3. Log Strain Difference $\varepsilon - \varepsilon^0$ Versus Log Time t for 90,000 hr. of Creep of Two Laminates at 77°F and 50 percent R.H. From [8.8], courtesy ASTM.

(a) Canvas laminate

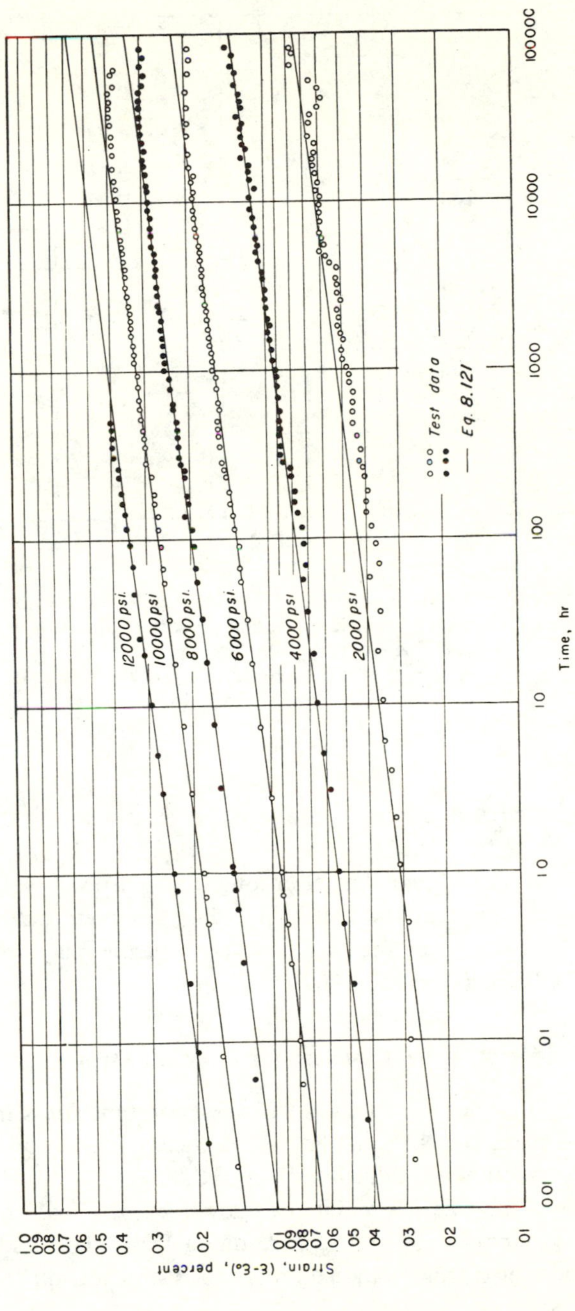

(b) Paper laminate

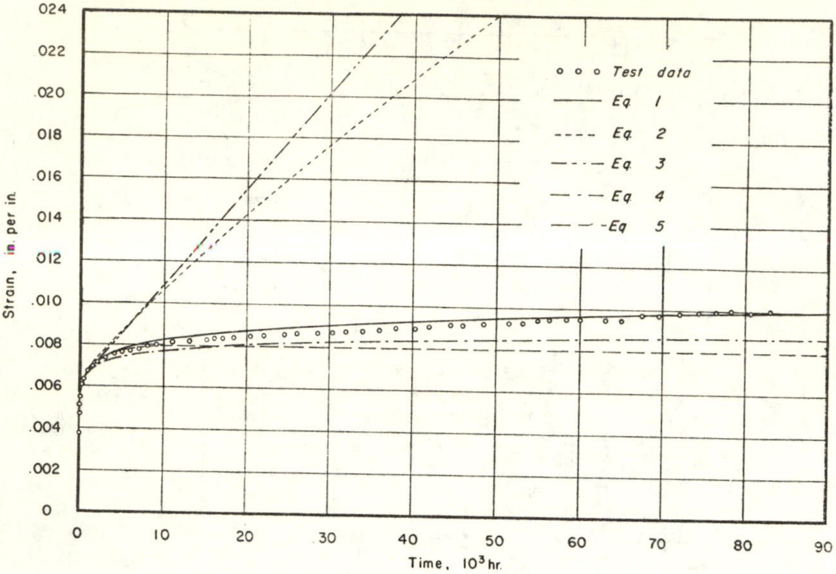

Fig. 8.4. Long-time Creep of Canvas Laminate at 3800 psi Together with Prediction of Several Equations Based on First 2000 hr. of Data. From [8.8], courtesy ASTM.
Eq. 1: $\varepsilon = \varepsilon^0 + \varepsilon^+ t^n$, (2.16),
Eq. 2: $\varepsilon = \varepsilon^0 + A \log t + Bt$, (2.13),
Eq. 3: $\varepsilon = \varepsilon^0 + A(1 - e^{-ct}) + Bt$, (2.15),
Eq. 4: $\varepsilon = \varepsilon^0 + A \log t$, (2.14),
Eq. 5: $\varepsilon = \varepsilon^0 + A(1 - e^{-ct})$.

the data may have resulted from aging as shown by a creep test of the same material after aging for 16 years at no stress, see Fig. 8.5. It was also shown in [8.9] that recovery and reloading after long time creep was well described by the Boltzmann superposition principle. For complete unloading and reloading to the same stress the superposition principle may be employed even in the nonlinear range as in [8.9].

8.12 Determination of Kernel Functions for Constant Stress Creep

As of this writing not all the twelve K's for constant stress involved in general three-dimensional stressing have been determined for any material, because of the experimental difficulties involved.

Biaxial Creep. The nine K's required to describe the biaxial strain components ε_{11}, ε_{22}, and ε_{12} (but not ε_{33}) resulting from biaxial stress states have been determined from combined axial load and torsion (axial stress

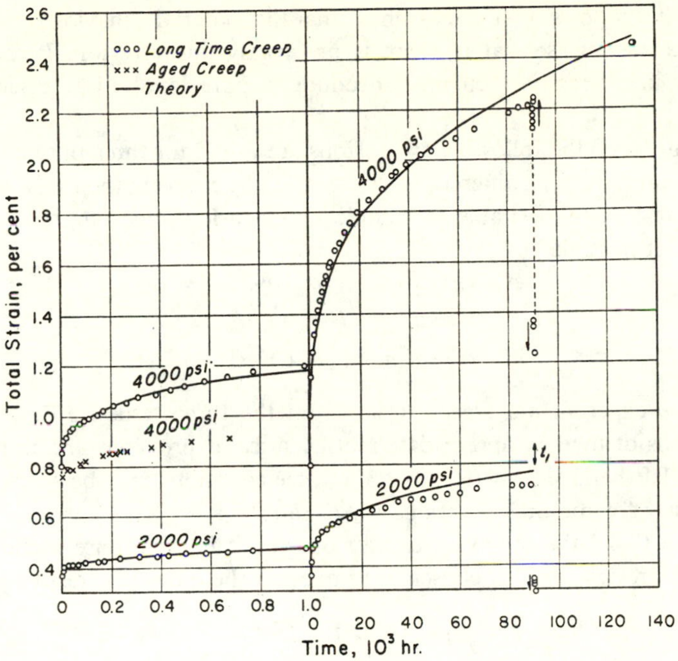

Fig. 8.5. Long-time Creep of Poly [Vinyl Chloride] (PVC) at 75°F, 50 percent R.H. From [8.9], courtesy Polymer Engineering and Science.

and shear stress) in several combinations for several polymers. The strain component ε_{33} normal to the biaxial plane is not described by these nine components. Fortunately many situations of interest in engineering practice are covered adequately by the biaxial state of stress and corresponding biaxial strain components. It is the author's opinion that combined axial and torsion tests are perhaps the easiest way to determine the nine kernel functions in the general biaxial stress state for the third order multiple integral representation. Accordingly the determination of the kernel functions for combined axial load and torsion will be discussed in detail. Either numerical or graphical methods may be employed.

Numerical. Since nine time functions, F_1, F_2, F_3, F_4, F_5, G_1, G_2, G_3, G_4 are required in (8.43), (8.44), and (8.46) nine different simultaneous equations involving the experimental results from creep tests at different states of stress are needed. Fortunately not all nine equations need be solved simultaneously. Also it is only necessary to employ (8.43) and (8.46); (8.44) need not be used in determining the kernel functions.

In what follows it will be assumed that the material behavior is nonlinear at all stress levels; that is, there is no strictly linear region. If there is a strictly linear region then the procedure is more involved as discussed in [8.1].

From (8.43) the following expressions are found for three pure tension or compression tests at different stresses σ_a, σ_b, σ_c covering the range of stress of interest. (It is desirable to include both tension and compression creep results, if possible.)

$$\begin{aligned} \varepsilon_{11a}(t) &= F_1\sigma_a + F_2\sigma_a^2 + F_3\sigma_a^3, \\ \varepsilon_{11b}(t) &= F_1\sigma_b + F_2\sigma_b^2 + F_3\sigma_b^3, \\ \varepsilon_{11c}(t) &= F_1\sigma_c + F_2\sigma_c^2 + F_3\sigma_c^3. \end{aligned} \tag{8.122}$$

If the experimental creep strains from the three tests are described as functions of time by appropriate mathematical expressions and introduced on the left-hand side of (8.122), these three equations may be solved simultaneously for the time functions F_1, F_2, and F_3.

From (8.46) the following expressions result for two pure torsion creep tests at stresses τ_a and τ_b chosen so as to cover the range of stress of interest:

$$\begin{aligned} \varepsilon_{12a}(t) &= G_1\tau_a + G_2\tau_a^3, \\ \varepsilon_{12b}(t) &= G_1\tau_b + G_2\tau_b^3, \end{aligned} \tag{8.123}$$

Again, describing the experimental strains from these creep tests by appropriate mathematical expressions and introducting these expressions on the left-hand side of (8.123), the pair of simulatenous equations (8.123) may be solved for G_1 and G_2 as functions of time.

From a pair of creep tests under different combinations of tension (or compression) and torsion σ_c, τ_c, and σ_d, τ_d using stresses in the upper range of interest, the following expressions are obtained from (8.43).

$$\begin{aligned} \varepsilon_{11c}(t) &= F_1\sigma_c + F_2\sigma_c^2 + F_3\sigma_c{}^3 + F_4\sigma_c\tau_c^2 + F_5\tau_c^2, \\ \varepsilon_{11d}(t) &= F_1\sigma_d + F_2\sigma_d^2 + F_3\sigma_d^3 + F_4\sigma_d\tau_d^2 + F_5\tau_d^2, \end{aligned} \tag{8.124}$$

and from (8.46)

$$\begin{aligned} \varepsilon_{12c}(t) &= G_1\tau_c + G_2\tau_c^3 + G_3\sigma_c\tau_c + G_4\sigma_c^2\tau_c, \\ \varepsilon_{12d}(t) &= G_1\tau_d + G_2\tau_d^3 + G_3\sigma_d\tau_d + G_4\sigma_d^2\tau_d. \end{aligned} \tag{8.125}$$

Using an appropriate mathematical description of the axial experimental creep strains from the two combined axial-torsion creep tests and substituting for the left-hand side of (8.124), a pair of equations results which may be solved for F_4 and F_5 since F_1, F_2 and F_3 are known from (8.122). Similarly, the shear strains corresponding to torsion from the two combined axial-torsion tests may be employed in (8.125) to solve the pair of simultaneous equations (8.125) for G_3 and G_4, since G_1 and G_2 are known from (8.123). Thus, all nine time functions may be determined from the results of three axial, two torsion and two combined axial-torsion creep tests under constant stress.

An alternative method of determining F_4 and F_5 is to use the axial creep strain resulting from one of the pure torsion creep tests τ_a to solve for F_5 directly since for pure torsion an axial strain is predicted from (8.43) when σ is zero,

$$\varepsilon_{11}(t) = F_5\tau_a^2. \tag{8.126}$$

Then F_4 may be determined directly from the axial strain from one of the combined axial-torsion creep tests. This method requires high precision in the measurement of the axial strain in pure torsion. Sufficient accuracy is difficult to obtain. Of course, the above operations may all be programmed for computer.

The above method is most suitable when the time-independent strain ε^0 is separable from the time-dependent strain $\varepsilon-\varepsilon^0$ and $\varepsilon-\varepsilon^0$ contains a time function such as t^n which is the same for different stresses and states of stress. Fortunately the behavior of many plastics and some other materials can be described within these limitations. Thus, each of the functions F_α, G_α in (8.43) and (8.46) has a time-independent part F_α^0, G_α^0 and a coefficient of the time-dependent part F_α^+, G_α^+. The time-independent terms F_α^0, G_α^0 may be determined separately by means of the procedure described above using (8.122) to (8.126). The time-dependent coefficients may be determined separately in a similar manner.

Many plastics investigated to date show that the time-independent strains ε^0 are very nearly linear functions, see Fig. 8.6a, b, for example, even when the time-dependent coefficient is quite nonlinear (see Fig. 8.6c, d). When it is appropriate to take the time-independent strains as linear only F_1^0 and G_1^0 of the nine values F_α^0, G_α^0 are non-zero. Thus many computations are considerably simplified.

Obviously, difficulties will arise with certain forms of time functions or if different forms are used for each creep test. However, a numerical method can always be used to solve (8.122) through (8.125). For example in solving for the time functions F_1, F_2 and F_3 of (8.122), instead of introducing the mathematical expressions for the creep strains of the entire time range into

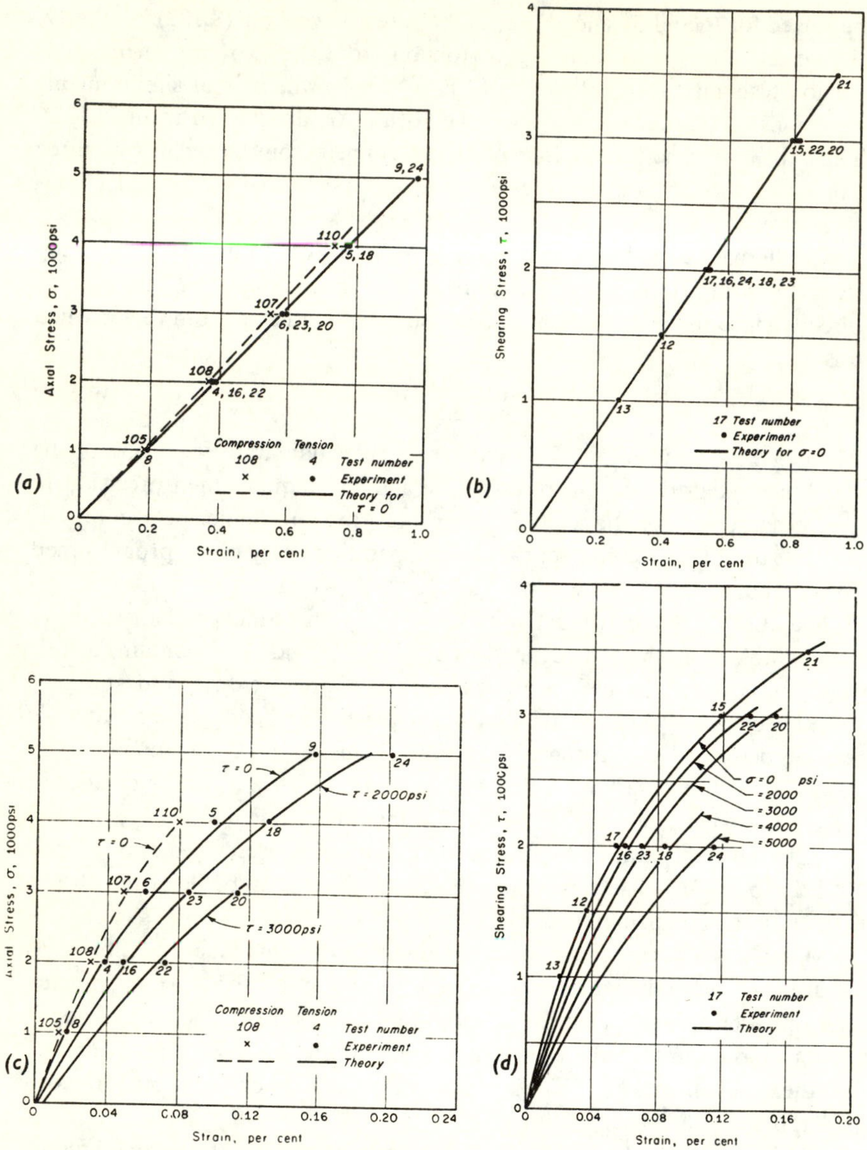

Fig. 8.6. Time-independent Strain ε^0 and Time-dependent Coefficient ε^+ Versus Stress for Creep of Polyurethane under Combined Tension and Torsion at 75°F and 50 per cent R.H.

(a) Axial component of strain, ε_{11}^0. From [8.5], courtesy ASME.
(b) Shearing component of strain, ε_{12}^0.
(c) Axial component ε_{11}^+. From [8.5], courtesy ASME.
(d) Shearing component ε_{12}^+.

the left-hand side of (8.122), the measured creep strains ε_{11_a}, ε_{11_b}, and ε_{11_c} at each time instance are introduced successively into (8.122). Thus, the time functions F_1, F_2, and F_3 at each time instance can be obtained.

When more than the minimum number of experiments for determining the time functions are available, a least square method together with the numerical method discussed above can be used to evaluate the kernel functions.

Graphical. Besides the numerical method discussed above, a graphical method can be used to determine the time functions. The graphical method in some instances may prove to be the most desirable when more than the minimum number of experiments for determining the time functions are available. To simplify the discussion only the coefficients of the time-dependent terms will be treated in the following. Consider, for example, that the time-dependent strain can be separated from the time-independent strain and that the former is describable by a power function of time as in (8.121) with the same value of n for all states of stress. Then the time-dependent coefficients F_α^+, G_α^+ corresponding to F_α, G_α in (8.43) and (8.46) may be determined graphically as follows.

Choosing one or more pure tension or compression and a pure torsion creep experiment at sufficiently low stresses the behavior will be nearly linear. Thus F_1^+ and G_1^+ may be determined directly from these experiments. Now making use of pure tension creep experiments in the nonlinear range the function $A(\sigma)$ given below may be plotted versus stress σ to yield a straight line (if three terms of the multiple integral representation are sufficient),

$$A = \frac{\varepsilon_{11}^+(\sigma) - F_1^+\sigma}{\sigma^2} = F_2^+ + F_3^+\sigma. \tag{8.127}$$

Equation (8.127) was found from (8.122) for pure tension by rearranging terms. Since F_1^+ has been determined above and $\varepsilon_{11}^+(\sigma)$ is the experimentally determined time-dependent coefficient for a given experiment under a pure tension stress σ the term A is known for each experiment. Thus if A vs. σ is a straight line F_2^+ and F_3^+ may be found as the intercept and slope of the straight line, respectively. The value of F_1^+ may be adjusted to achieve the best straight line.

Similarly, from pure torsion creep tests in the nonlinear range the function B, given below, may be evaluated for each torsion experiment and plotted versus stress τ,

$$B = \frac{\varepsilon_{12}^+(\tau) - G_1^+\tau}{\tau^2} = G_2^+\tau, \tag{8.128}$$

where G_1^+ has been previously determined and $\varepsilon_{12}^+(\tau)$ is the experimentally determined coefficient of the time-independent term in each pure torsion test τ. If a straight line through the origin results then G_2^+ may be determined as the slope of the straight line. If a straight line does not result, adjustment of $G_1{}^+$ may make it so.

After F_1^+, F_2^+, F_3^+ are found F_4^+ and F_5^+ may be determined by forming the quantity C from (8.43),

$$C = [\varepsilon_{11}^+(\sigma,\tau) - F_1^+\sigma - F_2^+\sigma^2 - F_3^+\sigma^3]/\tau^2 = F_4^+\sigma + F_5^+, \tag{8.129}$$

where $\varepsilon_{11}^+(\sigma, \tau)$ represents the values of ε_{11}^+ from combined tension-torsion creep experiments. A plot of C versus σ should yield a straight line with F_5^+ as intercept and F_4^+ as slope. If several data points are available the method of least squares may be used to determine F_4^+ and F_5^+.

Similarly, after determining G_1^+ and G_2^+ above, G_3^+ and G_4^+ may be obtained by forming the quantity D from (8.46),

$$D = [\varepsilon_{12}^+(\sigma, \tau) - G_1^+\tau - G_2^+\tau^3]/\sigma\tau = G_3^+ + G_4^+\sigma, \tag{8.130}$$

where $\varepsilon_{12}^+(\sigma, \tau)$ represents the values of ε_{12}^+ obtained from combined tension-torsion creep tests. Plotting D versus σ should yield a straight line of intercept G_3^+ and slope G_4^+.

Determining Remaining Kernel Functions. To describe ε_{33} under biaxial stress states requires determining F_6 and F_7 in (8.45) or (8.37). Thus it is necessary to measure ε_{33} under a biaxial stress state such as (8.42) or (8.34). With instrumentation suitable for this measurement determination of F_6 and F_7 may be made from two tests at different stress levels, having previously determined the other kernel functions involved, F_1, F_2, F_3, G_1, G_3, G_4, in (8.45) for example.

Since the most suitable specimen for determining the first nine kernel functions is a thin-walled tube for which ε_{33} is in the direction of the thickness of the tube it is very difficult to provide instrumentation with sufficient sensitivity and accuracy to yield meaningful values of ε_{33}. The gage length for strain measurement is the wall thickness, and this is too short. Thus no determinations of F_6, F_7 have been made to date.

If the other eleven kernel functions were known from determinations such as above then the last K_{12} required for describing creep behavior under constant triaxial stress could be determined from a single triaxial or hydrostatic creep test by measuring any one of the principal strains, see (8.17) and (8.23).

8.13 Determination of Kernel Functions for Constant-Strain Stress-Relaxation

In principle the kernel functions for constant strain relaxation can be determined by the same methods described above for creep since the mathematical relations are of exactly the same form. However, experimentally it is quite a different matter. For example, if as noted before normal stress is imposed along one axis of an unstressed solid there are no other stresses induced. However, if an axial strain ε is imposed along an axis of unstrained solid unprescribed strains are induced in the transverse directions due to the Poisson effect. In other words if one component of a stress tensor is imposed the resulting stress state is defined by that component. But in general if one normal strain component of a strain tensor is imposed the resulting strain state will include other induced strain components unless stresses are also imposed to suppress the induced strain components. Thus duplication of the methods used for creep to determine the kernel functions for relaxation introduce experimental complexities that are very difficult if not impossible to cope with.

Pure-Axial Strain. Consider equations (8.80) through (8.83). Kernel functions $\mathcal{F}_1, \mathcal{F}_2, \mathcal{F}_3$ can be determined from results of relaxation tests at three different values of ε, at least two of which are in the nonlinear range. However, the state of stress associated with the pure axial strain ε is triaxial.

There are two ways that can be used to achieve a pure axial strain state. One way is to restrict to zero the radial deflection of a specimen of circular cross-section and measure the radial stress $\sigma_r = \sigma_{22} = \sigma_{33}$ when the specimen is subjected to an axial strain. This type of experiment can be performed in principle when the axial strain is compressive if the confining means is sufficiently rigid and the friction between the specimen and the confining means can be eliminated. Another way of achieving the objective in principle is to use a servo system in which the radial strain ε_r serves as the input signal to control the confining pressure (along the radial direction only) such that the radial strain $\varepsilon_r = \varepsilon_{22} = \varepsilon_{33} = 0$ and measure the corresponding radial stress $\sigma_r(t) = \sigma_{22} = \sigma_{33}$. Both of the methods are extremely difficult to achieve with acceptable accuracy.

It may also be noted that a pure uniaxial strain is incompatible with an assumption of incompressibility.

Unconfined Axial Strain. The usual technique for performing a simple relaxation test is to impose a constant axial strain ε and no transverse stresses. Thus the transverse strains $\varkappa$ are not zero, but are time dependent even for constant axial strain. This strain state is given by

$$\boldsymbol{\epsilon} = \begin{vmatrix} \varepsilon & 0 & 0 \\ 0 & \varkappa(t) & 0 \\ 0 & 0 & \varkappa(t) \end{vmatrix}. \tag{8.131}$$

In view of the time dependence of $\varkappa$, (7.97) through (7.99) must be considered instead of (8.57) through (8.64). Since the transverse stresses σ_{22} and σ_{33} are zero (7.98) may be equated to zero and in principle solved for $\varkappa$ as a function of ε and t [8.10, 8.11]. This function may then be substituted in (7.97) to yield σ_{11} as a function of ε and three functions of the kernels ψ_a, ψ_b, ψ_c

$$\sigma_{11}(t) = \int_0^t \psi_a \dot{\varepsilon}\, d\xi_1 + \int_0^t \int_0^t \psi_b \dot{\varepsilon}^2\, d\xi_1\, d\xi_2 + \int_0^t \int_0^t \int_0^t \psi_c \dot{\varepsilon}^3\, d\xi_1\, d\xi_2\, d\xi_3. \tag{8.132}$$

For constant strain ε (8.132) becomes

$$\sigma_{11}(t) = \psi_a(t)\varepsilon + \psi_b(t)\varepsilon^2 + \psi_c(t)\varepsilon^3. \tag{8.133}$$

Then from three relaxation tests at different values of ε covering the nonlinear range the nature of ψ_a, ψ_b, ψ_c at constant strain can be determined. It must be noted, however, that ψ_a, ψ_b, ψ_c are not the same as $\mathcal{F}_1, \mathcal{F}_2, \mathcal{F}_3$ in (8.81). They involve other ψ's besides those given in (7.91). Another way of looking at (8.133) is that this equation represents the relation between axial stress σ_{11} and constant axial strain ε in an unconfined relaxation test in which the effect of the transverse strains on the stress σ_{11} has been lumped into the kernel functions ψ_a, ψ_b, ψ_c. This method of determining kernel functions for relaxation is quite satisfactory for use in computing behavior of structures for which the state of strain closely approximates that of the unconfined axial strain (8.131) as in bending for example. It does not, however, contribute directly to determination of the kernel functions for a biaxial or general state of strain.

Pure Shear. If experimental techniques permit testing under pure shear then kernel functions $\mathcal{G}_1$, $\mathcal{G}_2$, and $\mathcal{F}_5$ and $\mathcal{F}_6$ are easily determined from the results. $\mathcal{G}_1$ and $\mathcal{G}_2$ may be determined by methods described for creep from the results of two relaxation tests at two constant values of ε_{12} covering the nonlinear range by using (8.78). $\mathcal{F}_5$ and $\mathcal{F}_6$ may be obtained from measured values of σ_{11} and σ_{33} respectively during one test, using (8.76) and (8.77). However, the required experimental technique involves holding ε_{12} constant and holding ε_{11}, ε_{22} and ε_{33} zero while measuring σ_{12}, σ_{11} (or σ_{22}) and σ_{33}. The difficulties involved may be surmountable, but it is doubtful whether the results would be worth the effort even if successful.

Unconfined Shear. From an experimental standpoint an unconfined shear test is a simpler experiment than pure shear. Such a test may be accomplished by subjecting a thin-walled tube to a constant angle of twist and observing the torque (or shear stress). In such an experiment the normal stresses σ_{11}, σ_{22}, σ_{33} are zero but the corresponding strains ε_{11}, ε_{22}, ε_{33} are not zero and are time-dependent as indicated by

$$\boldsymbol{\epsilon} = \left|\begin{matrix} \varepsilon_{11}(t) & \varepsilon_{12} & 0 \\ \varepsilon_{12} & \varepsilon_{22}(t) & 0 \\ 0 & 0 & \varepsilon_{33}(t) \end{matrix}\right|. \tag{8.134}$$

Thus the state of strain is triaxial and involves many more than the desired two kernel functions $\mathcal{G}_1$, $\mathcal{G}_2$. Since the normal strains are expected to be small one approach is to lump their effect on the shear stress σ_{12} into the kernel functions in a similar manner to that discussed above for unconfined axial stress. This amounts to assuming that the state of strain is given by (8.75). Since the normal stresses are zero, (8.76) and (8.77) imply that $\mathcal{F}_5$ and $\mathcal{F}_6$ are zero. The equivalent of (8.78) is

$$\sigma_{12}(t) = \mathcal{G}_1^* \varepsilon_{12} + \mathcal{G}_2^* \varepsilon_{12}^3, \tag{8.135}$$

Where $\mathcal{G}_1^*$ and $\mathcal{G}_2^*$ are the lumped kernel functions for constant shear strain and are functions of time. Use of $\mathcal{G}_1^*$ and $\mathcal{G}_2^*$ should be satisfactory in stress analysis of structures such as circular bars and tubes in which an assumption that the normal stresses σ_{11}, σ_{22}, σ_{33} are negligibly small is acceptable.

General Biaxial Strain. If a suitable specimen form and instrumentation are devised to permit control of strain components ε_{12}, ε_{11}, ε_{22}, ε_{33} and measurement of stress components σ_{11} and σ_{12} at least, then the nine kernel functions required to express the biaxial stress components may be determined by the same methods described for the corresponding biaxial creep behavior. If in addition σ_{33} is measured, then two additional kernel functions may be determined from tests at two different levels of strain by using (8.68). From strain state (8.80) $\mathcal{F}_1$, $\mathcal{F}_2$, $\mathcal{F}_3$ may be determined by three tests at three strain levels; from strain state (8.75) $\mathcal{G}_1$, $\mathcal{G}_2$, and $\mathcal{F}_5$ may be obtained by tests at two strain levels; and from strain state (8.74) $\mathcal{F}_4$, $\mathcal{G}_3$, $\mathcal{G}_4$ may be found. Also from strain state (8.74) $\mathcal{F}_6$ and $\mathcal{F}_7$ may be obtained if σ_{33} is measured. As indicated before, however, devising a suitable specimen and loading frame and instrumentation to provide a known constant ε_{12}, maintaining ε_{11}, ε_{22} and ε_{33} equal to zero and measuring σ_{12}, σ_{11}, and σ_{33} is very difficult, has not been achieved to date and seems not worth the effort.

Triaxial Strain. If apparatus and specimen suitable for determining the kernel functions for biaxial strain states are available presumably it would also be possible with this apparatus to provide non-zero values of ε_{22} and ε_{33} as well as ε_{12} and ε_{11}. In this case the remaining kernel function for the triaxial state ψ_{12} can be determined from a single test under a hydrostatic or triaxial strain state using any one of equations (8.58) through (8.60).

8.14 Experimental Results of Creep

Tension and Compression. Representative tension creep curves for a laminated plastic were shown in Fig. 8.1. Similar curves are shown in Fig. 8.7(a, b, c, d) for tension creep of polyethylene [8.4], unplasticized poly (vinyl chloride) [8.1], monochlorotrifluoroethylene [8.12], and polyurethane (full density) [8.5]. Also shown in Fig. 8.7(a) are compression creep curves for the same material and temperature indicated for the corresponding tension curves. All of these curves show the same trends and all are describable with reasonable accuracy by (8.121). Typical values of the constants in (8.121) for various materials are shown in Table 8.1. It is remarkable how similar are the values of n for several different thermoplastic materials.

Experimental constants for nonlinear viscoelastic constitutive relations of other materials using the multiple integral representation have been reported by many researchers. For example, Ward and Onat [8.13] studied the behavior of oriented polypropylene monofilament in tension. Leaderman, McCrackin and Nakada [8.14] studied the behavior of a rubberlike network polymer in tension. Hadley and Ward [8.15], Ward and Wolfe [8.16] studied the creep behavior of several polypropylene fibers of different molecular structures. Lifshitz and Kolsky [8.17] studied the creep of polyethylene. Drescher and Kwaszczynska [8.18] studied the creep of poly (vinyl chloride) in which the strains were considered to be dependent on the stress up to the fifth order. Gottenberg, Bird and Agrawal [8.19] studied the relaxational behavior of plasticized cellulose acetate magnetic recording tape under uniaxial tension. Adeyeri, Krizek and Achenback [8.20] investigated the creep and stress relaxation behavior of a remolded clay under uniaxial compression.

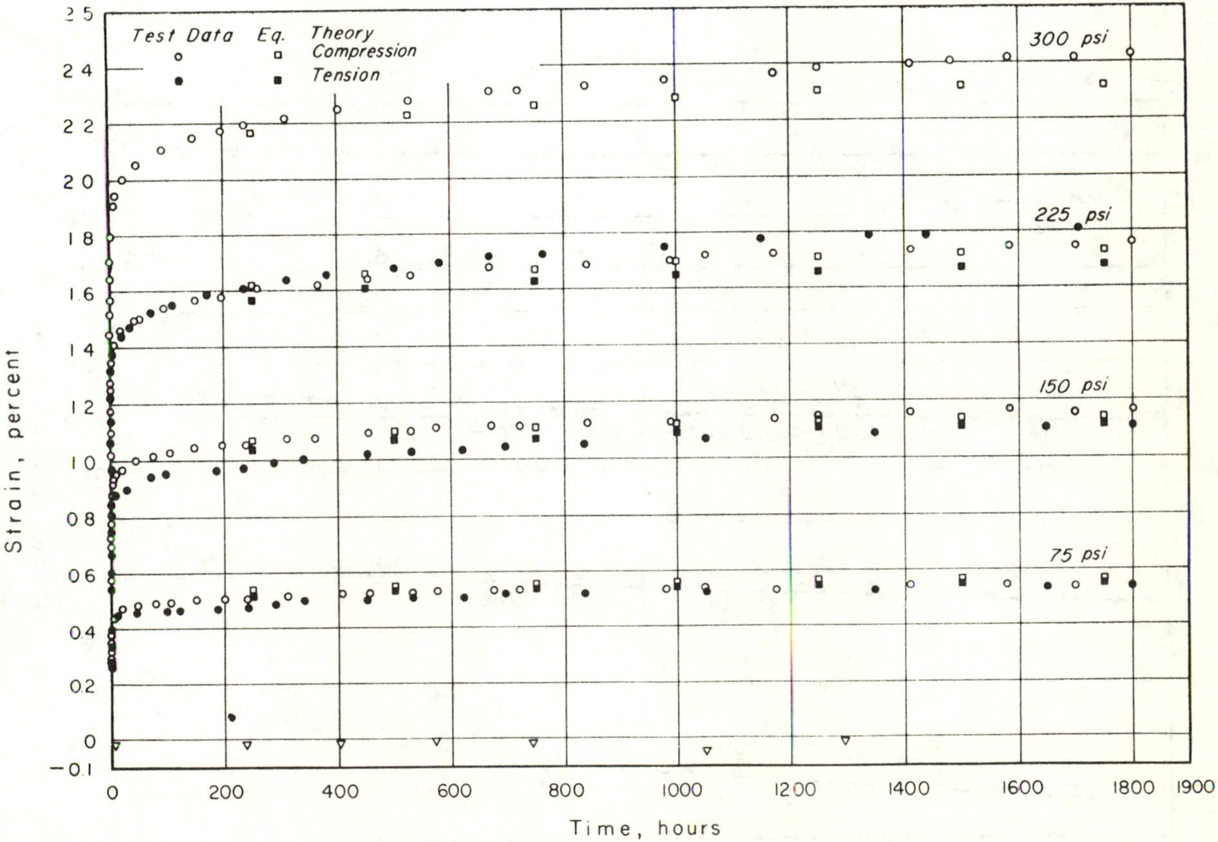

Fig. 8.7. Creep Curves for Several Plastics at 75°F and 50 per cent R.H.

Fig. 8.7*a*. Polyethylene in Tension and Compression. From [8.4], courtesy ASME. Theory: $\varepsilon = \varepsilon_0' \sinh \sigma/\sigma_\varepsilon + m't^n \sinh \sigma/\sigma_m$.

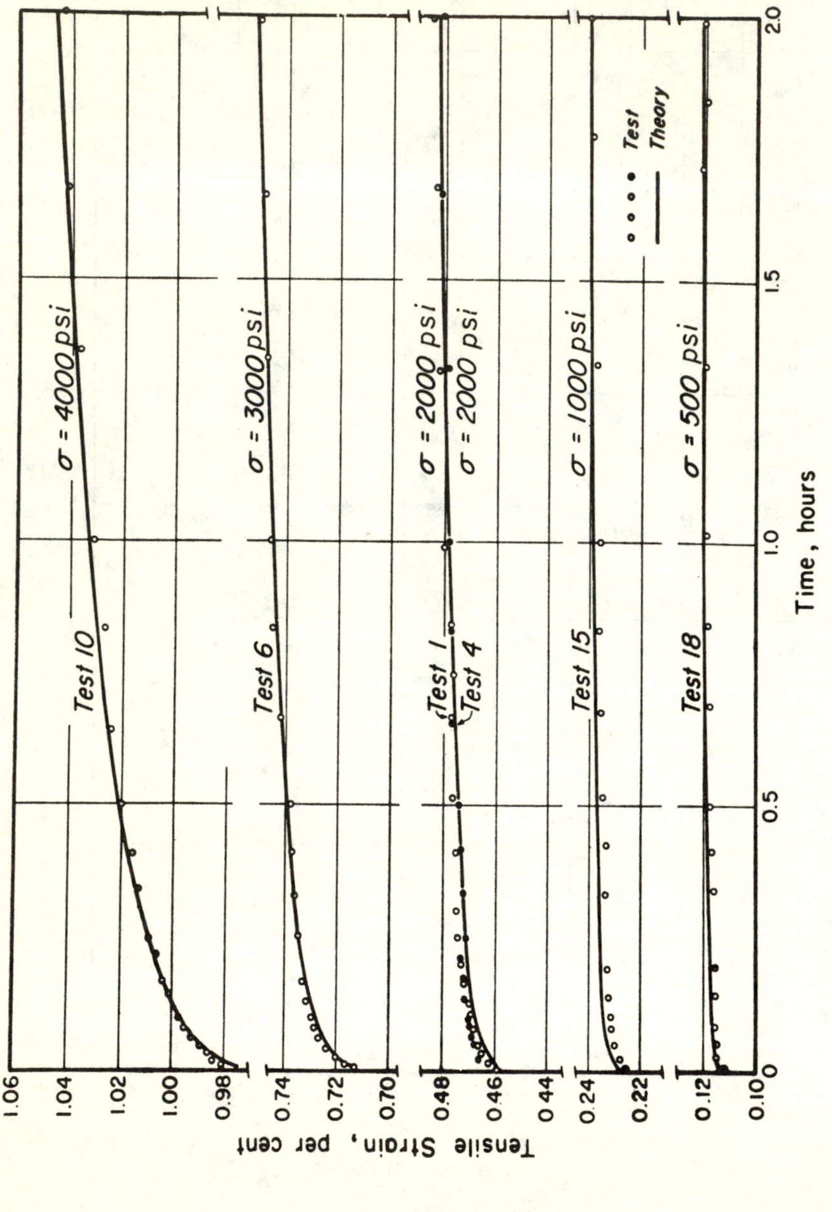
σ = 4000 psi
Test 10
σ = 3000 psi
Test 6
σ = 2000 psi
σ = 2000 psi
Test 1
Test 4
σ = 1000 psi
Test 15
σ = 500 psi
Test 18
Test
Theory
Time, hours
Tensile Strain, per cent
1.06
1.04
1.02
1.00
0.98
0.74
0.72
0.70
0.48
0.46
0.44
0.24
0.22
0.12
0.10
0
0.5
1.0
1.5
2.0

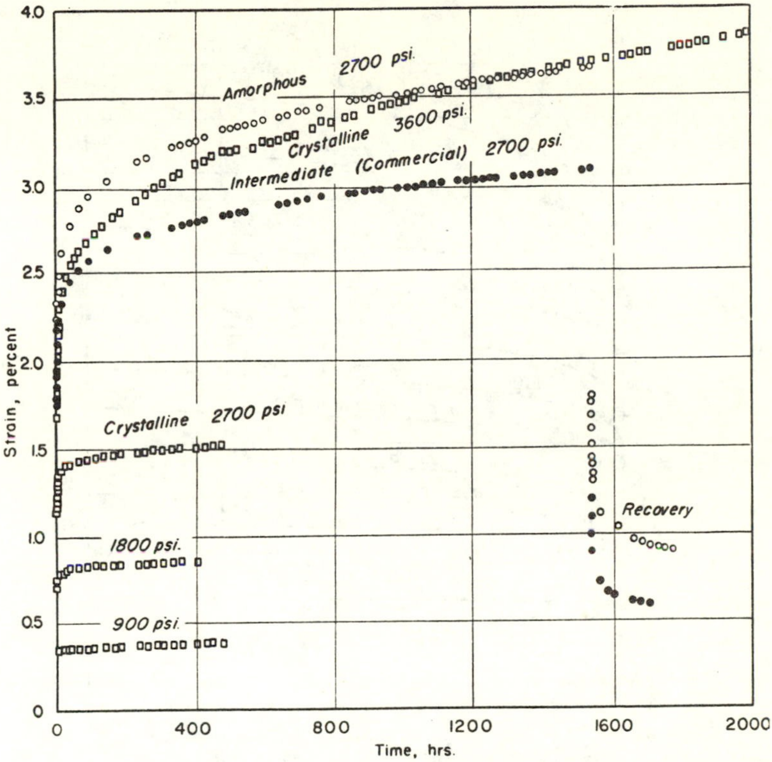

Fig. 8.7c. Monochlorotrifluoroethylene in Tension at Different Degrees of Crystallization. From [8.12], courtesy ASTM.

In Fig. 8.8 are shown tension or compression creep curves for red oak, copper, 2618-T61 aluminum alloy and asphalt concrete for various temperatures and times, together with representation of the data by (8.121). Creep of 304 stainless steel is shown in Fig. 2.1. The constants in (8.121) for some of the curves in Fig. 8.8 and 2.1 are given in Table 8.1. It should not be concluded that (8.121) can be used to describe the time dependence of all materials. For small strains, however, (8.121) is rather versatile. Also, it should be noted that even though creep at constant stress can be described by the same form of equation for dissimilar materials it does not follow that changes in stress (recovery for example) can be described by similar expressions. The recovery of metals is quite different from plastics, for example.

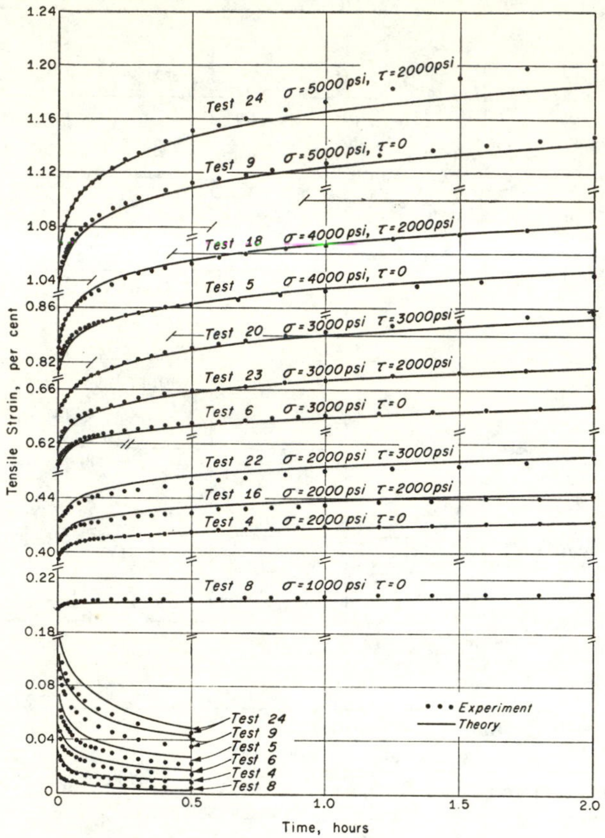

Fig. 8.7*d*. Tension Creep and Recovery of Full Density Polyurethane under Combined Tension and Torsion, From [8.5], courtesy ASME. Theory: (8.43) and (8.121).

Torsion and Combined Tension and Torsion. Examples of creep curves in torsion for polycarbonate, foam polyurethane (0.3 density) and unplasticized poly (vinyl chloride) are shown in Fig. 5.6, 8.9a and 8.9b respectively. The axial and shearing component of strains under combined tension and torsion for polyurethane are shown in Fig. 8.7d and 8.7e, respectively. These show similar characteristics to that observed in tension creep tests. The solid lines shown in Fig. 8.7(d, e) indicate the representation of creep of polyurethane by the nonlinear equations (8.43) and (8.46), where the time dependence of the F_α, G_α terms are as given by (8.121).

Time-independent Strain. The time-independent strain ε_{ij}^0 was determined from the creep tests for some of the materials, such as shown in Fig. 8.6a, b

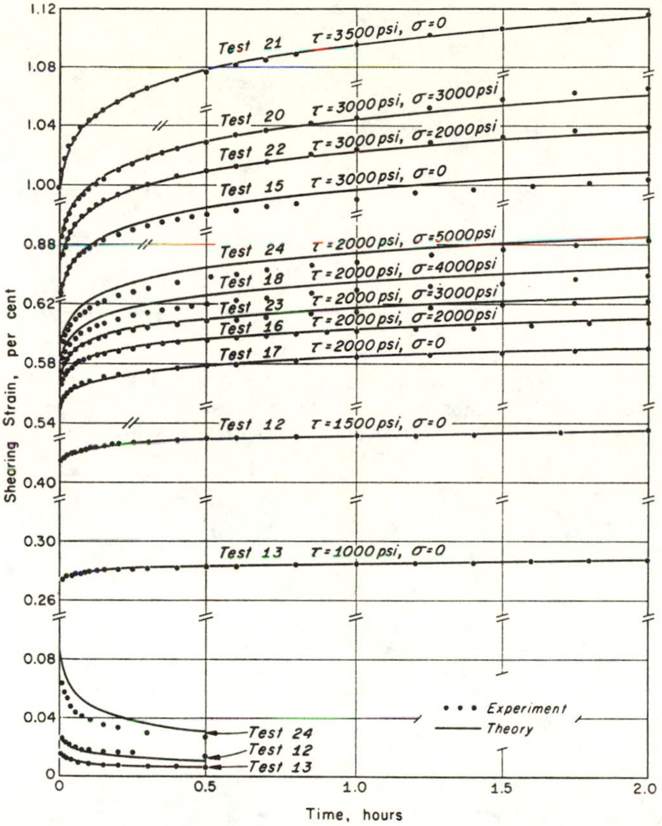

Fig. 8.7*e*. Shearing Creep and Recovery of Full Density Polyurethane under Combined Tension and Torsion. From [8.5], courtesy ASME. Theory: (8.46) and (8.121).

as described in a prior section. This was an indirect determination obtained by fitting the creep test data to (8.121). The time-independent strain cannot be determined directly because the strain is too time sensitive immediately after loading. The variation of ε_{ij}^0 with stress was found to be very nearly linear for all plastics which have been investigated as shown in Fig. 8.6a and b for combined tension and torsion of polyurethane. Some nonlinearity was observed. However it is small enough to be neglected in many applications.

Time-dependent Strain. The coefficient of the time-dependent strain ε_{ij}^+ was found to be a nonlinear function of stress for sufficiently large stresses in plastics. Poly (vinyl chloride) showed an essentially linear behavior of

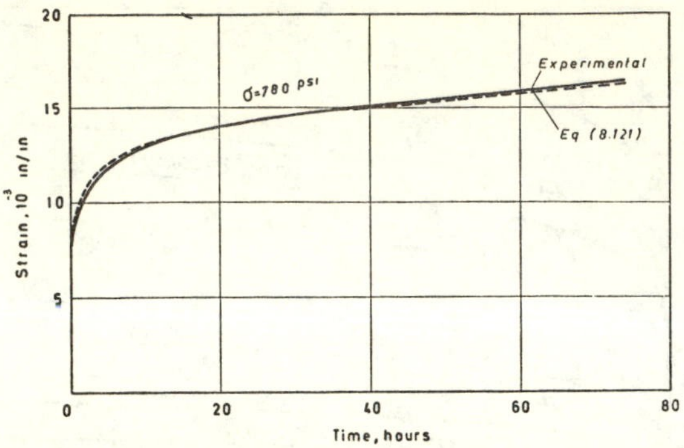

Fig. 8.8. Creep of Metals, Concrete and Wood.
Fig. 8.8*a*. Red Oak in Tension Perpendicular to Grain at 80°F (27°C) and 12 per cent Moisture Content. From [8.21], courtesy Forest Prod. Lab.

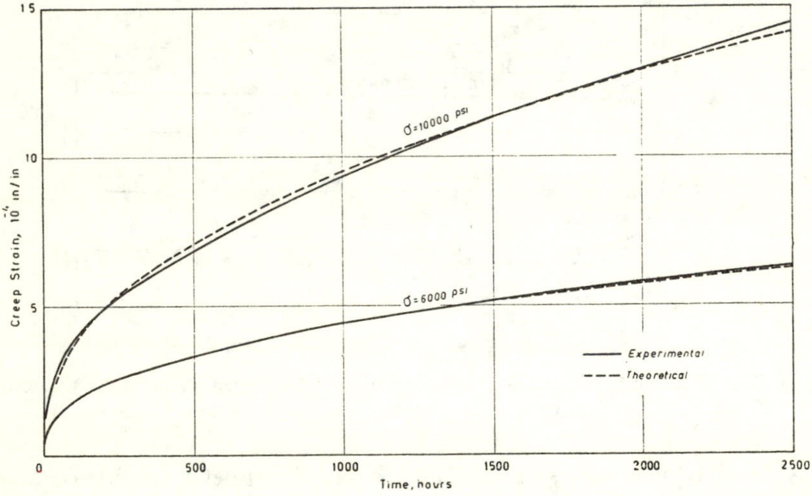

Fig. 8.8*b*. Oxygen-free Copper in Tension at 329°F (165°C). From [8.22], courtesy ASME.

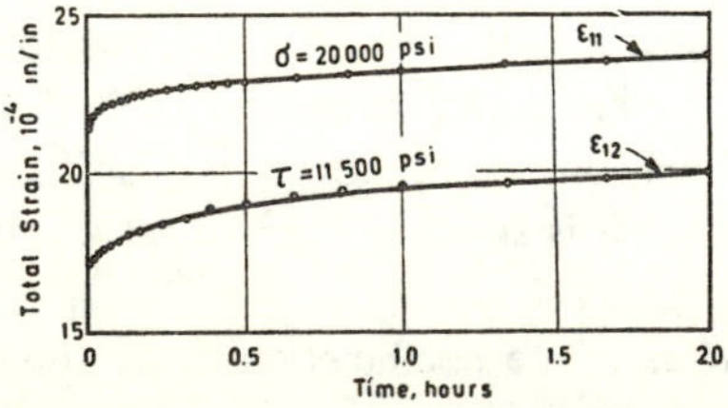

Fig. 8.8*c*. Aluminum Alloy 2618-T61 in Tension and in Torsion at 392°F (200°C). From [8.23], courtesy ASME.

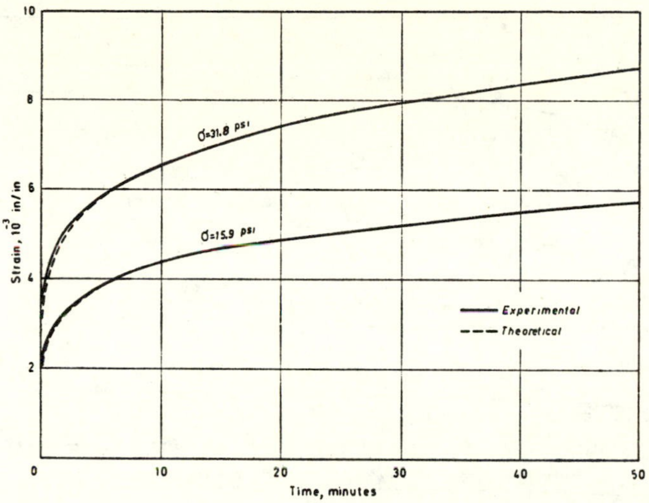

Fig. 8.8*d*. Asphalt Concrete in Compression at Room Temperature. From [8.24], courtesy Univ. of Utah.

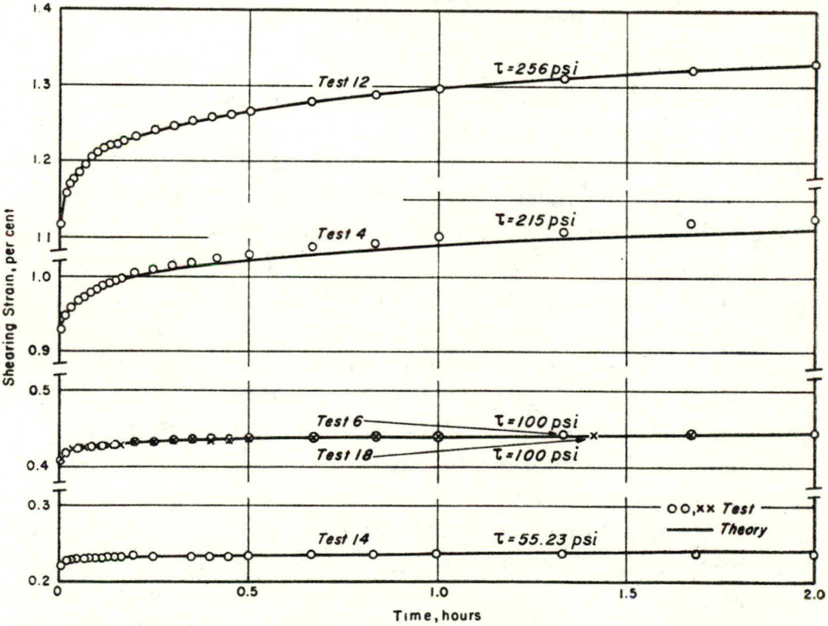

Fig. 8.9. Torsion Creep Curves for Several Plastics at 75°F and 50 per cent R.H.
Fig. 8.9*a*. 0.3 Density Foam Polyurethane in Torsion. From [8 3], courtesy ASTM. Theory: (8.51) and (8.121).

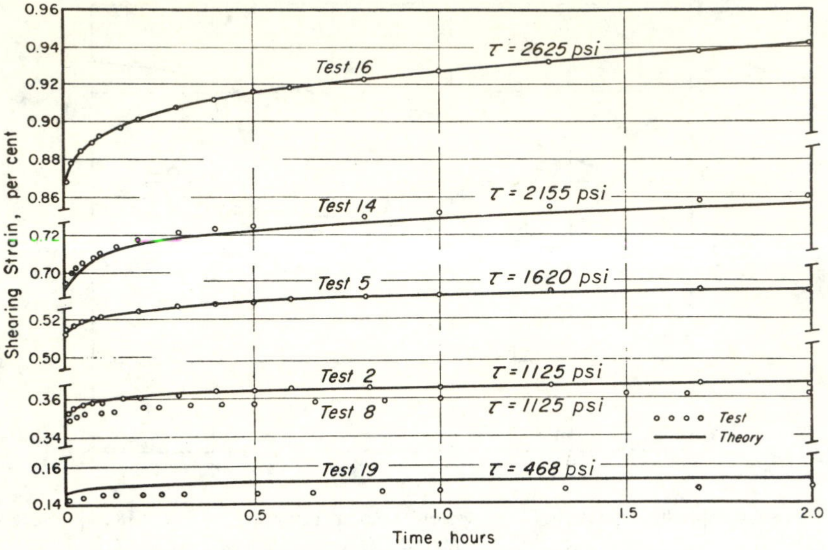

Fig. 8.9*b*. Unplasticized Poly (Vinyl Chloride) in Torsion. From [8.1], courtesy Soc. of Rheology. Theory: (8.51) and (8.121).

ε_{ij}^{+} at stresses below a principal shearing stress of about 1000 psi (or total strain of about 1/2 percent) but a strongly nonlinear behavior at higher stresses, as shown in Fig. 8.10 for ε_{12}^{+}. Polyurethane, on the other hand, showed a nonlinear behavior at all stress levels as nearly as could be determined both in the solid (full density) and foam, as shown in Fig. 8.6c and d.

High Pressure. Creep experiments in which polyethylene, poly (vinyl chloride) (PVC), methyl methacrylate (PMMA), polyurethane and epoxy were subjected to constant hydrostatic pressure up to 50,000 psi were reported by Findley, Reed and Stern [8.30]. In these experiments the volumetric strain was measured by means of a resistance strain gage. Results showed a temperature spike on pressurizing or decompressing which was dissipated in 0.2 hr. Following the return to constant temperature after the temperature spike a continuous decrease in volumetric strain was observed under constant pressure, see results for PMMA in Fig. 8.11. This strain could be described by (8.121). On release of pressure the recovery was essentially complete and instantaneous, except for the effect of the short duration temperature spike, see Fig. 8.11. This recovery is in marked contrast to the gradual recovery following tension or torsion creep of these materials. This difference appeared to result from the fact that under tension

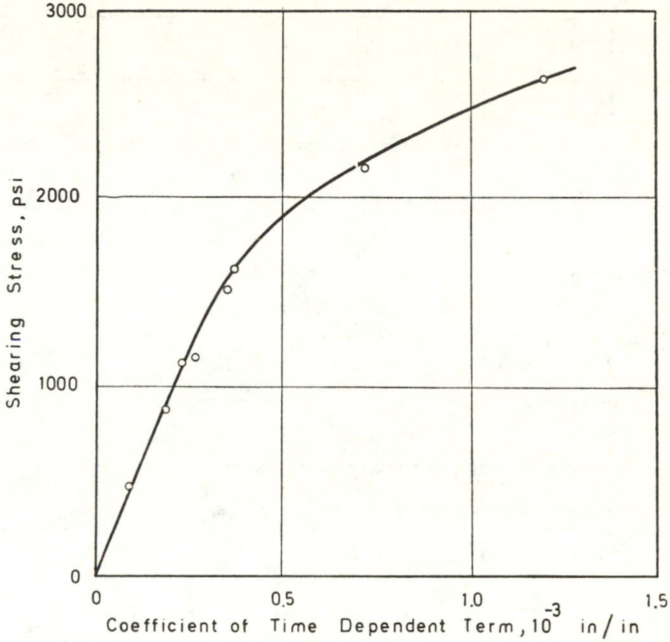

Fig. 8.10. Shearing Component of Coefficient of Time-dependent Strain ε_{12}^{+} for Pure Torsion Creep of Poly (Vinyl Chloride) at 75°F and 50 per cent R.H. From [8.1], courtesy Soc. of Rheology.

or torsion both creep and recovery took place in a material which was in essentially the same state; whereas volume creep under high hydrostatic pressure occurred in a compressed polymer structure while recovery following release of pressure occurred in an expanded polymer structure. The opening up of the structure on release of pressure permitted almost immediate return of molecular segments to their original configuration.

Diffusion of Gases. The effect on polycarbonate of simultaneous creep under biaxial stress and diffusion of gases under pressure was reported by Hojo and Findley [8.31]. It was observed that the rate of creep was considerably increased by absorption of solubility controlled gases, carbon dioxide and freon, because of the plasticizing action of their dissolved gases, as shown in Fig. 8.12. On the other hand, a diffusion controlled gas (helium) had no detectible effect on the creep rate, see Fig. 8.12. Air and nitrogen, which have a much smaller diffusivity than helium and smaller solubility than carbon dioxide, also had a negligible effect on creep rate, as shown in Fig. 8.12.

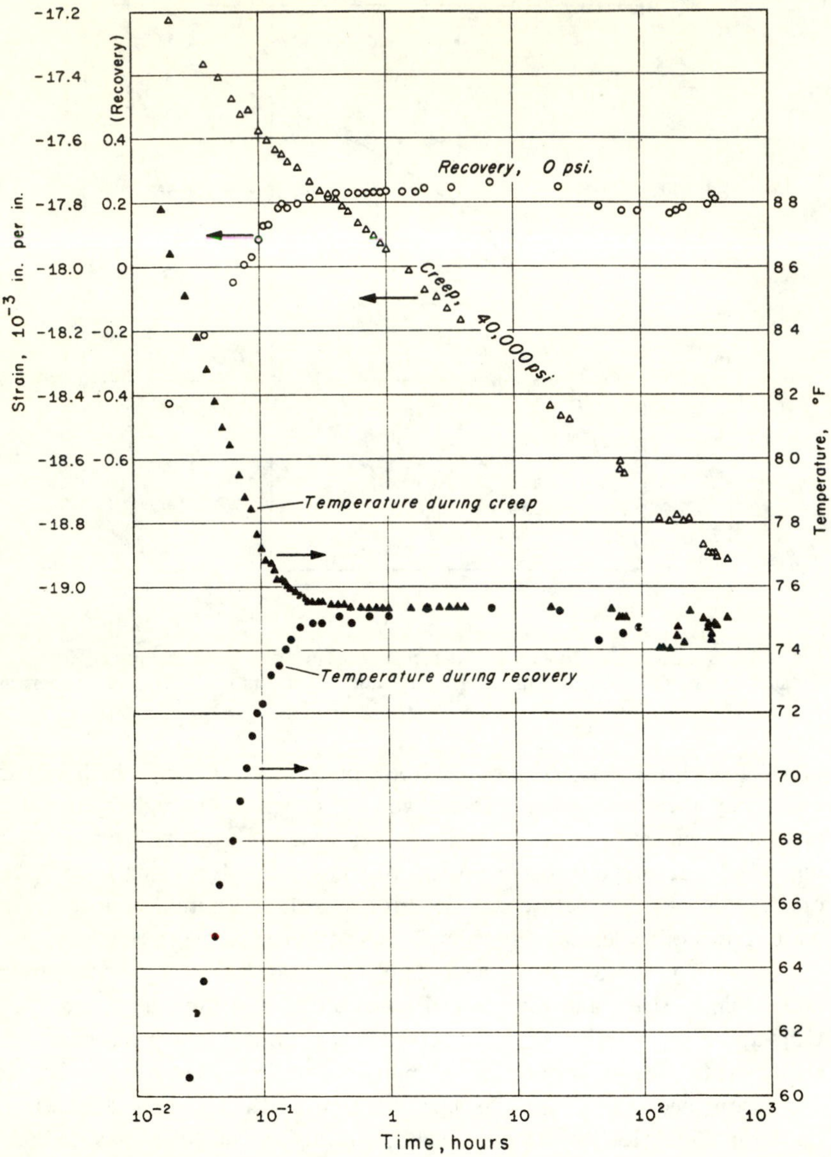

Fig. 8.11. Creep Under Hydrostatic Pressure, Recovery and Temperature Response for PMMA. From [8.30], courtesy ASME.

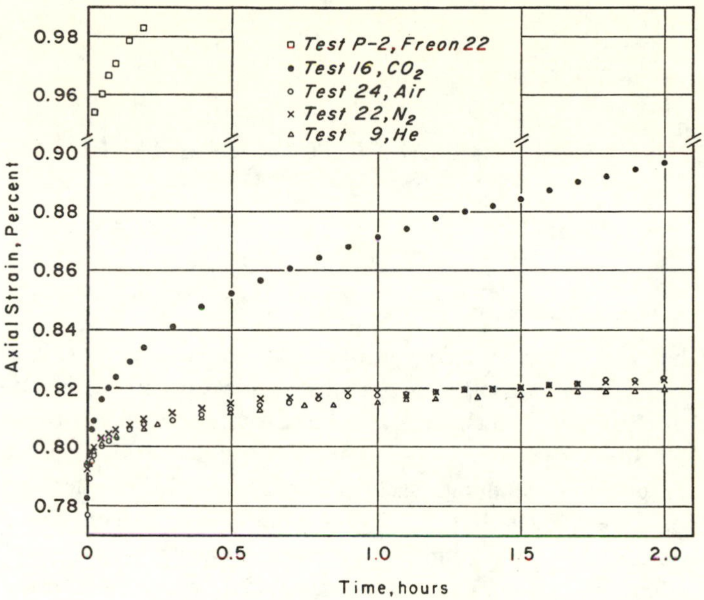

Fig. 8.12. Effect of Diffusion of Gases on Creep of a Polycarbonate Tube Subjected to Axial Tension and Internal Pressure (Axial stress $\sigma_1 = 3000$ psi, circumferential stress $\sigma_2 = 1000$ psi, internal pressure 85.4 psi) at 75°F. From [8.31], courtesy Polymer Engineering and Science.

Biaxial Tension-Tension. The time dependence of the biaxial creep reported in [8.31] due to tension and internal pressure in a thin-walled tube was well described by (8.121). Using material constants obtained from combined tension and torsion creep of a different specimen in [8.26] the axial strain under tension and internal pressure (biaxial tension-tension) was computed from (8.35). These computed values compared only fairly well with results of creep under tension and internal pressure reported in [8.31].

CHAPTER 9

NONLINEAR CREEP (OR RELAXATION) UNDER VARIABLE STRESS (OR STRAIN)

9.1 Introduction

In Chapter 7, the general constitutive equation was formulated using the multiple integral representation. The characterization of the equations required for constant inputs, such as constant stress for creep or constant strain for stress relaxation, were treated in Chapter 8. A variable input (variable loading or straining) such as discussed in this chapter requires a more complete characterization of the equation. A more complete characterization of the kernel functions requires experimental data from multiple step loading (or straining) and the number of tests needed is in the order of a hundred or more. This creates a great problem for both experimentation and analysis. Besides, the constitutive equation thus obtained is quite complicated. This creates considerable mathematical difficulty in solving boundary value problems. Therefore, several kinds of approximations have been proposed to provide a less time-consuming and less involved means of dealing with nonlinear viscoelastic behavior under variable input programs. Some of these approximate methods are described in subsequent sections of this chapter. In these methods the kernel functions determined from constant input programs are used to describe the nonlinear viscoelastic behavior under a variable input program. A method of direct experimental determination of the kernel functions of the constitutive equation under variable loading is discussed first in this chapter followed by several approximations.

9.2 Direct Determination of Kernel Functions

Determination of kernel functions for mixed time parameters, such as $K_7, \ldots, K_{12}$; $\psi_7, \ldots, \psi_{12}$ of equations (7.9) and (7.12), directly from experiments will be illustrated for the kernel functions $L(t, t, t-t_1)$, $L(t, t-t_1, t-t_1)$, and $L(t, t-t_1, t-t_2)$ which appear in the formulation for pure torsion, as follows. The general formulation of the constitutive relation of creep under pure torsion, can be expressed as follows, see (7.80),

$$\varepsilon_{12}(t) = \int_0^t R(t-\xi_1)\,\dot{\tau}(\xi_1)\,d\xi_1 + \int_0^t\int_0^t\int_0^t L(t-\xi_1, t-\xi_2, t-\xi_3)\,\dot{\tau}(\xi_1)\,\dot{\tau}(\xi_2)\,\dot{\tau}(\xi_3)\,d\xi_1\,d\xi_2\,d\xi_3, \tag{9.1}$$

where terms higher than the third order have been neglected and R and L have been substituted for G_1 and G_2 respectively. A complete characterization of these two kernel functions $R(t-\xi_1)$ and $L(t-\xi_1, t-\xi_2, t-\xi_3)$ requires creep tests with one step, two steps and three steps of constant

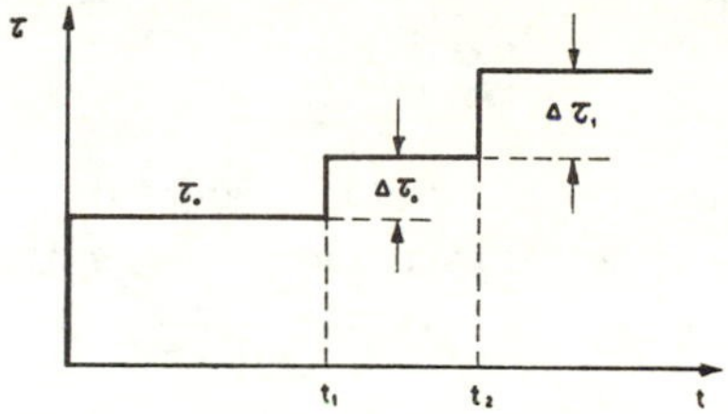

Fig. 9.1. Three-step Loading.

loading. A three step pure torsional stress history, as shown in Figure 9.1, can be expressed by the following equation

$$\tau(t) = \tau_0 H(t) + \Delta\tau_0 H(t-t_1) + \Delta\tau_1 H(t-t_2), \tag{9.2}$$

where τ_0 is the stress of the first step, $\Delta\tau_0$ the first change in stress and $\Delta\tau_1$ the second change in stress, and $H(t)$ is the Heaviside unit function. The shearing strain response under (9.2) as input can be obtained as follows by inserting (9.2) into (9.1) and performing the integration for the three different time intervals,

$$\varepsilon_{12}(t) = R(t)\tau_0 + L(t, t, t)\tau_0^3; \quad 0 < t \leq t_1, \tag{9.3}$$

$$\begin{aligned}\varepsilon_{12}(t) = {} & R(t)\tau_0 + L(t, t, t)\tau_0^3 + R(t-t_1)(\Delta\tau_0) + L(t-t_1, t-t_1, t-t_1)(\Delta\tau_0)^3 \\ & + 3L(t, t, t-t_1)\tau_0^2(\Delta\tau_0) + 3L(t, t-t_1, t-t_1)\tau_0(\Delta\tau_0)^2; \\ & t_1 < t \leq t_2,\end{aligned} \tag{9.4}$$

and

$$
\begin{aligned}
\varepsilon_{12}(t) = {} & R(t)\tau_0 + L(t, t, t)\tau_0^3 \\
& + R(t-t_1)(\Delta\tau_0) + L(t-t_1, t-t_1, t-t_1)(\Delta\tau_0)^3 \\
& + R(t-t_2)(\Delta\tau_1) + L(t-t_2, t-t_2, t-t_2)(\Delta\tau_1)^3 \\
& + 3[L(t, t, t-t_1)\tau_0^2(\Delta\tau_0) + L(t, t-t_1, t-t_1)\tau_0(\Delta\tau_0)^2] \\
& + L(t, t, t-t_2)\tau_0^2(\Delta\tau_1) + L(t, t-t_2, t-t_2)\tau_0(\Delta\tau_1)^2 \\
& + L(t-t_1, t-t_1, t-t_2)(\Delta\tau_0)^2(\Delta\tau_1) + L(t-t_1, t-t_2, t-t_2)(\Delta\tau_0)(\Delta\tau_1)^2 \\
& + 6L(t, t-t_1, t-t_2)\tau_0(\Delta\tau_0)(\Delta\tau_1); \quad t > t_2.
\end{aligned} \tag{9.5}
$$

From (9.3), (9.4) and (9.5), it can be seen that a complete characterization of the two kernel functions in (9.1) requires determining the following time functions,

$$
\begin{aligned}
&\text{(A)} \quad R(t),\ L(t, t, t), \\
&\text{(B)} \quad L(t, t, t-t_1),\ L(t, t-t_1, t-t_1), \\
&\text{(C)} \quad L(t, t-t_1, t-t_2),
\end{aligned} \tag{9.6}
$$

under arbitrary values of t_1 and t_2. One way of determining the time functions in (9.6) will be discussed using the following notations:

$$
\begin{aligned}
R(t) &= R_0, \\
R(t-t_1) &= R_1, \\
R(t-t_2) &= R_2, \\
L(t, t, t) &= L_{000}, \\
L(t, t, t-t_1) &= L_{001}, \\
L(t, t-t_1, t-t_1) &= L_{011}, \\
L(t, t-t_1, t-t_2) &= L_{012}, \text{ etc.}
\end{aligned} \tag{9.7}
$$

Part A. *Determination of $R(t)$ and $L(t, t, t)$*

From several constant stress creep tests at different shearing stresses in the nonlinear range, a set of creep curves of shearing strain versus time can be obtained. The ratio of the time dependent part of the shearing strain $\varepsilon_{12}(t_i)$ at given time t_i divided by shearing stress $(\varepsilon_{12}(t_i)/\tau_0)$ may be plotted versus τ_0 to form a series of isochronous curves corresponding to the different times t_i. According to (9.3) all of these curves should have parabolic form, namely,

$$
\frac{\varepsilon_{12}(t_i)}{\tau_0} = R(t_i) + L(t_i)\,\tau_0^2. \tag{9.8}
$$

From these curves R_0 and L_{000} can be determined at each chosen time t_i. Thus, the variation of R_0 and L_{000} with respect to time can be obtained. Although in principle, two creep tests at two different shearing stresses will be sufficient to determine R_0 and L_{000}, because only two points are needed to determine a parabola, three or four creep tests under different shearing stress are desirable.

Part B. *Determination of* $L(t, t, t-t_1)$ *and* $L(t, t-t_1, t-t_1)$

Two-step creep test data are required for the determination of L_{001} and L_{011}. Rewriting (9.4) in the following form

$$\varepsilon_{12}(t) = \varepsilon_a(t)+\varepsilon_b(t-t_1)+\varepsilon_c(t, t-t_1), \tag{9.9}$$

where

$$\varepsilon_a(t) = R(t)\tau_0+L(t, t, t)\tau_0^3,$$

$$\varepsilon_b(t-t_1) = R(t-t_1)(\Delta\tau_0)+L(t-t_1, t-t_1, t-t_1)(\Delta\tau_0)^3,$$

$$\varepsilon_c(t, t-t_1) = 3L(t, t, t-t_1)\tau_0^2(\Delta\tau_0)+3L(t, t-t_1, t-t_1)\tau_0(\Delta\tau_0)^2,$$

where $\varepsilon_{12}(t)$ are the measured quantities from two-step creep tests. Using (9.3) ε_a can be evaluated since $R(t)$, $L(t, t, t)$ have been determined from Part A. Also ε_b can be calculated from (9.3) and Part (A), since $R(t-t_1)$ and $L(t-t_1, t-t_1, t-t_1)$ are the same functions as $R(t)$ and $L(t, t, t)$, respectively, except for a shift in the time axis. Thus, from (9.9) $\varepsilon_c(t, t-t_1)$ can be calculated for each two-step creep test. L_{001} and L_{011} at each instant of prior time, $t = \xi$, can be calculated for a fixed value of t_1, by solving the following two simultaneous linear equations

$$\varepsilon_c^{(1)}(\xi, \xi-t_1) = 3L_{001}(\xi, \xi-t_1)(\tau_0^{(1)})^2(\Delta\tau_0^{(1)})+3L_{011}(\xi, \xi-t_1)(\tau_0^{(1)})(\Delta\tau_0{}^{(1)})^2,$$

$$\varepsilon_c^{(2)}(\xi, \xi-\xi_1) = 3L_{001}(\xi, \xi-t_1)(\tau_0^{(2)})^2(\Delta\tau_0^{(2)})+3L_{011}(\xi, \xi-t_1)$$
$$(\tau_0{}^{(2)})(\Delta\tau_0{}^{(2)})^2, \tag{9.10}$$

where $\varepsilon_c^{(1)}$, $\varepsilon_c^{(2)}$ are calculated, respectively, from two two-step creep tests having different stresses $\tau_0^{(1)}$, $\Delta\tau_0^{(1)}$ and $\tau_0^{(2)}$, $\Delta t_0^{(2)}$ but the same t_1. Therefore the variation of $L(t, t, t-t_1)$ and $L(t, t-t_1, t-t_1)$ versus t at the fixed value of t_1 can be obtained. By performing two-step creep tests at different values of t_1 and performing the same analysis, a series of curves for $L(t, t, t-t_1)$ and $L(t, t-t_1, t-t_1)$ versus t at different values of t_1 can be obtained. These curves may then be described by suitable functions.

Part C. *Determination of* $L(t, t-t_1, t-t_2)$

Three-step creep test data are required for the determination of $L(t, t-t_1, t-t_2)$. In (9.5) using notation of (9.7), R_0, L_{000} can be obtained from Part (A), L_{001} and L_{011} can be obtained from Part (B) and the other terms except $L(t, t-t_1, t-t_2)$ can be obtained from either Part (A) or Part (B) with a proper time shift. Hence the term $L(t, t-t_1, t-t_2)$ at each instance of arbitrary time, $t = \xi$, can be obtained from three-step creep data with fixed t_1 and t_2. By varying t_1 and t_2 in the three-step creep tests and using an analysis similar to that just described, the variation of the term $L(t, t-t_1, t-t_2)$ with respect to time t and with different values of t_1 and t_2 can be obtained. Therefore complete information on $L(t, t-t_1, t-t_2)$ can be displayed by α sets of curves of $L(t, t-t_1, t-t_2)$ versus time t where each set of curves has a different constant value of t_2 and has β curves with different values of t_1. These curves also may be described by suitable functions.

For a third order theory, Parts (A), (B), (C) complete the determination of the kernel functions for pure shear. With the kernels of (9.6) determined, the creep strain under arbitrary pure shear stress history can be predicted from (9.1).

The number of different values of t_1 and t_2 required for L_{001}, L_{011} and L_{012} such that a reasonable prediction of the creep behavior can be expected depends on the material behavior and the smoothness of the creep strain response to the application of the load. It is also strongly dependent on the type of stress input from which the creep strain will be predicted. The smoother the material response and the smoother the stress input, the fewer values of t_1 and t_2 are required. On the other hand if the stress input is less smooth, more t_1 and t_2 are required if better prediction is to be expected.

A rough estimation of how many tests are required in order that the kernel functions in the third order theory under pure shear can be characterized is as follows. Assume that five different values of t_1 and t_2 as shown in the following are sufficient,

$$t_1 = 1,\ 10,\ 10^2,\ 10^3,\ 10^4 \text{ seconds},$$
$$t_2 = 3\times10^0,\ 3\times10^1,\ 3\times10^2,\ 3\times10^3,\ 3\times10^4 \text{ seconds},$$

or

$$t_1 = 3\times10^{-4},\ 3\times10^{-3},\ 3\times10^{-2},\ 3\times10^{-1},\ 3 \text{ hours},$$
$$t_2 = 10^{-3},\ 10^{-2},\ 10^{-1},\ 1,\ 10 \text{ hours}.$$

The theoretical number of tests required is then as follows: two one-step tests for Part (A), ten two-step tests with five different values of t_1 for

Part (B), fifteen three-step tests with five different values of t_1 and t_2 with $t_2 > t_1$ for Part (C). Thus, the total number is 27. However, in order to minimize the possible experimental errors, twice this number of tests would be desirable.

The authors [9.1] explored the feasibility of using the method just described to determine some of the kernel functions of an unplasticized PVC copolymer using two-step uniaxial tensile and pure torsional loading. The analysis for pure torsion will be discussed in the following.

The tests were conducted using a single specimen at a temperature of $75 \pm 1/2$°F (24°C) and $50 \pm 1/2$ percent relative humidity. The specimen was first subjected to a constant shear stress τ_0 for a period of t_1; then the stress was changed to τ_1 and kept constant for one to two hours after which the stress was removed. Following each experiment, the specimen was allowed to recover at zero stress until the strain had returned to less than 25×10^{-6} in. per in. before proceeding to the next experiment.

The method of analysis of the test data for the first step was exactly the same as discussed in Chapter 8. Shearing strain was represented by

$$\varepsilon_{12}(t) = \varepsilon_{12}^{0} + \varepsilon_{12}^{+} t^n, \tag{9.11}$$

where the value of ε_{12}^{0}, ε_{12}^{+}, n for each test together with the compliances $\varepsilon_{12}^{0}/\tau_0$, $\varepsilon_{12}^{+}/\tau_0$ are given in Table 9.1.

Comparison of (9.11) with (9.3) implies that

$$\begin{aligned} R(t) &= R^0 + R^+ t^n, \\ L(t, t, t) &= L^0 + L^+ t^n. \end{aligned} \tag{9.12}$$

By using the numerical values of compliances $\varepsilon_{12}^{0}/\tau_0$ and $\varepsilon_{12}^{+}/\tau_0$ from Table 9.1 and employing a least square method, the constants R^0, R^+, L^0 and L^+ given in Table 9.2 were determined.

The method described in Part (B) of this section was employed to determine the kernels $L(t, t, t-t_1)$ and $L(t, t-t_1, t-t_1)$ from the second step creep data for each given value of t_1. The experimental results showed that for step-up tests the strains for the second step were nearly the same for all values of t_1. This may result from the fact that the memory effect on the strains in the second period due to the small stress in the first step was so small as to be nearly obscured by the scatter of the test results. On the other hand, for step-down tests the effect of the first loading on the strains in the second period was very significant. Therefore, the experimental data from step-down tests were used to determine the kernel functions L_{001} and L_{011}.

Table 9.1

Constants in $\varepsilon_{12} = \varepsilon_{12}^0 + \varepsilon_{12}^+ t^n$ for Creep of Unplasticized PVC at Different Stresses

Stress, τ psi	Constants of (9.11)		Constants of Creep Compliance	
	ε_{12}^0, 10^{-6} in/in	ε_{12}^+, 10^{-6} in/in	ε_{12}^0/τ, 10^{-6} in/in psi	ε_{12}^+/τ, 10^{-6} in/in psi
439	1293	90	2.938	0.204
878	2666	159	3.036	0.182
1755	5358	392	3.053	0.223
2457	7702	750	3.135	0.305
3072	9511	1569	3.096	0.516

$n = 0.155$

Table 9.2

Kernel Functions $R(t)$, $L(t, t, t)$ for Unplasticized PVC

$$R^0 = 2.993 \times 10^{-6} \text{ in/in psi}$$
$$R^+ = 0.156 \times 10^{-6} \text{ in/in psi hr}^n$$
$$L^0 = 15.10 \times 10^{-15} \text{ in/in (psi)}^3$$
$$L^+ = 32.80 \times 10^{-15} \text{ in/in (psi)}^3 \text{ hr}^n$$

It was found from these experiments that $L(t, t, t-t_1)$ and $L(t, t-t_1, t-t_1)$ could be described as a function of $(t-t_1)$ over the range of observation of $(t-t_1)$ by the following relationships,

$$L(t, t, t-t_1) = m_1(t-t_1)^{n_1},$$
$$L(t, t-t_1, t-t_1) = m_2(t-t_1)^{n_2}, \tag{9.13}$$

for each given value of t_1. It was observed by comparison of m_1, m_2, n_1, n_2 for different values of t_1, that n_1 and n_2 were independent of t_1, and m_1 and m_2 were approximately linear functions of t_1. Therefore, the following equations for the kernel functions $L(t, t, t-t_1)$ and $L(t, t-t_1, t-t_1)$ were obtained

$$L_{001} = L(t, t, t-t_1) = (50-15t_1)(t-t_1)^{0.24}\, 10^{-15} \text{ in/in (psi)}^3,$$
$$L_{011} = L(t, t-t_1, t-t_1) = (53-23t_1)(t-t_1)^{0.38}\, 10^{-15} \text{ in/in (psi)}^3. \tag{9.14}$$

No attempt was made in [9.1] to determine $L(t, t-t_1, t-t_2)$.

A similar approach was taken by Neis and Sackman [9.2] in the determination of the first-, second- and third-order creep kernel functions under a uniaxial stress state from the creep data of low density polyethylene under single-step, two-step and three-step uniaxial tensile and compressibe loading histories.

9.3 Product-Form Approximation of Kernel Functions

The results presented in the previous section indicate that even under pure shear stress direct determination of the kernel functions in the multiple integral representation of the constitutive relation requires a large number of well-conducted experiments. This limits the practical value of this method for describing variable stress situations. Hence, several approximate methods have been proposed to simplify the multiple integral representation when applying it to variable stress situations with nonlinear viscoelastic materials. Methods to be discussed in this and the following sections describe nonlinear viscoelastic behavior for variable stress by using the kernel functions determined only from one-step creep tests.

The method employed in this section is based on an assumption that all the kernel functions which depend on time arguments have the explicit form of the product of their time argument. This assumption was originally suggested by Nakada [9.3] and was further explored by the authors [9.4, 9.5]. Under this assumption the kernel functions of the higher order term under pure torsion, see (9.1), can be written as follows:

$$L(t-t_1, t-t_2, t-t_3) = [L(t-t_1)]^{1/3}\,[L(t-t_2)]^{1/3}[L(t-t_3)]^{1/3} \tag{9.15}$$

where $L(t-t_1)$, $L(t-t_2)$, $L(t-t_3)$ can be determined from one-step constant-stress creep tests as described in Chapter 8. This equation satisfies the symmetry requirement of the kernel function $L(t-t_1,\ t-t_2,\ t-t_3)$ with respect to its time arguments. Inserting (9.15) into (9.1) yields

$$\begin{aligned}\varepsilon_{12}(t) = &\int_0^t R(t-\xi_1)\,\dot{\tau}(\xi_1)\,d\xi_1 \\ &+ \int_0^t\int_0^t\int_0^t [L(t-\xi_1)]^{1/3}[L(t-\xi_2)]^{1/3}[L(t-\xi_3)]^{1/3}\,\dot{\tau}(\xi_1)\,\dot{\tau}(\xi_2)\,\dot{\tau}(\xi_3) \\ &d\xi_1\,d\xi_2\,d\xi_3\end{aligned} \tag{9.16}$$

or

$$\begin{aligned}\varepsilon_{12}(t) = &\int_0^t R(t-\xi_1)\,\dot{\tau}(\xi_1)\,d\xi_1 + \left\{\int_0^t [L(t-\xi_1)]^{1/3}\,\dot{\tau}(\xi_1)\,d\xi_1\right\} \\ &\left\{\int_0^t [L(t-\xi_2)]^{1/3}\,\dot{\tau}(\xi_2)\,d\xi_2\right\}\left\{\int_0^t [L(t-\xi_3)]^{1/3}\,\dot{\tau}(\xi_3)\,d\xi_3\right\}.\end{aligned} \tag{9.16a}$$

This equation, with $R(t)$, $L(t)$ determined from constant-stress creep tests, can be used to describe nonlinear creep behavior under an arbitrary stress history. Application of (9.16) and its equivalent for other states of stress to predict creep behavior under two different stress histories will be presented in Section 9.7.

9.4 Additive Forms of Approximation of Kernel Functions

Gottenberg, Bird and Agrawal [9.6] utilized a similar approach to describe the nonlinear viscoelastic behavior of plasticized cellulose acetate magnetic recording tape under consecutive constant strain rate programs from single-step stress relaxation tests. In their approach, an additive form was first assumed for kernel functions, such as in (9.1), as follows,

$$L(t-\xi_1, t-\xi_2, t-\xi_3) = L(t-\xi_1+t-\xi_2+t-\xi_3) = L(3t-\xi_1-\xi_2-\xi_3). \tag{9.17}$$

The form of this kernel function can be determined from tests having a single step input. However a product form such as the one shown in (9.16) was used in [9.6] instead of (9.17) in order to simplify the computations under constant strain rate inputs.

Cheung [9.7] utilized an additive form to describe the nonlinear viscoelastic behavior of blood vessels. In his approach the following additive form was assumed.

$$L(t-\xi_1, t-\xi_2, t-\xi_3) = \tfrac{1}{3}L(t-\xi_1)+\tfrac{1}{3}L(t-\xi_2)+\tfrac{1}{3}L(t-\xi_3), \qquad (9.18)$$

where $L(t, t, t)$ can be determined from a single step input, since when $\xi_1 = \xi_2 = \xi_3 = 0$, (9.18) reduces to $L(t, t, t) = L(t)$.

9.5 Modified Superposition Method

Again, creep behavior under pure torsion will be used in this section to illustrate the concept of the "modified superposition method" employed by the authors [9.8, 9.9, 9.10, 9.11]. Under constant shear stress τ, the nonlinear viscoelastic stress-strain-time relation can be represented by

$$\varepsilon_{12}(t) = f(\tau, t), \qquad (9.19)$$

where $f(\tau, t)$ is a nonlinear function of stress, τ. For instance

$$f(\tau, t) = R(t)\tau+L(t)\tau^3$$

results from the third-order multiple integral representation for pure shear stress, see Chapter 8.

The modified superposition method suggests that for N stepwise changes of stress input from τ_{i-1} to τ_i at time $t = t_i$, the corresponding creep strain at $t > t_N$ is represented by the following form

$$\varepsilon_{12}(t) = \sum_{i=0}^{N} [f(\tau_i, t-t_i)-f(\tau_{i-1}, t-t_i)], \quad t > t_N. \qquad (9.20)$$

For the case of $N = 1$, the stress is changed from τ_0 to τ_1 at $t = t_1$, see Fig. 9.2A. Thus (9.20) becomes

$$\varepsilon_{12}(t) = f(\tau_0, t-t_0)- f(\tau_0, t-t_1)+f(\tau_1, t-t_1), \quad t > t_1. \qquad (9.21)$$

Equation (9.21) can be interpreted as follows. At the instant of time t_1, the stress τ_0 is considered to be removed and at the same time τ_1 is applied to the specimen, both being considered as independent actions. The recovery strain ε^r resulting from removal of τ_0 at t_1, see Fig. 9.2B, is given by

$$\varepsilon^r(t) = f(\tau_0, t-t_0) - f(\tau_0, t-t_1), \quad t > t_1. \tag{9.22}$$

The creep strain ε^c due to application of τ_1 at time t_1, see Fig. 9.2C, is given by

$$\varepsilon^c(t) = f(\tau_1, \ t-t_1), \quad t > t_1. \tag{9.23}$$

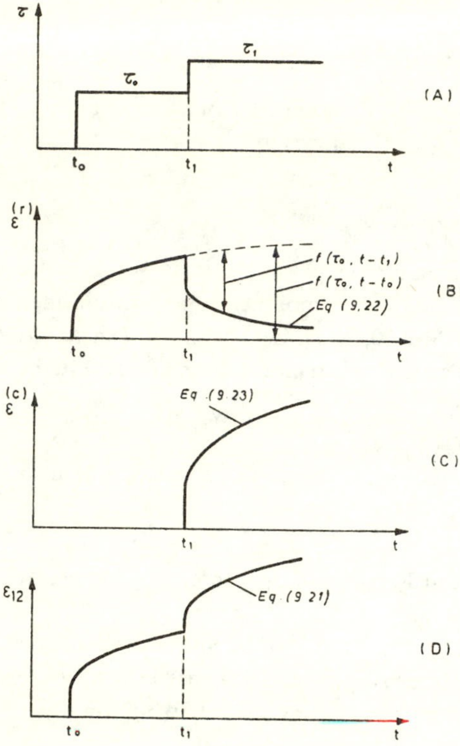

Fig. 9.2. Illustration of Modified Superposition Principle for Two-step Loading.

The total strain will be the sum of (9.22) and (9.23) and equal to (9.21), see Fig. 9.2D.

The modified superposition method cannot describe all the features of creep behavior describable by the multiple integral representation. It does satisfy several essential requirements. For instance, if $\tau_1 = \tau_0$, there is no stress change at $t = t_1$, and the strain response is independent of t_1 as shown by (9.19). The strain response predicted from (9.21) is $\varepsilon_{12}(t) = f(\tau_0, t-t_0)$, which is the same as (9.19) with a shift of time origin. Also, if $\tau_0 = 0$ the strain response should be independent of τ_0 and t_0. Equation (9.21) predicts that the strain response is

$$\varepsilon_{12}(t) = f(\tau_1, t-t_1),$$

which again reduces to (9.19) with a time translation. A discussion of (9.21) has been made by Pipkin and Rogers [9.12].

Arbitrary stress history can be considered as the limiting case as the number of steps tends to an infinite number of infinitesimal steps of stress. Thus, (9.20) becomes

$$\varepsilon_{12}(t) = \int_0^t \frac{\partial f[\tau(\xi), t-\xi]}{\partial \tau(\xi)} \, \mathrm{d}\tau(\xi). \tag{9.24}$$

If the stress history is differentiable (9.24) can be written as

$$\varepsilon_{12}(t) = \int_0^t \frac{\partial f[\tau(\xi), t-\xi]}{\partial \tau(\xi)} \frac{\mathrm{d}\tau(\xi)}{\mathrm{d}\xi} \, \mathrm{d}\xi, \tag{9.25}$$

thus changing the variable of integration to time ξ. This form has been mentioned by Turner [9.13] and Arutyunyan [9.14].

Equation (9.25) also can be derived directly from the multiple integral representation by applying certain restrictions to the time arguments in the kernel functions so that the kernels are functions of their smallest time argument only [9.15]. Appendix A5 gives a formal procedure for the derivation. For example, in the third-order kernel function restricting $L(t,-\xi_1\, t-\xi_2, t-\xi_3)$ to a function of the smallest time argument, or the most recent occurrence $(t-\xi_3)$,

$$L(t-\xi_1, t-\xi_2, t-\xi_3) \to L(t-\xi_3) \tag{9.26}$$

or

$$\int_0^t \int_0^t \int_0^t L(t-\xi_1, t-\xi_2, t-\xi_3)\, \dot{\tau}(\xi_1)\, \dot{\tau}(\xi_2)\, \dot{\tau}(\xi_3)\, \mathrm{d}\xi_1\, \mathrm{d}\xi_2\, \mathrm{d}\xi_3$$

$$\to \int_0^t L(t-\xi) \frac{\mathrm{d}}{\mathrm{d}\xi} [\tau(\xi)\, \tau(\xi)\, \tau(\xi)]\, \mathrm{d}\xi = \int_0^t L(t-\xi) \frac{\mathrm{d}}{\mathrm{d}\xi} [\tau^3(\xi)]\, \mathrm{d}\xi \tag{9.26a}$$

where

$$L(t-\xi) = L(t-\xi, t-\xi, t-\xi).$$

Furthermore, by retaining the two smaller time arguments $(t-\xi_2)$ and $(t-\xi_3)$ only a "second-order memory representation" [9.15] can also be obtained.

Using the principles involved in (9.26) the following expression is found for (9.1), for example, when the modified superposition principle is employed

$$\varepsilon_{12}(t) = \int_0^t [R(t-\xi)+3\tau^2(\xi)\, L(t-\xi)]\, \dot{\tau}(\xi)\, d\xi. \tag{9.27}$$

When the forms of $R(t)$ and $L(t)$ are known, such as the ones given in Table 9.2, where

$$R(t) = R_0+R^+t^n,$$
$$L(t) = L_0+L^+t^n,$$

then (9.27) can be rewritten in the following more explicit form

$$\varepsilon_{12}(t) = R_0\, \tau(t) + L_0\, \tau^3(t) + \int_0^t (t-\xi)^n\, [R^+ + 3L^+\tau^2(\xi)]\, \dot{\tau}(\xi)\, d\xi. \tag{9.28}$$

For applications to other special states of stress and the general stress state see Appendix A5.3.

The modified superposition method is an approximate method which utilizes the kernel functions determined from one-step creep tests to describe nonlinear creep behavior under arbitrary stress history. The accuracy of the description of the creep behavior depends on the material and type of stress history. The accuracy is higher for smooth changes of stress than for abrupt change of stress. An improvement of this method using N-step creep test data was proposed by Pipkin and Rogers [9.12] in which the modified superposition method discussed in this section was the first approximation. However, in order to obtain a further approximation beyond the first one (modified superposition method), required a large amount of multi-step creep test data to determine the corresponding kernel functions. The analysis of the test data is expected to be cumbersome and the constitutive relation will be complicated. Thus, it may not be justifiable to use this refinement.

The following is an illustration of a type of material behavior which is predicted by the general form of the multiple integral representation or by the product form of kernel function but not predicted by the modified superposition principle. Consider a combined tension-torsion creep experiment in the nonlinear range in which the torque is removed after a period of time so that recovery of the shearing strain begins. Then at a later period the axial load is removed. The modified superposition principle predicts no change in the recovery pattern for shear strain when the tensile load is removed; whereas a small effect is predicted by the general form and is found in experiments on rigid polyurethane foam as shown by the shearing strains in period IV in Fig. 9.3.

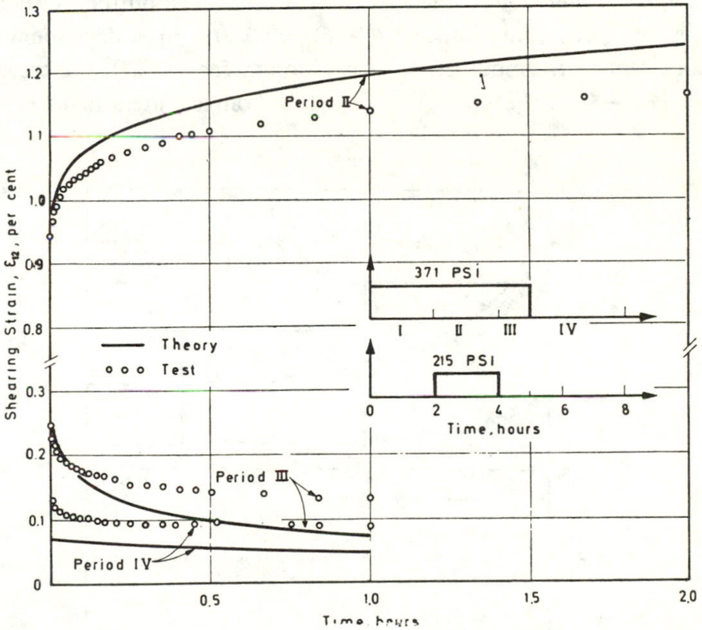

Fig. 9.3. Creep of Rigid Polyurethane Foam (0.3 Density) under a Complex History of Tension and Torsion with Prediction by the Modified Superposition Principle. From [9.10], courtesy ASTM.

Application of the modified superposition principle to predict creep behavior under two different stress histories will be presented in Section 9.7.

9.6 Physical Linearity Approximation of Kernel Functions

The "physical linearity" assumption suggests that the response of a "physically linear" viscoelastic material is linearly dependent on the history of the input, but is nonlinearly dependent on the present value of the input.

This approximate form first proposed by Coleman and Noll [9.16] and considered by Stafford [9.17] is based on the principle of fading memory applied to the theory of finite linear viscoelasticity. Using this assumption, (9.1) may be shown to become

$$\varepsilon_{12}(t) = \int_0^t [R(t-\xi)+\tau^2(t)L(t-\xi)]\,\dot{\tau}(\xi)\,d\xi. \tag{9.29}$$

In (9.29) the nonlinearity comes from the dependence on the present input. That is, the creep complicance $[R(t-\xi)+\tau^2(t)L(t-\xi)]$ is dependent on the present stress. In the modified superposition method (9.27), the creep compliance $[R(t-\xi)+3\tau^2(\xi)L(t-\xi)]$ is dependent on the stress history.

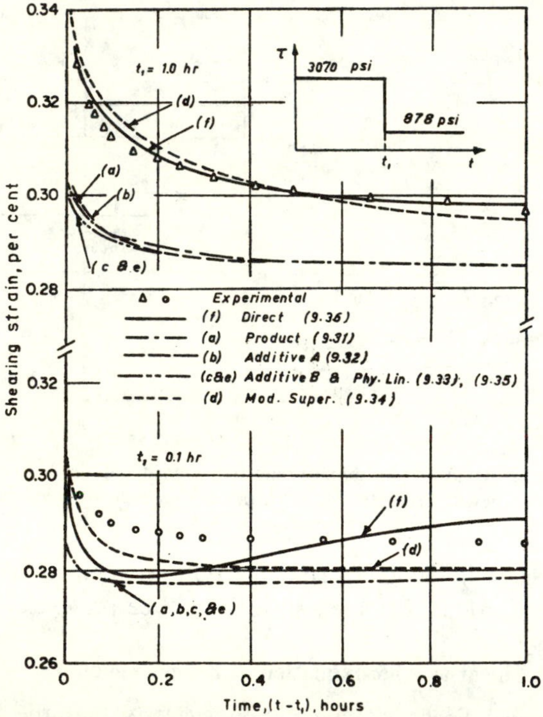

Fig. 9.4. Torsion Creep for the Second Step of Stress (σ = 878 psi) in Step-down Tests of Poly (Vinyl Chloride) Copolymer, Unplasticized, Compared with Prediction of Several Theories. Data from [9.1], courtesy ASME.

9.7 Comparison

In this section, the direct determination and the approximate methods discussed in Sections 9.3 through 9.6 are compared. The methods include:

(a) product form by Nakada, (9.16),
(b) additive form A by Gottenberg, Bird, and Agrawall, (9.17),
(c) additive form B by Cheung, (9.18),
(d) modified superposition method by Findley, Khosla and Lai, (9.25),
(e) physical linearity by Coleman and Noll, (9.29),
(f) direct method, see section 9.2.

These methods were applied to predict the creep behavior of an unplasticized PVC copolymer under two-step shearing stress as shown in Fig. 9.4 and Fig. 9.5. From one-step creep tests, the kernel functions $R(t)$ and $L(t) = L(t, t, t,)$ for the material were determined in [9.1] as shown in (9.12) and Table 9.2.

A two-step pure torsion stress history can be represented by

$$\tau(t) = \tau H(t) + \Delta\tau H(t-t_1), \tag{9.30}$$

where τ is the constant stress from $t = 0+$ to $t = t_1$, $\Delta\tau$ is the change in stress occurring at $t = t_1$. Inserting (9.30) into (9.1) and utilizing the various approximate methods yields the following creep strains during the second step, $t > t_1$,

(a) *Product Form from* (9.16a)

$$\begin{aligned} \varepsilon_{12}(t) &= R(t)\tau + R(t-t_1)\Delta\tau + \{[L(t)]^{1/3}\tau + [L(t-t_1)]^{1/3}\,\Delta\tau\}^3 \\ &= R(t)\tau + R(t-t_1)\Delta\tau + L(t)\tau^3 + 3[L(t)]^{2/3}[L(t-t_1)]^{1/3}\tau^2\,\Delta\tau \\ &\quad + 3[L(t)]^{1/3}[L(t-t_1)]^{2/3}\,\tau(\Delta\tau)^2 + L(t-t_1)\,(\Delta\tau)^3, \end{aligned} \tag{9.31}$$

$$\begin{aligned} \varepsilon_{12}(t) &= 2.993(\tau+\Delta\tau)\;10^{-6} + 15.1(\tau+\Delta\tau)^3\;10^{-15} \\ &\quad + 0.156[\tau t^n + \Delta\tau(t-t_1)^n]10^{-6} + 32.8[\tau^3 t^n + 3\tau^2\Delta\tau t^{2n/3}(t-t_1)^{n/3} \\ &\quad + 3\tau(\Delta\tau)^2 t^{n/3}(t-t_1)^{2n/3} + (\Delta\tau)^3\,(t-t_1)^n]\;10^{-15} \end{aligned} \tag{9.31a}$$

(b) *Additive Form A from* (9.17)

$$\begin{aligned} \varepsilon_{12}(t) &= R(t)\tau + R(t-t_1)\Delta\tau + \tau^3 L(3t) + 3\tau^2\Delta\tau L(3t-t_1) + 3\tau(\Delta\tau)^2 L(3t-2t_1) \\ &\quad + (\Delta\tau)^3 L(3t-3t_1), \end{aligned} \tag{9.32}$$

$$\begin{aligned} \varepsilon_{12}(t) &= 2.993(\tau+\Delta\tau)\;10^{-6} + 15.1(\tau+\Delta\tau)^3\;10^{-15} + 0.156[\tau t^n + \Delta\tau(t-t_1)^n]\;10^{-6} \\ &\quad + 32.8[\tau^3(3t)^n + 3\tau^2\,\Delta\tau(3t-t_1)^n + 3\tau(\Delta\tau)^2\,(3t-2t_1)^n \\ &\quad + (\Delta\tau)^3\,(3t-3t_1)^n]10^{-15}. \end{aligned} \tag{9.32a}$$

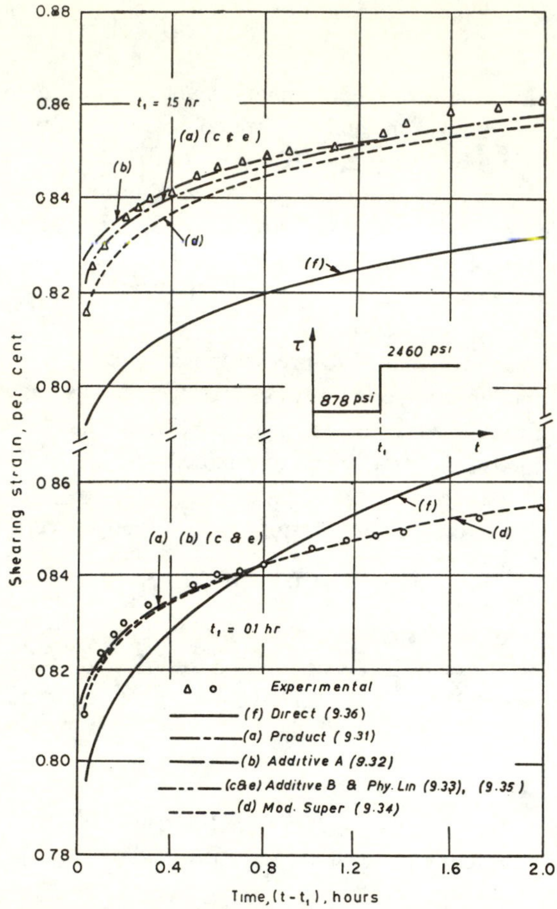

Fig. 9.5. Torsion Creep for the Second Step of Stress ($\sigma = 2460$ psi) in Step-up Tests of Poly (Vinyl Chloride) Copolymer, Unplasticized, Compared with Prediction of Several Theories. Data from [9.1], courtesy ASME.

(c) *Additive Form B from* (9.18)

$$\varepsilon_{12}(t) = R(t)\tau + R(t-t_1)\Delta\tau + L(t)(\tau+\Delta\tau)^2\tau + L(t-t_1)(\tau+\Delta\tau)^2\,\Delta\tau, \tag{9.33}$$

$$\varepsilon_{12}(t) = 2.993(\tau+\Delta\tau)10^{-6} + 15.1(\tau+\Delta\tau)^3 10^{-15} + 0.156[\tau t^n + \Delta\tau(t-t_1)^n]$$
$$10^{-6} + 32.8[(\tau+\Delta\tau)^2\,\tau t^n + (\tau+\Delta\tau)^2\,\Delta\tau(t-t_1)^n]10^{-15}. \tag{9.33a}$$

(d) *Modified Superposition Method from* (9.27)

$$\varepsilon_{12}(t) = R(t)\tau + R(t-t_1)\Delta\tau + L(t)\tau^3 + L(t-t_1)\,[(\tau+\Delta\tau)^3 - \tau^3], \tag{9.34}$$

$$\begin{aligned}\varepsilon_{12}(t) = {} & 2.993(\tau+\Delta\tau)10^{-6}+15.1(\tau+\Delta\tau)^3 10^{-15} \\ & +0.156\,[\tau t^n+\Delta\tau(t-t_1)^n]\,10^{-6} \\ & +32.8\{\tau^3 t^n+[(\tau+\Delta\tau)^3-\tau^3]\,(t-t_1)^n\}\,10^{-15}.\end{aligned} \tag{9.34a}$$

(e) *Physical Linearity from* (9.29)

$$\varepsilon_{12}(t) = R(t)\tau+R(t-t_1)\,(\Delta\tau)+L(t)\,(\tau+\Delta\tau)^2\tau+L(t-t_1)\,(\tau+\Delta\tau)^2\,(\Delta\tau). \tag{9.35}$$

Notice that (9.35) is identical to (9.33). Also when $t_1 = 0$, equations (9.31)–(9.35) all reduce to

$$\varepsilon_{12}(t) = R(t)\,(\tau+\Delta\tau)+L(t)\,(\tau+\Delta\tau)^3,$$

which represents the creep response under a constant stress of $(\tau+\Delta\tau)$.

(f) *Direct Method, Section* 9.2

The prediction of creep strains in the second step from the direct method as discussed in Section 9.2 is given by (9.4)

$$\begin{aligned}\varepsilon_{12}(t) = {} & R(t)\tau+R(t-t_1)\Delta\tau+L(t,\,t,\,t)\tau^3+3L(t,\,t,\,t-t_1)\tau^2\,\Delta\tau \\ & +3L(t,\,t-t_1,\,t-t_1)\,\tau(\Delta\tau)^2+L(t-t_1,\,t-t_1,\,t-t_1)\,(\Delta\tau)^3.\end{aligned} \tag{9.36}$$

Inserting (9.12) and (9.14) into (9.36) yields

$$\begin{aligned}\varepsilon_{12}(t) = {} & 2.993(\tau+\Delta\tau)\,10^{-6}+15.1(\tau+\Delta\tau)^3 10^{-15} \\ & +0.156[\tau t^n+\Delta\tau(t-t_1)^n]10^{-6} \\ & +32.8[\tau^3 t^n+(\Delta\tau)^3(t-t_1)^n]10^{-15} \\ & +3(50-15t_1)\,(t-t_1)^{0.24}\,\tau^2\,\Delta\tau 10^{-15} \\ & +3(53-23t_1)\,(t-t_1)^{0.38}\tau(\Delta\tau)^2 10^{-15}.\end{aligned} \tag{9.36a}$$

Note that as would be expected the linear terms and the time-independent part of the nonlinear terms of all six methods as expressed by (9.31) to (9.36) are the same.

Data from two-step pure torsion creep tests [9.1] of an unplasticized PVC copolymer in which the second step was smaller than the first were used to obtain the kernel functions by the direct method (9.36a). The same kernel functions corresponding to single step loadings were used in the five other approximate methods. These six methods were then used to predict the behavior of the same material for different experiments which included the following cases; the second step is smaller than the first step (Fig. 9.4), and the second step is greater than the first step (Fig. 9.5). It would appear that the simplest form (modified superposition principle) gives as accurate a prediction as any, at least for 2-step loadings.

In Fig. 9.4 for step-down tests, the prediction by the direct method shows the best agreement with the experimental results because these data were used in determining the kernel functions for the direct method. These results were obtained using equipment having a strain sensitivity of 2 to 3×10^{-6} in. per in; the same specimen was used for all tests; and a step-down type of two-step creep test was employed. Even so, the accuracy available in the test data was insufficient to determine the form of the two kernel functions with the desired precision, as shown in Fig. 9.5 by the fact that the prediction by the direct method was worse than by the other methods considered. Evaluation of other kernel functions using three-step loadings is even less likely to yield the desired accuracy.

CHAPTER 10

CONVERSION AND MIXING OF NONLINEAR CREEP AND RELAXATION

10.1 Introduction

In previous chapters creep behavior and stress relaxation behavior in the nonlinear range have been discussed separately where methods of determining the creep function (creep compliance) and stress relaxation function (relaxation modulus) from corresponding creep tests and stress relaxation tests have also been shown.

Since creep and stress relaxation behaviors are two aspects of the time sensitive mechanical behavior of materials, one behavior should be predictable if the other behavior is known. For a linear viscoelastic material, the relationship between and the interconversion of creep and stress relaxation functions are relatively simple and have been discussed in Chapter 5. In this chapter some aspects of the relationship between and the interconversion of creep and stress relaxation functions of a nonlinear viscoelastic material are discussed. Since the basic principle involved in prediction of creep from stress relaxation and prediction of relaxation from creep are essentially the same, in the following only the method of prediction of stress relaxation from given creep behavior is described in detail.

Furthermore, the method of interconversion between creep and stress relaxation in the linear range as discussed in Chapter 5 only for the simple stress state, can be extended quite straightforwardly to the multi-stress state since the linear superposition principle applies in the linear case. In the nonlinear range the extension from a simple stress state to the multi-stress state is more involved. The difficulty is mainly due to the nonlinear relationship between stress and strain and also due to the existence of the product terms of different stress components. In this chapter a method of interconversion of nonlinear creep and stress relaxation in the multi-stress state is discussed.

10.2 Relation Between Creep and Stress Relaxation for Uniaxial Nonlinear Viscoelasticity

In Chapter 7, the multiple integral representations of creep and stress relaxation of nonlinear viscoelastic materials were derived. The creep formulation employed is permissible only under small deformation as pointed out in Appendix A2. This restriction is implied in the rest of this chapter in discussion of the interconversion between creep and stress relaxation. In Chapter 9, simplifications of the multiple integral representation were made by assuming the product form of the kernel functions or by using the modified superposition method, etc. The merit of using the modified superposition method in comparison with the other methods was also shown in Chapter 9. In the following creep and stress relaxation formulations of the nonlinear viscoelastic constitutive relations derived either from the modified superposition method or the product form are utilized to demonstrate the principles involved in the interconversion. The method discussed in the following can be applied equally well to constitutive relations derived from other methods.

When the modified superposition method is used the following is obtained from (9.25), for creep under uniaxial stress, see also Appendix A5,

$$\varepsilon(t) = \int_0^t F[\sigma(\xi),\, t-\xi]\, \dot{\sigma}(\xi)\, d\xi, \tag{10.1}$$

where

$$F[\sigma(\xi),\, t-\xi] = F_1(t-\xi)+2F_2(t-\xi)\,\sigma(\xi)+3F_3(t-\xi)\,\sigma^2(\xi). \tag{10.2}$$

Equation (10.2) is obtained by differentiating (8.54) with respect to σ in accordance with the equivalent of (9.25) for uniaxial stress, and taking σ to be a function of time, $\sigma(t)$.

For stress relaxation under prescribed uniaxial strain the following form can be obtained similarly by utilizing (8.133) and the modified superposition method

$$\sigma(t) = \int_0^t \psi[\varepsilon(\xi),\, t-\xi]\, \dot{\varepsilon}(\xi)\, d\xi, \tag{10.3}$$

where

$$\psi[\varepsilon(\xi),\, t-\xi] = \psi_a(t-\xi)+2\psi_b(t-\xi)\,\varepsilon(\xi)+3\psi_c(t-\xi)\,\varepsilon^2(\xi). \tag{10.4}$$

By "prescribed uniaxial strain" is meant a state in which the strain in one direction is prescribed and the strains in the orthogonal directions are not prescribed and generally not zero. Thus the strain state is triaxial with only one strain prescribed, but the stress state is uniaxial in the direction of the prescribed strain. Comparing (10.1), (10.3) with (5.103) and (5.105) suggests that $F[\sigma, t]$ and $\psi[\varepsilon, t]$ could be called the creep compliance and relaxation modulus respectively of nonlinear viscoelasticity. Due to nonlinearity, the relationship between F and ψ is more complicated than that between $J(t)$ and $E(t)$ of creep compliance and relaxation modulus, respectively, of linear viscoelasticity and cannot be reduced to a form like (5.111), (5.112).

The interconversion or prediction of relaxational stress from a given creep function in the nonlinear range such as F in (10.2) by the method derived in Chapter 5 is inapplicable. A different method will be developed in the following. Consider a creep test performed under a prescribed stress program $\sigma = \sigma(t)$, for which the corresponding creep strain response is $\varepsilon = \varepsilon(t)$. Then in a relaxation test performed under a prescribed axial strain program equal to the strain response $\varepsilon = \varepsilon(t)$ resulting from the prescribed creep test program, the observed stress response in the relaxation test would be equal to the prescribed stress program $\sigma = \sigma(t)$ in the creep test. Thus the input is stress $\sigma = \sigma(t)$ and the response is strain $\varepsilon = \varepsilon(t)$ in the creep test, or the input is strain $\varepsilon = \varepsilon(t)$ and the output is stress $\sigma = \sigma(t)$ in the relaxation test.

Consider now a variable-stress creep test program where $\sigma = \sigma(t)$, for which the resulting strain response, such as described by (10.1), is a constant strain $\varepsilon = \varepsilon_0$. Then from the previous conclusion the stress response from a relaxation test, at prescribed constant strain $\varepsilon(t) = \varepsilon_0 H(t)$, will be $\sigma = \sigma(t)$. This is the same function which served as input to the variable-stress creep situation just described. The question remaining is how to find a function $\sigma = \sigma(t)$ such that when inserted into (10.1), the corresponding creep strain $\varepsilon(t)$ will be equal to the desired constant strain ε_0.

The following example shows one way to find this function $\sigma = \sigma(t)$. The method developed in the following example is equally applicable to the prediction of creep from stress relaxation as to the prediction of stress relaxation from creep. It is also applicable regardless of what method is used to determine the kernel functions whether by the direct method, the product form, modified superposition, or others.

A somewhat different approach to the inversion of (10.1) was made by Arutyunyan [10.1] and Bychawski and Fox [10.2]. They considered an integral representation, such as (10.1) as a solution of a nonhomogeneous differential equation. Therefore a nonlinear differential equation was first constructed from the integral representation and the inversion was then made in the differential representation.

10.3 Example: Prediction of Uniaxial Stress Relaxation from Creep of Nonlinear Viscoelastic Material

In [10.3] the product form of the kernel function, (9.15), was utilized where

$$\varepsilon(t) = \int_0^t K_1(t-\xi_1)\dot{\sigma}(\xi_1)\,d\xi_1$$
$$+\int_0^t\int_0^t [K_2(t-\xi_1)]^{1/2}[K_2(t-\xi_1)]^{1/2}\,\dot{\sigma}(\xi_1)\,\dot{\sigma}(\xi_2)\,d\xi_1\,d\xi_2$$
$$+\int_0^t\int_0^t\int_0^t [K_3(t-\xi_1)]^{1/3}[K_3(t-\xi_2)]^{1/3}[K_3(t-\xi_3)]^{1/3}\,\dot{\sigma}(\xi_1)\,\dot{\sigma}(\xi_2)\,\dot{\sigma}(\xi_3)$$
$$d\xi_1\,d\xi_2\,d\xi_3. \tag{10.5}$$

Creep tests were conducted at several different constant tensile stresses on a polyurethane specimen. Results from these tests were used to determine the three kernel functions K_1, K_2, K_3 of (10.5). A method of determining those kernel functions from creep test data was discussed in Chapter 8. Using the product form assumption (10.5) may be employed to describe the strain-time relationship under various prescribed stress histories. For a constant stress input (10.5) yields the following

$$\varepsilon(t) = K_1(t)\sigma + K_2(t)\sigma^2 + K_3(t)\sigma^3, \tag{10.6}$$

where K_1, K_2, K_3 were predetermined from creep test data. Inversion of (10.6), to yield σ as functions of ε and K_α, yields the following form [10.4]

$$\sigma(t) = f_1(t)\varepsilon + f_2(t)\varepsilon^2 + f_3(t)\varepsilon^3 + \ldots, \tag{10.7}$$

where

$$f_1(t) = K_1^{-1}(t),$$
$$f_2(t) = -K_2(t)\,K_1^{-3}(t),$$
$$f_3(t) = [2K_2^2(t) - K_1(t)\,K_3(t)]\,K_1^{-5}(t),$$
etc.

(10.6) and (10.7) are identical but inverted; that is, they express $\varepsilon(t)$ as a function of constant stress and time. If a constant strain ε_0 is desired to predict stress relaxation from creep $\varepsilon = \varepsilon_0$ is inserted in (10.7) as an approximation as follows

$$\sigma_1(t) = f_1(t)\varepsilon_0 + f_2(t)\varepsilon_0^2 + f_3(t)\varepsilon_0^3, \tag{10.8}$$

where terms higher than the third order have been neglected. Equation (10.8) is a first approximation of the desired function. Inserting (10.8) into (10.5) and performing the integration a corresponding creep strain $\varepsilon(t) = \varepsilon_1(t)$ was obtained. The deviation between $\varepsilon_1(t)$ and the desired constant value ε_0 then served as the input into (10.7) to obtain a second approximation $\sigma_2(t)$ of the stress function,

$$\sigma_2(t) - \sigma_1(t) = f_1(t)[\varepsilon_1(t) - \varepsilon_0] + f_2(t)[\varepsilon_1(t) - \varepsilon_0]^2 + \dots \tag{10.9}$$

If the quantity $[\varepsilon_1(t) - \varepsilon_0]$ is small, then higher order terms in (10.9) can be omitted so that

$$\sigma(t) = \sigma_2(t) = \sigma_1(t) + f_1(t)[\varepsilon_1(t) - \varepsilon_0]. \tag{10.10}$$

Inserting (10.8) into (10.10) yields

$$\sigma(t) = \sigma_2(t) = f_1(t)\, \varepsilon_1(t) + f_2(t)\varepsilon_0^2 + f_3(t)\varepsilon_0^3.$$

This second approximation of stress was employed in (10.5) to find the corresponding creep strain $\varepsilon(t) = \varepsilon_2(t)$. This process can be repeated until at, say the lth iteration, the deviation between $\varepsilon_l(t)$ and ε_0 is negligibly small. Then the lth approximation of stress $\sigma(t) = \sigma_l(t)$ is considered to be the desired variable stress which will produce the constant creep strain $\varepsilon(t) = \varepsilon_0$. The argument toward the end of Section 10.2 then implies that $\sigma(t) = \sigma_l(t)$ is the relaxational stress due to a constant prescribed uniaxial strain $\varepsilon(t) = \varepsilon_0$.

The numerical procedure described above was used to compute the relaxational stress for several different constant tensile strains extending into the nonlinear range. The resulting relaxation behavior predicted from creep data showed good agreement with the actual stress relaxational behavior as shown in Figure 10.1.

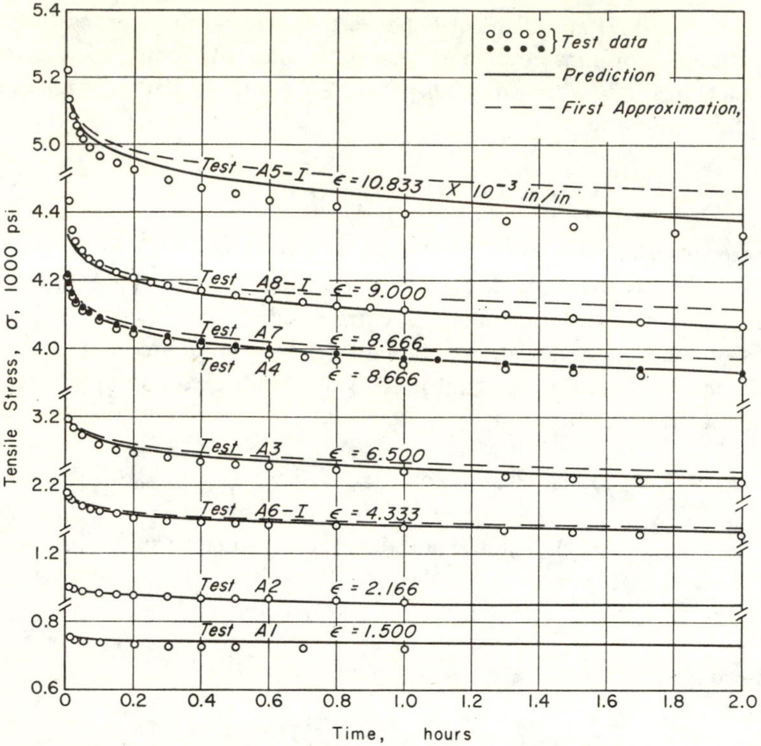

Fig. 10.1. Relaxation of Solid Polyurethane under Constant Tensile Strain at 75°F and 50 per cent R.H., with Prediction from Creep Tests Data. From [10.3], courtesy Soc. of Rheology.

10.4 Relation Between Creep and Relaxation for Biaxial Nonlinear Viscoelasticity

When the stress state is biaxial or triaxial, the interconversion of creep and relaxation is more complicated than in the uniaxial stress case. For instance, when the stress state is combined tension and torsion the constitutive equations for creep, using the modified superposition method, are as follows for axial and shearing strain components

$$\varepsilon_{11}(t) = \int_0^t \varphi_{1111}[\sigma, \tau, t-\xi]\dot{\sigma}(\xi)\,\mathrm{d}\xi + \int_0^t \varphi_{1112}[\sigma, \tau, t-\xi]\,\dot{\tau}(\xi)\,\mathrm{d}\xi, \tag{10.11}$$

$$\varepsilon_{12}(t) = \int_0^t \varphi_{1211}[\sigma, \tau, t-\xi]\dot{\sigma}(\xi)\,\mathrm{d}\xi + \int_0^t \varphi_{1212}[\sigma, \tau, t-\xi]\,\dot{\tau}(\xi)\,\mathrm{d}\xi, \tag{10.12}$$

where the creep functions $\varphi_{1111}, \varphi_{1112}, \varphi_{1211}, \varphi_{1212}$ are approximated by polynomials of σ and τ with time-dependent coefficients, see (A5.12), (A5.13), (A5.14),

$$\begin{aligned}
\varphi_{1111}[\sigma, \tau, t-\xi) &= F_1(t-\xi)+2F_2(t-\xi)\,\sigma(\xi)+3F_3(t-\xi)\,\sigma^2(\xi)+F_4(t-\xi)\,\tau^2(\xi),\\
\varphi_{1112}[\sigma, \tau, t-\xi] &= 2F_4(t-\xi)\,\sigma(\xi)\,\tau(\xi)+2F_5(t-\xi)\,\tau(\xi),\\
\varphi_{1211}[\sigma, \tau, t-\xi] &= G_3(t-\xi)\,\tau(\xi)+2G_4(t-\xi)\,\sigma(\xi)\,\tau(\xi),\\
\varphi_{1212}[\sigma, \tau, t-\xi] &= G_1(t-\xi)+3G_2(t-\xi)\,\tau^2(\xi)+G_3(t-\xi)\,\sigma(\xi)+G_4(t-\xi)\,\sigma^2(\xi).
\end{aligned} \tag{10.13}$$

For example, consider that the prescribed quantities in (10.11), (10.12) are tensile strain ε_{11} and shear strain ε_{12} and no other strain components are prescribed* while the quantities σ and τ are the relaxational stress functions to be determined from (10.11), (10.12) with given ε_{11} and ε_{12}. The argument of the existence of the solution is the same as described in Section 10.2. Mathematically, this is equivalent to solving two simultaneous nonlinear integral equations. Numerical methods are seemingly the only way to obtain a solution. Details of the numerical method, which involves the proper choice of the initial functions of $\sigma = \sigma_0(\varepsilon_{11}, \varepsilon_{12}, t)$ and $\tau = \tau_0(\varepsilon_{11}, \varepsilon_{12}, t)$ and the iteration process, are dependent on the character of the material functions $\varphi_{1111}, \varphi_{1112}, \varphi_{1211}, \varphi_{1212}$.

10.5 Behavior of Nonlinear Viscoelastic Material under Simultaneous Stress Relaxation in Tension and Creep in Torsion

In the following section a combined tension-torsion problem is presented in which the prescribed quantities are ε_{11} and τ_{12}, relaxation in tension and creep in torsion respectively [10.5].

For constant stresses, (10.11), (10.12) and (10.13) become, as shown before in (8.43) and (8.46) respectively,

$$\varepsilon_{11}(t) = F_1(t)\,\sigma+F_2(t)\,\sigma^2+F_3(t)\,\sigma^3+F_4(t)\,\sigma\tau^2+F_5(t)\tau^2, \tag{10.14a}$$

$$\varepsilon_{12}(t) = G_1(t)\,\tau+G_2(t)\,\tau^3+G_3(t)\,\sigma\tau+G_4(t)\,\sigma^2\tau. \tag{10.14b}$$

The method of determining these kernel functions was discussed in Section 8.12.

---

*In this case ε_{22} and ε_{33} are not zero but ε_{23} and ε_{13} are zero generally.

The method described in Section 10.3 can be extended here to predict simultaneous stress relaxation in tension and creep in torsion when the material is subjected to a constant prescribed tensile strain and constant shearing stress in torsion. This situation can be considered as one of combined tension and torsion creep (prescribed stresses) with variable tensile stress and constant shearing stress in torsion. The desired variable tensile stress $\sigma(t)$ is unknown. This unknown quantity has to satisfy the condition that the tensile strain produced by the constant shearing stress τ_0 and this unknown varying tensile stress $\sigma(t)$ acting together on the material must produce the prescribed constant tensile strain ε_0. Once the varying tensile stress $\sigma = \sigma(t)$ has been determined, the corresponding shearing strain can be obtained by inserting $\sigma = \sigma(t)$ and $\tau = \tau_0$ into (10.12). Under this kind of input, iteration of a nonlinear equation, between (10.11) and (10.14a) where $\tau = \tau_0$ is a known quantity, is required instead of iteration of two simultaneous nonlinear equations in the case of predicting the stress relaxation under combined tension and torsion from creep as in Section 10.4.

A first approximation of the varying tensile stress $\sigma(t) = \sigma_1(t)$ may be obtained as follows. Rewriting (10.14a) as,

$$\varepsilon_{11} - F_5(t)\tau^2 = [F_1(t) + F_4(t)\tau^2]\sigma + F_2(t)\sigma^2 + F_3(t)\sigma^3, \tag{10.15}$$

equation (10.15) is equivalent to (10.6). Thus (10.15) may be inverted to solve for $\sigma(t)$, and ε_0 substituted for ε_{11} to yield the first approximation $\sigma_1(t)$ similar to (10.8). Then insertion of $\sigma_1(t)$ and τ_0 in (10.11) yields $\varepsilon_1(t)$, the first approximation of the desired value ε_0. This procedure may be

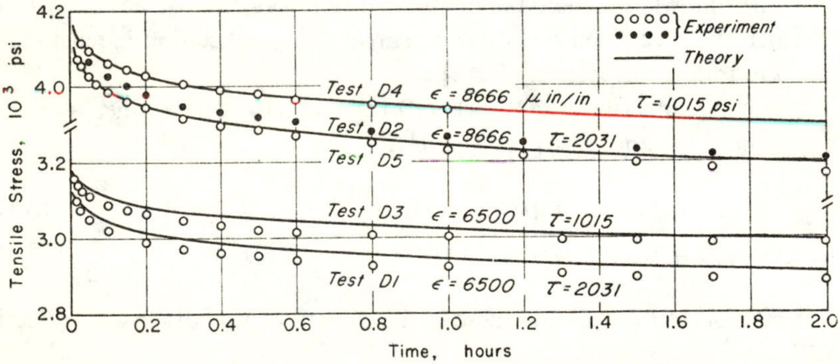

Fig. 10.2. Tensile Stress Versus Time under Simultaneous Stress Relaxation in Tension and Creep in Torsion of Solid Polyurethane at 75°F and 50 per cent R.H., with Prediction from Creep Data. From [10.5], courtesy ASME.

continued as described in Section 10.3 to obtain the tensile relaxational stress $\sigma(t)$.

Once $\sigma(t)$ is obtained, together with the prescribed shearing stress τ, (10.12) and (10.13) can be used to evaluate the shearing strain $\varepsilon_{12}(t)$.

The numerical procedure described was used to compute the tensile relaxational stress and the torsional creep strain under prescribed tensile strain and shearing stress using the kernel functions determined from combined tension-torsion creep tests at constant stress. Some of the results are shown in Figures 10.2 and 10.3.

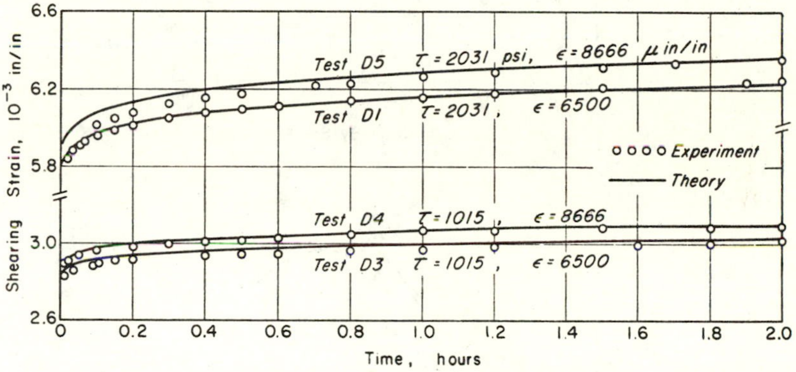

Fig. 10.3. Shearing Strain Versus Time under Simultaneous Stress Relaxation in Tension and Creep in Torsion of Solid Polyurethane at 75°F and 50 percent R.H., with Prediction from Creep Data. From [10.5], courtesy ASME.

10.6 **Prediction of Creep and Relaxation under Arbitrary Input**

The method described in the previous sections may be applied to arbitrary input although it is more applicable to a constant input such as constant stress or constant strain. This method would be much more elaborate for arbitrary input, though the basic method would be the same. Another approach [10.6] is as follows. Stress which varies with time in an irregular manner can be approximated either by intervals consisting of steps or by intervals of constant rate, as shown in Figure 10.4. Therefore the input, for example stress, during the ith interval beginning at t_i can be written as

$$\sigma(t) = \sigma(t_i)+\Delta\sigma_i, \tag{10.16a}$$

or

$$\sigma(t) = \sigma(t_i)+C_i(t-t_i), \quad t > t_i, \tag{10.16b}$$

for a constant step approximation or a constant rate approximation respectively, where C_i is the rate at the ith interval beginning at t_i and ending at t_{i+1}. The former approximation (10.16a) would result in the integrals becoming algebraic forms while the latter approximation (10.16b) would retain

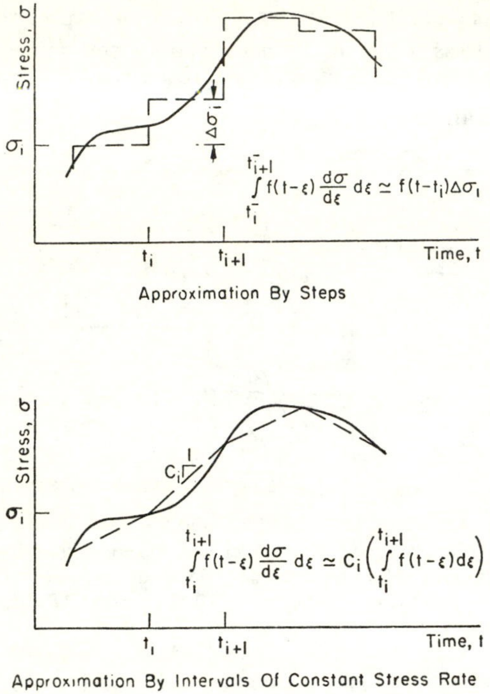

Fig. 10.4. Approximations of a Variable Stress History. From [10.6], courtesy Soc. of Rheology.

the integral form as shown in Figure 10.4. Perhaps the first approximation may be simpler to use, though it depends on each individual case.

The method of intervals of constant rate was employed by Nolte and Findley [10.6] to predict behavior under irregular loading using data from constant stress creep tests. It was found that good prediction resulted from the use of the product form of kernel functions, Section 9.3, but computational difficulties arose after a number of time intervals using the modified superposition method.

CHAPTER 11

EFFECT OF TEMPERATURE ON NONLINEAR VISCOELASTIC MATERIALS

11.1 Introduction

The mechanical properties of most viscoelastic materials are very sensitive to temperature. Therefore, in many structural applications of these materials, an understanding of the effect of temperature on the mechanical properties is necessary. For linear viscoelastic theory, it has been found that the time-temperature superposition principle satisfactorily accounts for the effect of temperature on the creep and stress relaxation behavior of most linear viscoelastic materials, as discussed in Chapter 5. However, there is little experimental information on the effect of temperature on the behavior of nonlinear viscoelastic materials. Previous investigations of nonlinear viscoelasticity discussed in the preceding chapters of this book were concerned mostly with constant temperature, usually with room temperature.

In this chapter, an extension of the multiple integral representation discussed in the previous chapters to account for the temperature effect is described. Although the most general equation described is capable of accounting for various loading and temperature conditions in principle, the experimental difficulty of characterizing the kernel functions may prevent it from being useful in practice. Therefore, several approximations are made in order to simplify the constitutive equation. In this chapter only creep behavior is discussed. However, stress relaxation behavior can be treated using the same concepts.

11.2 Nonlinear Creep Behavior at Elevated Temperatures

In Chapter 7 and Appendix A2, the nonlinear constitutive equation derived was for isothermal conditions. Consequently the kernel functions of the nonlinear constitutive equation, as shown in (7.9) and (7.10), are independent of temperature. In principle this isothermal constitutive equation is applicable for describing creep behavior under all isothermal states, but the kernel functions at each temperature are different. Therefore, the total strain

$\boldsymbol{\epsilon}$ under a given temperature, T, and given stress $\boldsymbol{\sigma}$ can be represented by

$$\boldsymbol{\epsilon} = \boldsymbol{\epsilon}^T + \boldsymbol{\epsilon}^t, \tag{11.1}$$

where $\boldsymbol{\epsilon}^t$ is the strain due to stress and time and $\boldsymbol{\epsilon}^T$ is the tensor of thermal expansion. In (11.1) the tensorial form of the strain is utilized. The tensor for thermal expansion $\boldsymbol{\epsilon}^T$ can be represented by the following due to the fact that thermal expansion only changes normal strains for homogeneous isotropic materials,

$$\boldsymbol{\epsilon}^T = \mathbf{I}\alpha(T), \tag{11.2}$$

where $\alpha(T)$ is the thermal expansion. $\alpha(T)$ in (11.2) may be a nonlinear function of temperature and stress, though for most materials $\alpha(T)$ is approximately a linear function of temperature over the usual range of temperature in structural applications. The creep strain $\boldsymbol{\epsilon}^t$ in (11.1) is represented by (7.9) provided the kernel functions as given in (7.10) are dependent on temperature as well as on time variables as follows:

$$\begin{aligned}\boldsymbol{\epsilon}^t(t) = & \int_0^t [\mathbf{I}K_1\,\bar{\dot{\boldsymbol{\sigma}}}(\xi_1)+K_2\,\dot{\boldsymbol{\sigma}}(\xi_1)]\mathrm{d}\xi_1+\int_0^t\int_0^t [\mathbf{I}K_3\,\bar{\dot{\boldsymbol{\sigma}}}(\xi_1)\,\bar{\dot{\boldsymbol{\sigma}}}(\xi_2)+\mathbf{I}\,K_4\overline{\dot{\boldsymbol{\sigma}}(\xi_1)\dot{\boldsymbol{\sigma}}(\xi_2)} \\ & +K_5\,\bar{\dot{\boldsymbol{\sigma}}}(\xi_1)\dot{\boldsymbol{\sigma}}(\xi_2)+K_6\,\dot{\boldsymbol{\sigma}}(\xi_1)\dot{\boldsymbol{\sigma}}(\xi_2)]\,\mathrm{d}\xi_1\,\mathrm{d}\xi_2 \\ & +\int_0^t\int_0^t\int_0^t [\mathbf{I}\,K_7\,\overline{\dot{\boldsymbol{\sigma}}(\xi_1)\,\dot{\boldsymbol{\sigma}}(\xi_2)\,\dot{\boldsymbol{\sigma}}}(\xi_3)+\mathbf{I}\,K_8\,\overline{\dot{\boldsymbol{\sigma}}(\xi_1)\,\dot{\boldsymbol{\sigma}}}(\xi_2)\,\bar{\dot{\boldsymbol{\sigma}}}(\xi_3) \\ & +K_9\,\bar{\dot{\boldsymbol{\sigma}}}(\xi_1)\,\bar{\dot{\boldsymbol{\sigma}}}(\xi_2)\,\dot{\boldsymbol{\sigma}}(\xi_3)+K_{10}\,\overline{\dot{\boldsymbol{\sigma}}(\xi_1)\,\dot{\boldsymbol{\sigma}}}(\xi_2)\dot{\boldsymbol{\sigma}}\,(\xi_3) \\ & +K_{11}\,\bar{\dot{\boldsymbol{\sigma}}}(\xi_1)\,\dot{\boldsymbol{\sigma}}(\xi_2)\,\dot{\boldsymbol{\sigma}}(\xi_3)+K_{12}\,\dot{\boldsymbol{\sigma}}(\xi_1)\,\dot{\boldsymbol{\sigma}}(\xi_2)\,\dot{\boldsymbol{\sigma}}(\xi_3)]\,\mathrm{d}\xi_1\,\mathrm{d}\xi_2\,\mathrm{d}\xi_3,\end{aligned} \tag{11.3}$$

where

$$\begin{aligned} K_j &= K_j(t-\xi_1;\,T), \quad j = 1, 2, \\ K_j &= K_j(t-\xi_1,\,t-\xi_2;\,T), \quad j = 3, 4, 5, 6, \\ K_j &= K_j(t-\xi_1,\,t-\xi_2,\,t-\xi_3;\,T), \quad j = 7, 8, 9, 10, 11, 12. \end{aligned} \tag{11.4}$$

In Chapter 8 and Chapter 9, methods were discussed for determining the kernel functions under isothermal conditions. In order to obtain an explicit form for the temperature dependence as well as time dependence of the kernel functions, creep experiments such as those suggested in Chapters 8 and 9 at several constant temperature levels instead of one temperature, are required. The number of the temperature levels required depends on the temperature range of interest in the application, the property of the material and the accuracy required. In principle, if the number of creep tests required for complete characterization at a given temperature is in the order of one hundred as discussed in Chapter 9, then the number of creep tests required is in the order of a thousand in order to completely determine the kernel functions in (11.3) as a function of temperature.

The discussion so far has been confined to constant temperature creep behavior. The creep behavior under varying temperature will be discussed in Section 11.4 and 11.5. In Chapter 9, the difficulty of determining all the kernel functions for different time variables has been shown. Thus several approximate constitutive equations were derived in Chapter 9 in order to reduce the effort in determining the kernel functions. Those approximations, such as the modified superposition method and the product form of kernel functions, can be applied equally well to the temperature problem discussed in this section. For those approximate constitutive equations it was shown in Chapter 9 that only constant stress creep tests are required to characterize the kernel functions. Therefore adapting these approximate methods in the temperature problem would also imply that only constant stress creep tests, though at different temperature levels, are required. In the following section, a method of determining the temperature dependent kernel functions under uniaxial stress and pure torsion stress states will be discussed.

11.3 Determination of Temperature Dependent Kernel Functions

A possible method of determining the temperature dependent kernel functions is discussed in this section. The discussion in this section is concentrated on the creep behavior under the simplest stress states, uniaxial stress and pure shear (torsional) stress, at elevated temperatures. The method can be used equally as well for more general stress states.

Under uniaxial stress, the total amount of strain due to the application of stress $\sigma(t)$ and temperature T can be represented by

$$\begin{aligned}\varepsilon_{11}(t) = \alpha(T) + \int_0^t F_1(t-\xi_1; T)\dot{\sigma}(\xi_1)\,\mathrm{d}\xi_1 \\ + \int_0^t\int_0^t F_2(t-\xi_1, t-\xi_2; T)\,\dot{\sigma}(\xi_1)\,\dot{\sigma}(\xi_2)\,\mathrm{d}\xi_1\,\mathrm{d}\xi_2 \\ + \int_0^t\int_0^t\int_0^t F_3(t-\xi_1, t-\xi_2, t-\xi_3; T)\,\dot{\sigma}(\xi_1)\,\dot{\sigma}(\xi_2)\,\dot{\sigma}(\xi_3)\,\mathrm{d}\xi_1\,\mathrm{d}\xi_2\,\mathrm{d}\xi_3,\end{aligned} \tag{11.5}$$

where the kernel functions F_1, F_2, F_3 depend on temperature and time. The isothermal values of kernel functions F_1, F_2, F_3 with equal time variables $(t-\xi_1 = t-\xi_2 = t-\xi_3 = t-\xi)$ can be determined from constant stress creep tests. Under constant stress and temperature (11.5) becomes

$$\varepsilon_{11}(t) = \alpha(T)+F_1(t, T)\,\sigma+F_2(t, T)\,\sigma^2+F_3(t, T)\,\sigma^3. \tag{11.6}$$

The methods described in Section 8.12 for determination of kernel functions for constant stress can be used to determine $F_1(t, T_i)$, $F_2(t, T_i)$ and $F_3(t, T_i)$ at each given temperature T_i. After that the relations of F_1, F_2, F_3 versus temperature can be obtained graphically by plotting F_1, F_2, F_3 versus T. An analytical form describing F_1, F_2 and F_3 as functions of the temperature parameter T can be obtained from the graph.

Under pure shear stress, the total amount of strain due to the application of stress $\tau(t)$ and temperature T can be represented by

$$\begin{aligned}\varepsilon_{12}(t) = &\int_0^t G_1(t-\xi_1; T)\,\dot{\tau}(\xi_1)\,d\xi_1 \\ &+\int_0^t\int_0^t\int_0^t G_3(t-\xi_1, t-\xi_2, t-\xi_3; T)\,\dot{\tau}(\xi_1)\,\dot{\tau}(\xi_2)\,\dot{\tau}(\xi_3)\,d\xi_1\,d\xi_2\,d\xi_3.\end{aligned} \tag{11.7}$$

Under constant stress, (11.7) becomes

$$\varepsilon_{12}(t) = G_1(t, T)\,\tau+G_2(t, T)\,\tau^3. \tag{11.8}$$

Lai and Findley [11.1, 11.2] conducted pure tension and pure torsional creep experiments under constant stress and at several temperatures from 75°F to 160°F on solid (full-density) polyurethane.

It was found from these tests for both tension and torsion that the experimental creep strain versus time at different stresses and different temperatures could be approximated rather well by a power law of time t with a constant time exponent n independent of stress and temperature. Using the power law representation for the time function and the method of determining kernel functions described in Section 8.12, it was found for tension that the time-independent strain was linear and nearly independent of temperature. Only the linear part of the time-dependent term $F_1^+(t, T)$ was a function of temperature, and the third order time-dependent term $F_3^+(t, T)$ was negligibly small. These observations greatly simplified the mathematical representation of the kernel functions. Thus for tension,

$$\varepsilon_{11}(\sigma, t, T) = \alpha(T) + F_1(t, T)\sigma + F_2(t, T)\sigma^2 + F_3(t, T)\sigma^3, \qquad (11.9)$$

where

$$\begin{aligned} \alpha(T) &= \alpha_0 T_1 = 35.2 \times T_1 \times 10^{-6} \text{ in/in/°F}, \\ F_1(t, T) &= [1.95 + (0.17 + 0.0038\, T_1 + 0.000045\, T_1^2)t^n] \times 10^{-6} \text{ in/in/(psi)}, \\ F_2(t, T) &= 0.04t^n \times 10^{-9} \text{ in/in/(psi)}^2, \\ F_3(t, T) &= 0, \\ n &= 0.143, \end{aligned}$$

where $T_1 = T - 75°\text{F}$, and where $\alpha_0 = 35.2 \times 10^{-6}$ in/in/°F was the observed thermal expansion coefficient determined by measuring the thermally induced strain at zero load. For torsion there is no shear strain due to thermal expansion and it was found that the time-independent term was nearly independent of temperature. The experimental results for torsion thus were represented by

$$\varepsilon_{12}(\tau, t, T) = G_1(t, T)\tau + G_2(t, T)\tau^3, \qquad (11.10)$$

where

$$\begin{aligned} G_1(t, T) &= [2.59 + (0.25 + 0.0074\, T_1 + 0.00007\, T_1^2)t^n] \times 10^{-6} \text{ in/in/(psi)}, \\ G_3(t, T) &= [0.065 + (0.04 - 0.0011\, T_1)t^n] \times 10^{-12} \text{ in/in/(psi)}^3, \\ n &= 0.143. \end{aligned}$$

Some of the experimental results and corresponding theory (11.9) and (11.10) are shown in Figures 11.1 and 11.2.

Another approach to the temperature problem in nonlinear creep of plastics as employed by Findley [11.3] was based on the activation energy theory as applied to creep of metals by Kauzmann [11.4]. According to this theory the frequency ν with which the movement (jump) of one portion of the molecular structure of the material past the energy barrier is given by

$$\nu = (\mathsf{K}T/\mathsf{h}) \exp(-\Delta f^*/\mathsf{K}T) \qquad (11.11)$$

if no external stress exists, where K and h are the Boltzmann and Planck constants, respectively, T is the absolute temperature and Δf^* is the free energy of the barrier. In the absence of stress the frequency of jumps in either direction is the same. But if a shear stress, τ, for example, is applied the frequency is increased in the direction of stressing and decreased in the

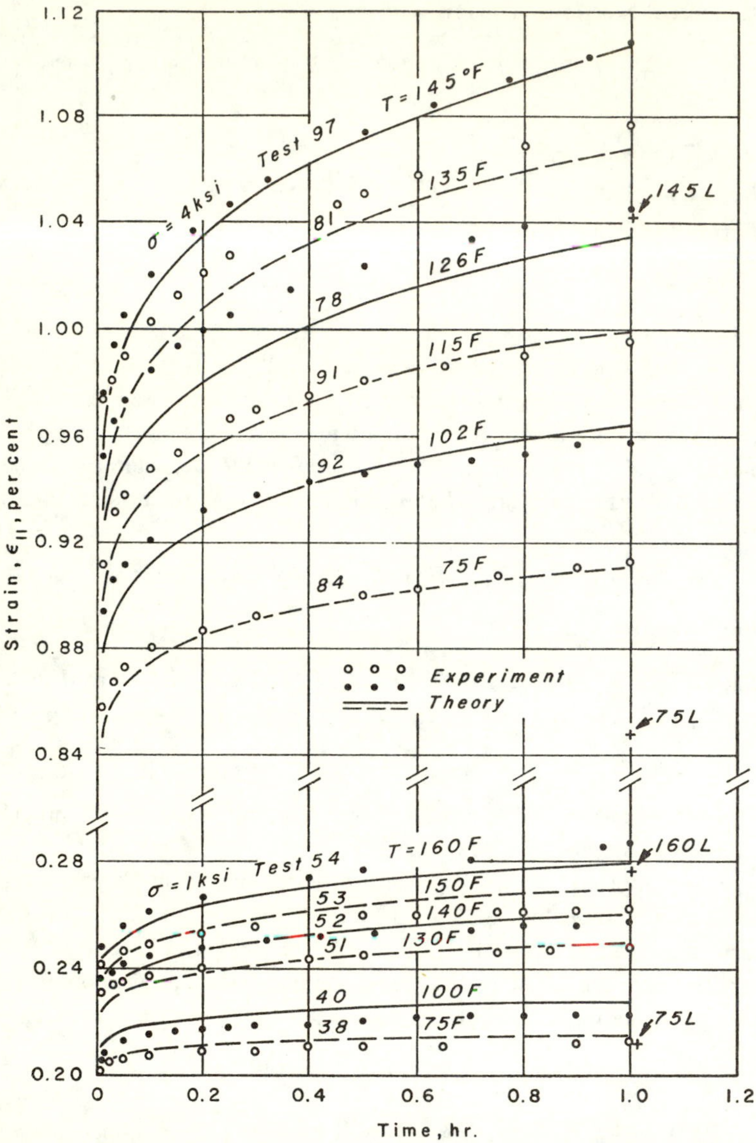

Fig. 11.1. Tensile Strain for Tension Creep of Solid Polyurethane at Various Temperatures. Numbers followed by letter *L* indicate the strain computed from the linear term only at the indicated temperature. From [11.1], courtesy Soc. of Rheology.

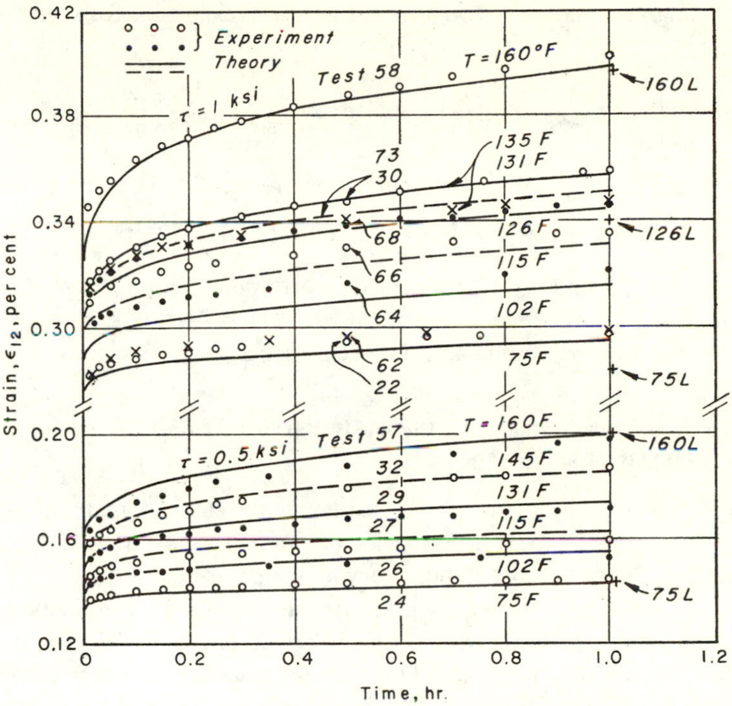

Fig. 11.2. Shearing Strain for Torsion Creep of Solid Polyurethane at Various Temperatures. Numbers followed by letter *L* indicate the strain computed from the linear term only at the indicated temperature. From [11.2], courtesy Soc. of Rheology.

opposite direction. Thus, the net frequency becomes

$$\nu = 2\left(\frac{\mathrm{K}T}{\mathrm{h}}\right)\exp\left(\frac{-\Delta f^*}{\mathrm{K}T}\right)\sin\mathrm{h}\left(\frac{\lambda\tau}{2\,\mathrm{K}T}\right), \tag{11.12}$$

where λ is the average distance, in the direction of stress, between equilibrium positions in successive jumps. The shear creep rate $\dot{\gamma}$ is proportional to the net frequency of jumps. Therefore

$$\dot{\gamma} = 2\dot{\varepsilon}_{12} = 2a\left(\frac{\mathrm{K}T}{\mathrm{h}}\right)\exp\left(\frac{-\Delta f^*}{\mathrm{K}T}\right)\sin\mathrm{h}\left(\frac{\lambda\tau}{2\,\mathrm{K}T}\right), \tag{11.13}$$

where a is a constant. Expansion of the hyperbolic sine function results in

$$\dot{\varepsilon}_{12} = a\left(\frac{\mathrm{K}T}{\mathrm{h}}\right)\exp\left(\frac{-\Delta f^*}{\mathrm{K}T}\right)\left[\frac{\lambda\tau}{2\,\mathrm{K}T}+\frac{1}{6}\left(\frac{\lambda\tau}{2\,\mathrm{K}T}\right)^3+\ldots\right]. \tag{11.14}$$

Neglecting all but the first two terms of the series expansion for sin h $\lambda\tau/2KT$ results in

$$\dot{\varepsilon}_{12} = a\left[\exp\left(\frac{-\Delta f^*}{KT}\right)\right]\left(\frac{\lambda}{2h}\right)\tau + a\left[\exp\left(\frac{-\Delta f^*}{KT}\right)\right]\left(\frac{\lambda^3}{48\,hK^2T^2}\right)\tau^3. \quad (11.15)$$

Differentiating (11.10) yields the following results for the strain rate:

$$\dot{\varepsilon}_{12} = n(a_0+a_1T_1+a_2T_1^2)t^{n-1}\tau + n(b_0+b_1T_1)t^{n-1}\tau^3, \quad (11.16)$$

where a_0, a_1, a_2 and b_0, b_1 represent the corresponding numerical constants given in (11.10). The stress dependence of (11.16) is the same as that of (11.15) but the temperature and time dependence are different.

11.4 Creep Behavior under Continuously Varying Temperature — Uniaxial Case

In Chapter 5, the thermorheologically simple class of material and the time-temperature superposition principle were employed to account for linear viscoelastic behavior under varying temperature.

A possible way of generalizing the time-temperature superposition principle for use in nonlinear viscoelastic behavior under varying temperature will be presented in this section. Different approaches for this same objective have been described also by Bernstein, Kearsly and Zapas [11.5], Lianis and McGuirt [11.6, 11.7] and by Schapery [11.8]. Again, for the sake of convenience, creep under a uniaxial stress state will be used in the following discussion. The method employed can be applied to more general stress states.

For many materials it has been shown that the strain during creep at constant stress is separable into a time-independent part and a time-dependent part with a separable time function for this part of the strain. For such materials the strain during creep at a constant uniaxial stress can be expressed in the following form

$$\varepsilon^t(\sigma, T_1, t) = \Phi^0(\sigma, T_1) + \Phi^+(\sigma, T_1)\,\mathsf{G}(t), \quad (11.17)$$

where ε^t is the strain due to stress, not including thermal expansion,

$$\Phi^0(\sigma, T_1) = \mathsf{a}_1(T_1)\sigma + \mathsf{a}_2(T_1)\sigma^2 + \mathsf{a}_3(T_1)\sigma^3, \quad (11.18a)$$

$$\Phi^+(\sigma, T_1) = \mathsf{b}_1(T_1)\sigma + \mathsf{b}_2(T_1)\sigma^2 + \mathsf{b}_3(T_1)\sigma^3, \quad (11.18b)$$

$\mathsf{G}(t)$ is a separable time function, and where $T_1 = T - T_0$ and T_0 is a constant reference temperature. After rearranging terms in (11.18b), as illustrated in an example below, (11.17) can be expressed in the following form

$$\varepsilon^t(\sigma, T_1, t) = \Phi^0(\sigma, T_1) + \mathsf{F}^+(\sigma)\, \mathsf{G}(\zeta), \tag{11.19}$$

where

$$\zeta = \frac{t}{a_T(T_1, \sigma)} \tag{11.20}$$

is the reduced time and

$$\mathsf{F}^+(\sigma) = \mu_1\sigma + \mu_2\sigma^2 + \mu_3\sigma^3 \tag{11.21}$$

is independent of temperature. Equation (11.19) can be interpreted as the time-temperature superposition principle for nonlinear viscoelasticity, provided that the shift factor a_T depends on both temperature and stress.

The following example is chosen to illustrate how (11.19) may be rewritten into the following form by considering that b_1, b_2, b_3 in (11.18b) may contain a temperature-independent term, as follows

$$\Phi^+(\sigma, T_1) = [\mu_1\sigma + \mu_2\sigma^2 + \mu_3\sigma^3 + \varrho_1(T_1)\sigma + \varrho_2(T_1)\sigma^2 + \varrho_3(T_1)\sigma^3]. \tag{11.22}$$

Rearranging by factoring out the temperature-independent terms in (11.22)

$$\Phi^+(\sigma, T_1) = [\mu_1\sigma + \mu_2\sigma^2 + \mu_3\sigma^3]\left[1 + \frac{\varrho_1(T_1) + \varrho_2(T_1)\sigma + \varrho_3(T_1)\sigma^2}{\mu_1 + \mu_2\sigma + \mu_3\sigma^2}\right], \tag{11.23}$$

where μ_1, μ_2, μ_3 are independent of temperature, while ϱ_1, ϱ_2, ϱ_3 are functions of temperature $T_1 = T - T_0$ and $\varrho_1 = \varrho_2 = \varrho_3 = 0$ at $T = T_0$ or $T_1 = 0$. Factor

$$f(\sigma, T_1) = 1 + \frac{\varrho_1(T_1) + \varrho_2(T_1)\sigma + \varrho_3(T_1)\sigma^2}{\mu_1 + \mu_2\sigma + \mu_3\sigma^2} \tag{11.24}$$

in (11.23) and $\mathsf{G}(t)$ in (11.17) can be combined into

$$\mathsf{G}(\zeta) = f(T_1, \sigma)\, \mathsf{G}(t) \tag{11.25}$$

with ζ defined by (11.20).

For example when $\mathsf{G}(t) = t^n$, (11.25) implies that

$$\zeta^n = f(T_1, \sigma) t^n,$$

or

$$\zeta = t/a_T(T_1, \sigma), \tag{11.26}$$

where the shift factor

$$a_T(T_1, \sigma) = [f(T_1, \sigma)]^{-1/n}$$

$$= \left[1 + \frac{\varrho_1(T_1) + \varrho_2(T_1)\sigma + \varrho_3(T_1)\sigma^2}{\mu_1 + \mu_2\sigma + \mu_3\sigma^2}\right]^{-1ln} \tag{11.27}$$

For continuously varying temperature, (11.20) together with (5.213) can be used to redefine the reduced time ζ as follows

$$\zeta = \int_0^t \frac{ds}{a_T[T_1(s), \sigma(t)]}, \tag{11.28}$$

where s is a dummy variable between 0 and t. Equations (11.19) and (11.28) can be used to describe creep behavior under varying temperature, provided that the temperature shift factor a_T has been determined from a series of constant temperature creep tests.

The modified superposition method described in Section 9.5 and Appendix A5 can be combined with (11.19) and (11.28) to account for creep behavior under continuously varying stress and temperature as follows

$$\varepsilon^t(\sigma, T_1, t) = \Phi^0(\sigma, T_1) + \int_0^\zeta [\mu_1 + 2\mu_2\sigma(\zeta') + 3\mu_3\sigma^2(\zeta')]G(\zeta - \zeta')\dot{\sigma}(\zeta')\, d\zeta', \tag{11.29}$$

where

$$\zeta = \int_0^t \frac{ds}{a_T[T_1(s), \sigma(t)]}, \tag{11.30}$$

$$\zeta' = \int_0^\xi \frac{ds}{a_T[T_1(s), \sigma(\xi)]}. \quad \begin{matrix} \zeta' \leq \zeta \\ \xi \leq t \end{matrix} \tag{11.31}$$

In (11.29) the terms in the bracket were obtained by differentiating (11.21) with respect to σ. ζ' and ξ in (11.31) are the generic reduced time and generic real time respectively.

For $T = T_0$, i.e., $T_1 = 0$, (11.26) reduces to a form similar to that in Section 9.5 since as defined in (11.24), at $T_1 = 0$, $f(\sigma, T_1) = 1$, $\zeta = t$, $\zeta' = \xi$.

For a constant temperature T but different from T_0, and a varying stress $\sigma = \sigma(t)$, the strain may be described by (11.29) provided that ζ and ζ' in (11.30) and (11.31) reduce to

$$\zeta = \frac{t}{a_T[T_1, \sigma(t)]}, \tag{11.32}$$

$$\zeta' = \frac{\xi}{a_T[T_1, \sigma(\xi)]}, \tag{11.33}$$

respectively.

For varying temperature but constant stress, $\sigma = \sigma_0 H(t)$, (11.29) and (11.30) reduce to the following equations which are identical to (11.19) and (11.28) for constant stress.

$$\varepsilon^t(t) = \Phi^0(\sigma_0, T_1) + [\mu_1\sigma_0 + \mu_2\sigma_0^2 + \mu_3\sigma_0^3]\mathrm{G}(\zeta), \tag{11.34}$$

$$\zeta = \int_0^t \frac{\mathrm{d}s}{a_T[T_1(s), \sigma_0]}. \tag{11.35}$$

Lai and Findley [11.1, 11.2] described a creep test under constant tensile stress with varying temperature, see Figure 11.3, and three creep tests under

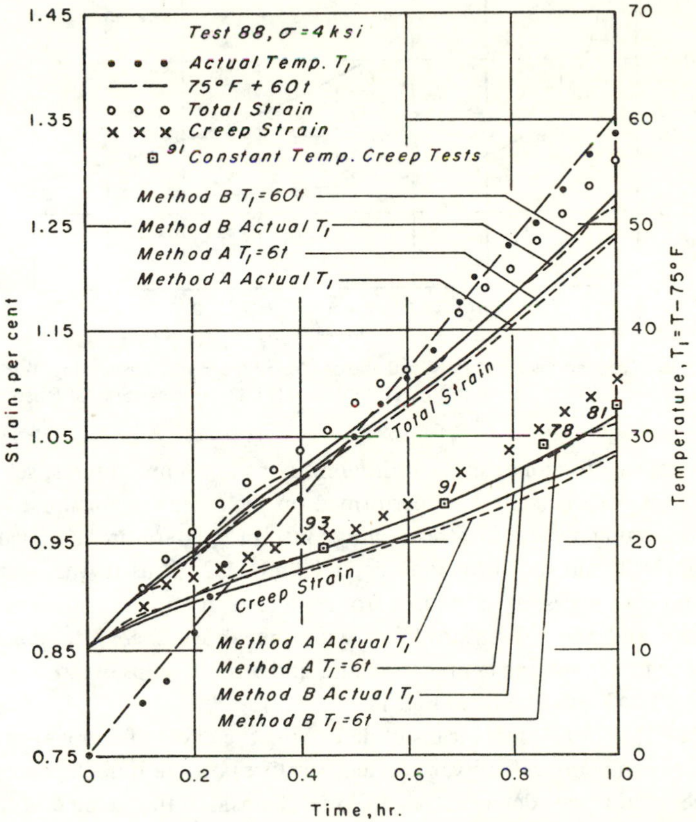

Fig. 11.3a. Tensile Creep of Solid Polyurethane at Constant Stress but Increasing Temperature. From [11.1], courtesy Soc. of Rheology.

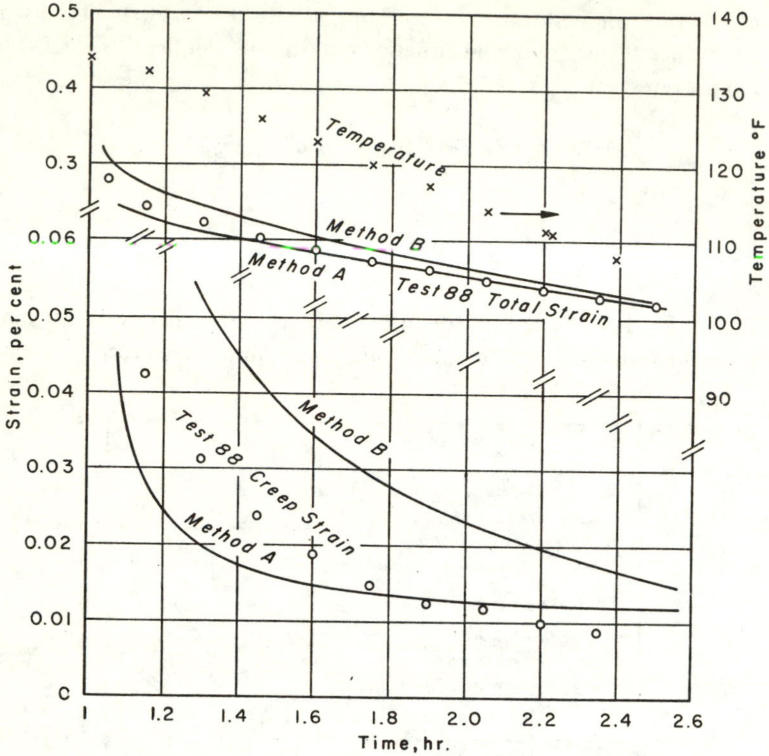

Fig. 11.3*b*. Recovery of Solid Polyurethane at Zero Stress and Decreasing Temperature Following Creep Shown in Fig. 11.3*a*. From [11.1], courtesy Soc. of Rheology.

multi-steps of torsional stress at different constant temperatures, see Figure 11.4. These experiments were performed on the same specimen used for the creep tests under constant stresses and constant temperatures as discussed in Section 11.3, and shown in Figures 11.1 and 11.2. Thus it was possible to determine all necessary constants from the tests at constant stress and temperature and use this information with the above theory to predict the behavior under varying temperature and under multi-steps of stress as shown in Fig. 11.3 (Method A) and Fig. 11.4.

In [11.1] a more direct means of describing the effect of temperature (both constant temperature and varying temperature) on the time-dependent part of the strain was described as follows. Consider that temperature only affects the coefficients $\mathsf{b}_1(T_1)$, $\mathsf{b}_2(T_1)$ and $\mathsf{b}_3(T_1)$ of (11.18). Employing the modified superposition method the creep behavior under varying load and

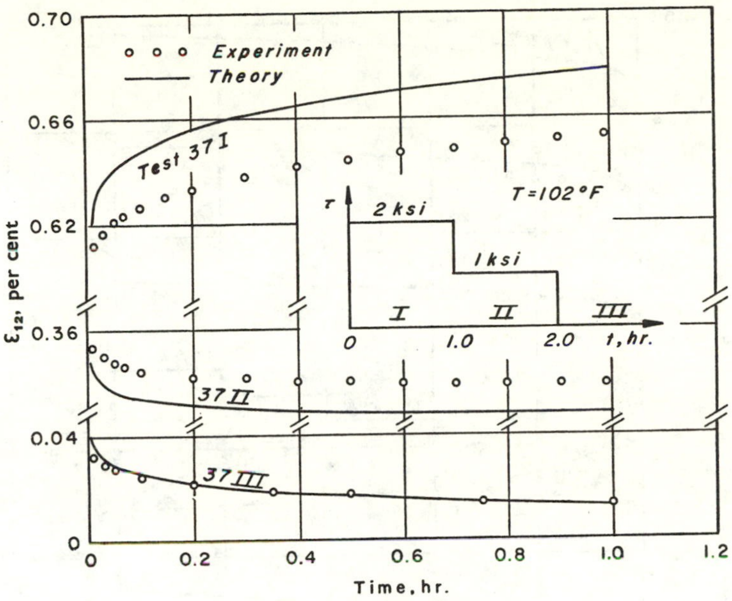

Fig. 11.4*a*. Step-loading Creep and Recovery of Solid Polyurethane at 102°F in Torsion. From [11.2], courtesy Soc. of Rheology.

temperature can be described from (11.17) and (11.18) as follows

$$\varepsilon^t(t) = \Phi^0(\sigma, T_1) + \int_0^t [b_1(T_1) + 2b_2(T_1)\sigma(\xi) + 3b_3(T_1)\sigma^2(\xi)]\, G(t-\xi)\dot{\sigma}(\xi)\, d\xi. \tag{11.36}$$

In (11.36), T_1 in the bracket as well as in the time-independent part is a function of the current time t only. The terms in brackets were obtained from (11.18b) by differentiating with respect to σ.

This method has also been employed to predict the creep behavior under varying temperature. In Figure 11.3, the prediction by this method is designated as Method B, while the prediction by the previous method is designated as Method A.

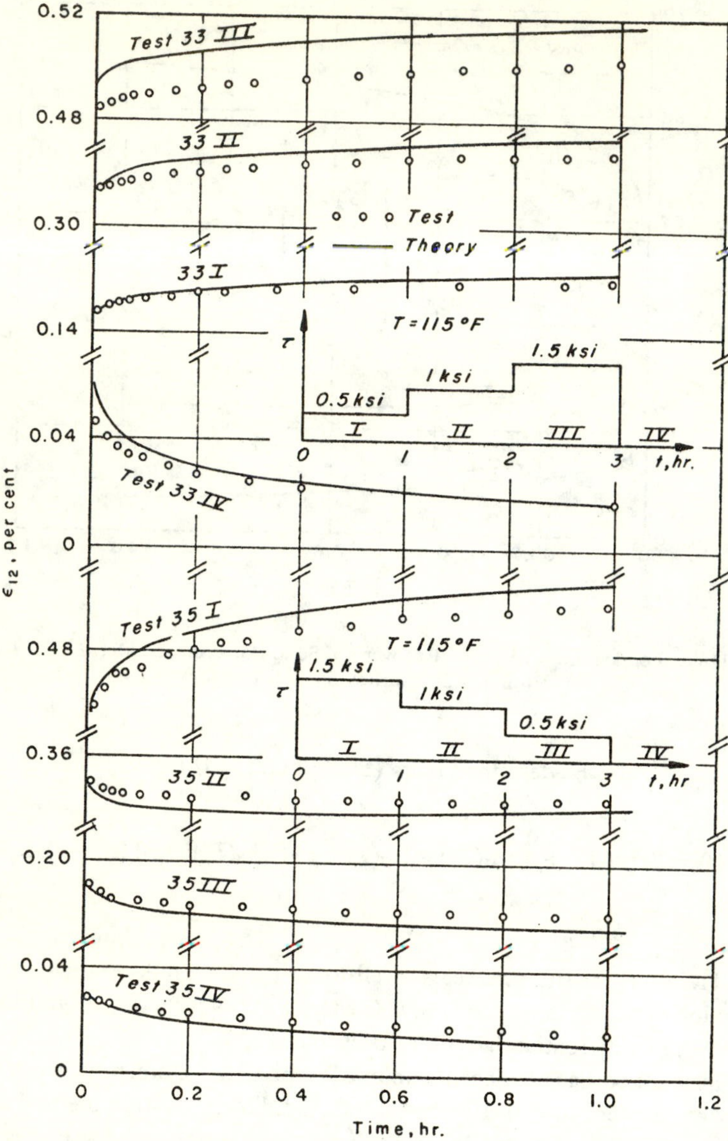

Fig. 11.4*b*. Step-loading Creep and Recovery of Solid Polyurethane at 115°F in Torsion From [11.2], courtesy Soc. of Rheology.

11.5 Creep Behavior under Continuously Varying Temperature for Combined Tension and Torsion

Results of creep experiments under combined tension and torsion at different temperatures were described by Mark and Findley [11.9] for polyurethane. This information, together with results from [11.1, 11.2], permitted describing the temperature dependence of all the terms in (8.43) and (8.46) for the tensile ε_{11} and shear ε_{12} components of strain under constant combined tension σ and torsion τ as follows,

$$\begin{aligned}\varepsilon_{11} = {} & 0.195\sigma+(0.0008+0.00031T_1-0.000{,}0112T_1^2)\sigma\tau^2 \\ & +(0.0011-0.00058T_1+0.000{,}0312T_1^2)\tau^2 \\ & +[(0.017+0.00038T_1+0.000{,}0045T_1^2)\sigma+0.004\sigma^2 \\ & +(0.0040-0.00017T_1+0.000{,}0094T_1^2)\sigma\tau^2 \\ & +(-0.0111+0.000{,}05T_1-0.000{,}0238T_1^2)\tau^2]t^{0.143},\end{aligned} \tag{11.37}$$

$$\begin{aligned}\varepsilon_{12} = {} & 0.259\tau+0.0065\tau^3+(-0.0149+0.00140T_1-0.000{,}0155T_1^2)\sigma\tau \\ & +(0.0041-0.00029T_1+0.000{,}0028T_1^2)\sigma^2\tau \\ & +[(0.025+0.00074T_1+0.000{,}007T_1^2)\tau+(0.004-0.00011T_1)\tau^3 \\ & +(-0.0157+0.00014T_1-0.000{,}0065T_1^2)\sigma\tau \\ & +(0.0050-0.000{,}08T_1+0.000{,}0028T_1^2)\sigma^2\tau]t^{0.143},\end{aligned} \tag{11.38}$$

where t is time in hr. and $T_1 = T-75°\text{F}$ is the temperature measured from room temperature (75°F).

Using (11.37) and (11.38) the strains $\varepsilon_{11}(t)$ and $\varepsilon_{12}(t)$ were predicted for a creep experiment [11.10] under constant combined tension and torsion and variable temperature. The stresses were applied at $t = 0$ and held constant during the experiment. The temperature increased linearly from $t = 0$ at 75°F to 135.5°F then decreased exponentially as shown in Fig. 11.5. Two approaches were used to predict the behavior: A temperature-history dependent theory similar to that described above using a reduced time given by (11.30); and a temperature-history independent theory as described above. In both theories a constant value was used for the coefficient of thermal expansion.

The results of these calculations are shown together with the experimental data in Fig. 11.6. During increasing temperature there is excellent agreement between the temperature history independent formulation and experimental data for both axial and shearing components of strain, but poor agreement during decreasing temperature. The strains predicted from the temperature

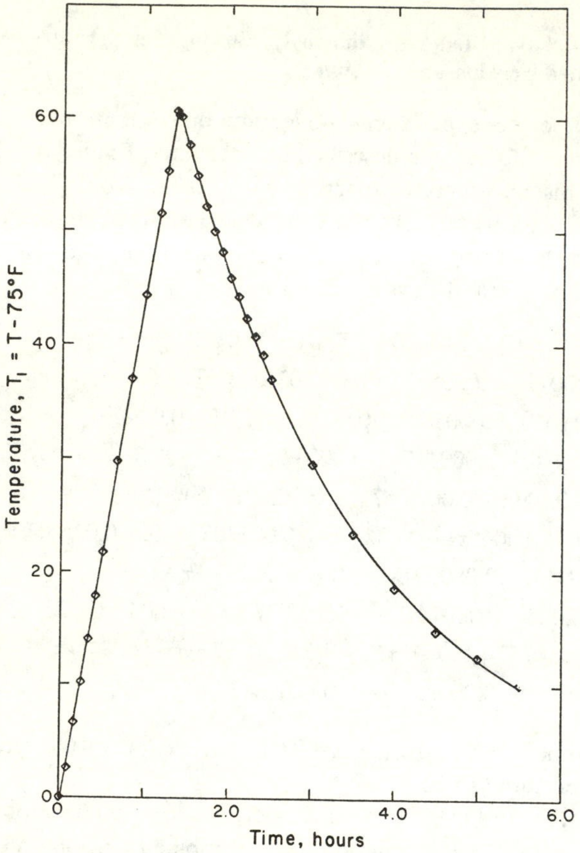

Fig. 11.5. Program of Temperature Change During Creep Test Shown in Fig. 11.6. From [11.10], courtesy Soc. of Rheology.

history dependent formulation show poor agreement with the data during rising temperature but good agreement with the trend during falling temperature. This suggests that creep may be independent of temperature history during rising temperature but dependent on temperature history during falling temperature. A prediction based on this assumption agrees well with the data as shown in Fig. 11.6.

11.6 Thermal Expansion Instability

It was shown by Spencer and Boyer [11.11] and Kovács [11.12] for amorphous thermoplastics that when the temperature was changed rapidly

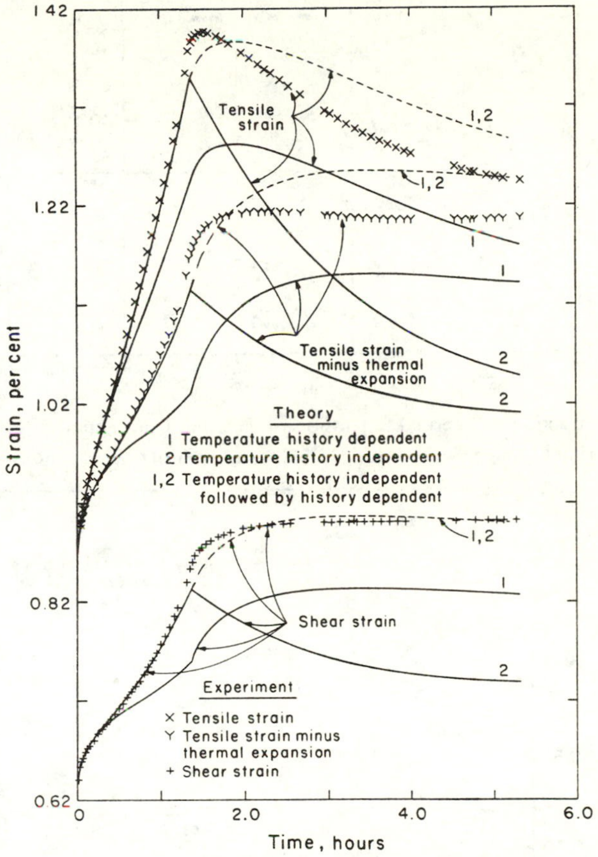

Fig. 11.6. Creep of Polyurethane under Constant Combined Tension ($\sigma = 4$ ksi) and Torsion ($\tau = 2$ ksi) During Temperature Rise and Fall as Shown in Fig. 11.5. From [11.10], courtesy Soc. of Rheology.

there was an almost instantaneous change in volume (thermal expansion) followed by an additional slow change of the same sign toward an equilibrium value. These results were explained in terms of molecular adjustments responsible for collapse of free volume [11.12].

On the other hand a highly cross-linked polyurethane was shown by Findley and Reed [11.13] to have an opposite type of thermal expansion instability. The thermal expansion resulting from heating to a new constant temperature gradually dissipated, as shown in Fig. 11.7. If the temperature was lowered after reaching equilibrium at the elevated temperature the

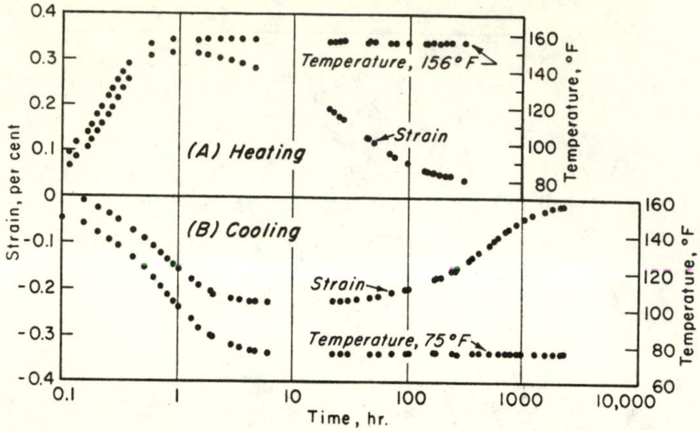

Fig. 11.7. Retraction of Thermal Expansion or Thermal Contraction of Polyurethane at Constant Temperature. From [11.13], courtesy Poly. Engr. and Science.

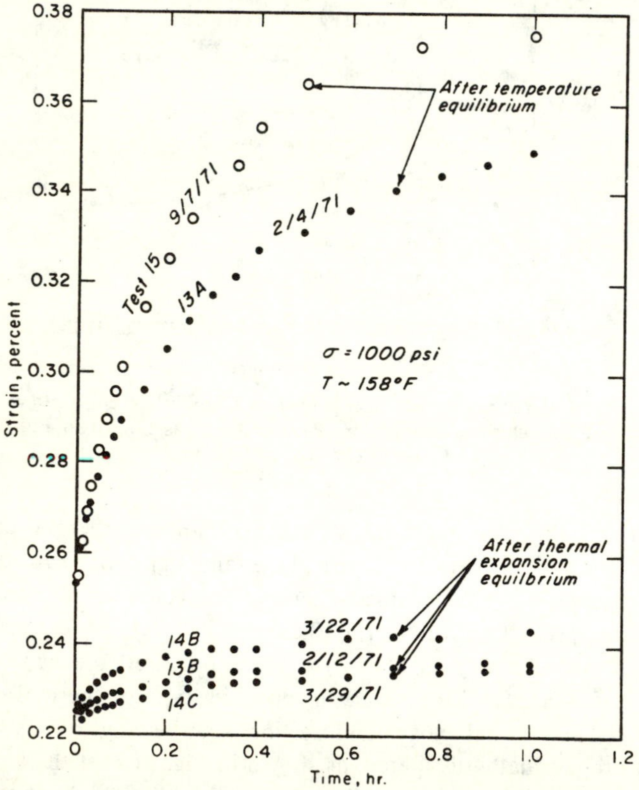

Fig. 11.8. Creep of Polyurethane After Temperature Equilibrium and After Nearing Thermal Expansion Equilibrium (see Fig. 11.7). All tests at about the same temperature of 158°F and all on the same specimen used in Fig. 11.7. From [11.13], courtesy Poly. Engr. and Science.

thermal contraction again was dissipated while at constant temperature, see Fig. 11.7.

It was also shown in [11.13] that the state of thermal expansion had a large influence on creep. Immediately after the temperature was raised so that thermal expansion was at its greatest value the creep rate was large; whereas when thermal expansion equilibrium had been nearly reached at the same temperature the creep rate was much smaller, as shown in Fig. 11.8. Other experiments reported in [11.13] suggest that creep correlates with the state of thermal expansion rather than temperature.

Added in proof: Unpublished experiments by R. Mark and and W. N. Findley have shown that a thermal expansion instability in epoxy such as described above for polyurethane was caused by small gradual changes in moisture content. These experiments also showed that creep behavior correlated with the state of "thermal" expansion rather than either temperature or moisture content. Other unpublished experiments by Mark and Findley on polyethylene showed no thermal expansion instability. Creep of polyethylene at constant stress with increasing and/or decreasing temperature was best described by a temperature history dependent relation.

CHAPTER 12

NONLINEAR VISCOELASTIC STRESS ANALYSIS

12.1 Introduction

As in the case of the theory of linear viscoelasticity discussed in Chapters 5 and 6, nonlinear viscoelastic theory consists of two major areas: characterization of the mechanical properties and solution of boundary value problems. Unfortunately, these two areas have not received equal attention, and have often been investigated independently by researchers in different disciplines. As a result, progress has been slow especially in the area of stress analysis.

Nonlinear viscoelastic stress analysis problems always result in a set of nonlinear differential or integral or differential-integral equations. Use of different types of nonlinear viscoelastic constitutive equations results in different kinds of differential or integral equations which require different techniques to solve them. To solve nonlinear viscoelastic boundary value problems a simplified nonlinear constitutive equation usually has been selected in order to facilitate solution of the problem.

In Chapters 7 through 11, the characterization of nonlinear viscoelastic behavior has been discussed. In Chapter 12 examples are given of nonlinear viscoelastic problems using the type of nonlinear viscolastic constitutive equations described in previous chapters.

Three problems are selected for consideration: a circular cross-section shaft under torsion, a rectangular beam under pure bending and a thick walled cylinder under axially symmetric loading. Similar problems have been treated by Findley, Poczatek and Mathur [12.1, 12.2] for beams in pure bending, by Huang and Lee [12.3] and by Ting [12.4, 12.5] for thick walled cylinders under axially symmetric loading and by Christensen [12.6] for a circular cross-section shaft under torsion.

12.2 Solid Circular Cross-section Shaft under Twisting

Consider a long solid circular cross-section shaft of length l and constant diameter d subjected to a twisting moment T, as shown in Fig. 12.1, so as to produce a prescribed angle of twist Ψ. This problem will be treated for a nonlinear material by an approximate method to solve for the torque and stress distribution as a function of time resulting from the prescribed angle of twist. No other external stresses are applied to the shaft except the torque.

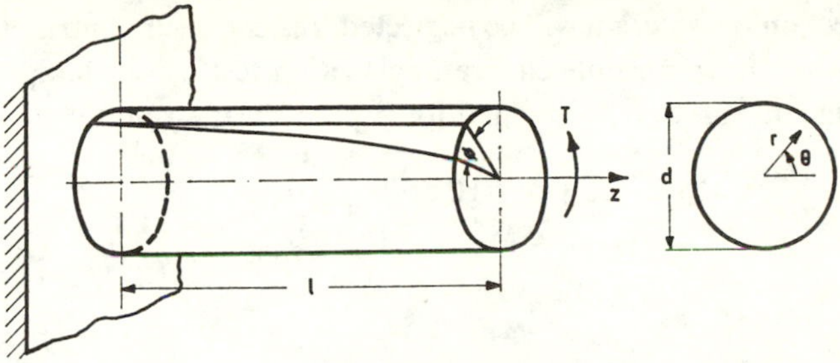

Fig. 12.1. Pure Torsion.

Thus the state of stress $\boldsymbol{\sigma}$ at the cylindrical surface is a pure shear,

$$\boldsymbol{\sigma} = \begin{vmatrix} 0 & \tau & 0 \\ \tau & 0 & 0 \\ 0 & 0 & 0 \end{vmatrix}. \tag{12.1}$$

where τ is the shearing stress. The state of stress in the interior is also approximately the same type as (12.1). The behavior of the material under this state of stress can be obtained experimentally by tests of thin-walled tubes under torque. The usual unconfined stress relaxation test in shear is performed in this manner by holding the angle of twist constant and measuring the torque versus time.

For a nonlinear material the strains resulting from (12.1) are given by (7.78) through (7.81). However, to solve this problem efficiently the inverse of (7.78) through (7.81) are needed, that is, expressions for stress versus strain and time.

It is clear from (7.78) through (7.81) that the state of strain corresponding to (12.1) is not a pure shear strain but a triaxial strain state including non-zero normal strains as given by

$$\boldsymbol{\epsilon} = \begin{vmatrix} \varkappa_{11}(t) & \varepsilon_{12} & 0 \\ \varepsilon_{12} & \varkappa_{22}(t) & 0 \\ 0 & 0 & \varkappa_{33}(t) \end{vmatrix}, \tag{12.2}$$

where in the present situation ε_{12} is the prescribed shearing strain and, $\varkappa_{11}$, $\varkappa_{22}$, $\varkappa_{33}$ are unprescribed time-dependent normal strains, and $\varkappa_{11} = \varkappa_{22}$. The $\varkappa$-strains must vary with time in such a way that the corresponding normal stresses are zero as in (12.1). As shown by (7.78), (7.79) these strains are functions of $\tau(t)$ only. Thus in principle they can be determined if kernel functions $\mathcal{F}_5$ and $\mathcal{F}_6$ are known completely. However, since this determination is quite difficult to do and since the normal strains in pure shear are small their effect on the solution will be neglected. Instead their contribution to the shear stress τ in an unconfined stress-relaxation test will be lumped into the kernel functions for $\tau(t)$ computed for a pure shear strain,

$$\boldsymbol{\epsilon} = \begin{vmatrix} 0 & \varepsilon_{12} & 0 \\ \varepsilon_{12} & 0 & 0 \\ 0 & 0 & 0 \end{vmatrix}. \tag{12.3}$$

That is,

$$\tau(t) = \int_0^t \mathcal{G}_1 \dot{\varepsilon}_{12}\, d\xi_1 + \int_0^t \int_0^t \int_0^t \mathcal{G}_2 \dot{\varepsilon}_{12}^3\, d\xi_1\, d\xi_2\, d\xi_3, \tag{12.4}$$

where (12.4) is obtained from the corresponding pure shear equation (7.80) by interchanging creep and relaxation symbols and making use of (7.91) as explained in Section 7.14.

The torsion problem will be solved by assuming the strain distribution and computing the stresses as follows. It may be shown from symmetry that concentric circles inscribed in a cross-section of a circular shaft in torsion do not warp. Also in a long shaft compatibility of strains requires that diameters remain straight lines.

Thus the displacement of a general point in the tangential direction u_θ expressed in cylindrical coordinates r, θ, z (see Fig. 12.1) is

$$u_\theta = \alpha r z, \tag{12.5}$$

where $\alpha = \varphi/l$ is the angle of twist per unit length. The shear strain $\varepsilon_{z\theta}$ in cylindrical coordinates is equivalent to ε_{12} in (12.3). The relation between strain $\varepsilon_{z\theta}$ and displacements in cylindrical coordinates, equivalent to (3.68), is

$$\varepsilon_{z\theta} = \frac{1}{2}\left(\frac{\partial u_\theta}{\partial z} + \frac{1}{r}\frac{\partial u_z}{\partial \theta}\right). \tag{12.6}$$

Substituting (12.5) into (12.6) and noting that because of symmetry there is no variation with respect to θ yields

$$\varepsilon_{z\theta} = \tfrac{1}{2}\alpha r. \tag{12.7}$$

Introducing (12.7) for ε_{12} in the expanded form of (12.4) yields

$$\tau(t) = \frac{1}{2} r \int_0^t \mathcal{G}_1(t-\xi_1)\,\dot{\alpha}(\xi_1)\,d\xi_1 + \frac{r^3}{8} \int_0^t \int_0^t \int_0^t \mathcal{G}_2(t-\xi_1, t-\xi_2, t-\xi_3)\,\dot{\alpha}(\xi_1)\,\dot{\alpha}(\xi_2)\,\dot{\alpha}(\xi_3)\,d\xi_1\,d\xi_2\,d\xi_3, \tag{12.8}$$

where $\mathcal{G}_1$, $\mathcal{G}_2$ are material functions determined from stress relaxation tests for twisting of an unconfined thin-walled tube or the equivalent.

The torque $\mathsf{T}(t)$ required to maintain a constant angle of twist α can be calculated as follows by summing the moment of tangential force increments $\sigma_{z\theta}\,dA$ over the cross-sectional area A (where $\sigma_{z\theta} = \tau(t)$)

$$\mathsf{T}(t) = \int_0^{d/2} \sigma_{z\theta}(2\pi r)\, r\, dr. \tag{12.9}$$

Inserting (12.8) into (12.9) yields after integration with respect to r

$$\mathsf{T}(t) = \frac{\mathcal{J}}{2} \int_0^t \mathcal{G}_1(t-\xi_1)\,\dot{\alpha}(\xi_1)\,d\xi_1 + \frac{d^2\mathcal{J}}{48} \int_0^t \int_0^t \int_0^t \mathcal{G}_2(t-\xi_1, t-\xi_2, t-\xi_3)\,\dot{\alpha}(\xi_1)\,\dot{\alpha}(\xi_2)\,\dot{\alpha}(\xi_3)\,d\xi_1\,d\xi_2\,d\xi_3, \tag{12.10}$$

where $\mathcal{J} = \int_0^{d/2} 2\pi r^3\,dr = \pi d^4/32$ is the polar moment of inertia of the cross-sectional area.

For a linear theory $\mathcal{G}_2 = 0$. Thus

$$\mathsf{T}(t) = \frac{\mathcal{J}}{2} \int_0^t \mathcal{G}_1(t-\xi_1)\,\dot{\alpha}(\xi_1)\,d\xi_1. \tag{12.11}$$

Equation (12.11) is the same as that obtained directly from the linear elastic solution by use of the correspondence principle.

For a constant angle of twist α_0 the torque $T(t)$ is obtained by inserting $\alpha = \alpha_0 H(t)$ into (12.10) with the result

$$\mathrm{T}(t) = \frac{\mathcal{J}}{2}\mathcal{G}_1(t)\,\alpha_0 + \frac{d^2\mathcal{J}}{48}\mathcal{G}_2(t)\alpha_0^3. \tag{12.12}$$

Equations (12.8) and (12.10) are capable of describing the stresses and the torque under any prescribed history of twisting $\alpha(t)$ provided the kernel functions $\mathcal{G}_1(t-t_1)$, $\mathcal{G}_2(t-t_1, t-t_2, \ t-t_3)$ are known completely. The procedures required and the difficulties of determining these kernel functions have been discussed in Section 9.2. Other simplified methods described in the subsequent sections of Chapter 9 could be used to describe the stresses and torque due to a varying angle of twist.

For example consider a two-step twisting, in which the angle of twisting $\alpha(t)$ is given by

$$\alpha(t) = \alpha_0 H(t) + \Delta\alpha_0 H(t-t_1), \tag{12.13}$$

where α_0 is the initial angle of twist and $\Delta\alpha_0$ is the change in angle of twist at time t_1. The resulting torque can be determined by using (12.10) and the modified superposition method of Section 9.5. Using the modified superposition method, see Appendix A5, (12.10) becomes

$$\mathrm{T}(t) = \frac{\mathcal{J}}{2}\int_0^t \mathcal{G}_1(t-\xi)\,\dot{\alpha}(\xi)\,d\xi + \frac{d^2\mathcal{J}}{48}\int_0^t \mathcal{G}_2(t-\xi)\frac{\partial}{\partial\xi}[\alpha(\xi)]^3\,d\xi. \tag{12.14}$$

Inserting (12.13) into (12.14) yields the following result for $t > t_1$,

$$\mathrm{T}(t) = \frac{\mathcal{J}}{2}[\mathcal{G}_1(t)\,\alpha_0 + \mathcal{G}_1(t-t_1)\,(\Delta\alpha_0)]$$
$$+\frac{d^2\mathcal{J}}{48}[\mathcal{G}_2(t)\,\alpha_0^3 - \mathcal{G}_2(t-t_1)\,\alpha_0^3 + \mathcal{G}_2(t-t_1)\,(\alpha_0+\Delta\alpha_0)^3]. \tag{12.15}$$

The torsion problem discussed above can be called the "relaxation formulation" for torsion since the angle of twist is prescribed and the torque required to hold the prescribed angle of twist is determined. If the angle of twist is constant the torque required decreases (relaxes) with time. The counterpart is the creep formulation for torsion in which the torque is prescribed and the corresponding angle of twist is sought. The formulation of this problem is the same as the relaxation problem except that in (12.10) or (12.14) $T(t)$ is prescribed and $\alpha(t)$ is sought. In this case (12.10) becomes a nonlinear multiple integral equation and (12.14) becomes a nonlinear single integral equation.

The above is based on the relaxation form of constitutive equation. It would seem more appropriate to use the creep form of constitutive equation to obtain the creep formulation for torsion. However, even under constant torque the stress distribution is not known and varies with time. Since stress appears under the integral signs in the creep form of constitutive equation this constitutive equation is difficult to employ in (12.9).

12.3 Beam under Pure Bending

The same simply supported beam treated in Chapter 6 is considered in this section, see Fig. 12.2, but the constitutive relation of the material of the beam is considered to be nonlinearly viscoelastic. For simplicity, small deforma-

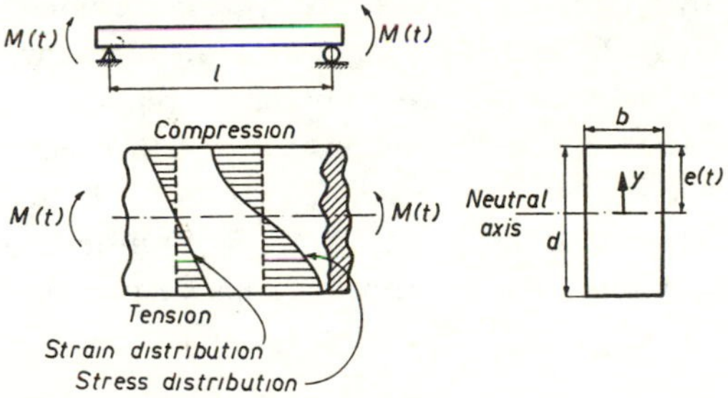

Fig. 12.2. Nonlinear Viscoelastic Beam under Pure Bending.

tions are assumed. For long beams of symmetrical cross-section under pure bending, plane sections before bending remain plane after bending. This fact has been demonstrated experimentally and analytically to be independent of the stress-strain behavior.

In the following discussion the relaxation formulation is employed for simplicity since stresses rather than strains appear in the equilibrium equations. In the first example the curvature $\Omega(t)$ of the beam under pure bending is prescribed; and the stress distribution along a cross-section and the external bending moment M required to maintain the prescribed curvature are to be determined. In the second example the problem of determining the curvature and the stress distribution resulting from a prescribed bending moment $M(t)$ will be discussed.

The fact that plane sections before bending remain plane after bending implies that strain is linearly proportional to the distance from the neutral axis, whose location $e(t)$ is to be determined. Therefore,

$$\varepsilon(t, y) = -\Omega(t)y, \tag{12.16}$$

where ε represents the axial strain, positive for tension and negative for compression, y is the distance measured positively upward from the neutral axis and Ω is the curvature of the neutral axis.

Since the stress state in a long straight beam in bending is essentially uniaxial (tension below the neutral axis and compression above the neutral axis for a positive bending moment), the nonlinear constitutive equation under a uniaxial stress state can be utilized.

A relaxation formulation for a uniaxial stress state can be obtained from Section 7.14 as follows. In the uniaxial stress state, as in the present example of bending, the transverse stresses are zero but the transverse strains are not zero. Thus the relaxation form of this stress state is an unconfined strain state as given by (7.96). The axial stress $\sigma_{11}(t)$ corresponding to this strain state is given by (7.97) and is a function of all three strain components, axial $\varepsilon_{11}(t)$ and transverse $\varkappa_{22}(t)$, $\varkappa_{33}(t)$. In the conventional relaxation test the transverse strains are not considered except that their effect is lumped into the functions relating the axial stress and strain, (7.93) for uniaxial strain, as though the state of strain was uniaxial (7.92) instead of triaxial (7.96). The axial stress σ versus axial strain ε is thus written as follows, similarly to (7.93),

$$\begin{aligned}\sigma(t) = &\int_0^t \mathcal{F}_1'(t-\xi_1)\,\dot{\varepsilon}(\xi_1)\,\mathrm{d}\xi_1\\ &+\int_0^t\int_0^t \mathcal{F}_2'(t-\xi_1, t-\xi_2)\,\dot{\varepsilon}(\xi_1)\,\dot{\varepsilon}(\xi_2)\,\mathrm{d}\xi_1\,\mathrm{d}\xi_2\\ &+\int_0^t\int_0^t\int_0^t \mathcal{F}_3'(t-\xi_1, t-\xi_2, t-\xi_3)\,\dot{\varepsilon}(\xi_1)\,\dot{\varepsilon}(\xi_2)\,\dot{\varepsilon}(\xi_3)\,\mathrm{d}\xi_1\,\mathrm{d}\xi_2\,\mathrm{d}\xi_3.\end{aligned} \tag{12.17}$$

The kernel functions in (12.17) must be obtained from relaxation tests under unconfined uniaxial strain, i.e. the state of strain given by (7.96). Hence these kernel functions $\mathcal{F}'_\alpha$ are not the same as $\mathcal{F}_\alpha$ and thus are not useful in describing any of the kernel functions in a general strain state. It is to be noted that (12.17) implies that tensile and compressive stress-strain relations are different if the second-order strain term in (12.17) is non-zero.

The conditions of equilibrium for the beam are expressed by the following equations

$$\iint_A \sigma \, \mathrm{d}A = 0, \tag{12.18}$$

$$\iint_A \sigma y \, \mathrm{d}A + M(t) = 0. \tag{12.19}$$

Equation (12.18) is the balance of the resultant force of the axial stress on the cross-section A of the beam and (12.19) is the balance of the internal moment of the axial stress on the cross-section of a portion of the beam with the external applied moment M on the same portion. For a rectangular beam of width b and depth d the above equations can be rewritten as follows

$$b \int_{-[d-e(t)]}^{e(t)} \sigma(t, y) \, \mathrm{d}y = 0, \tag{12.20}$$

$$b \int_{-[d-e(t)]}^{e(t)} \sigma(t, y) \, y \, \mathrm{d}y + M(t) = 0, \tag{12.21}$$

where $e(t)$ is the distance from the neutral axis to the top side of the beam.

Inserting (12.16) into (12.17) yields the stress distribution in terms of $\Omega(t)$,

$$\begin{aligned}\sigma(t) = &-y \int_0^t \mathcal{F}'_1(t-\xi_1) \, \dot{\Omega}(\xi_1) \, \mathrm{d}\xi_1 \\ &+ y^2 \int_0^t \int_0^t \mathcal{F}'_2(t-\xi_1, t-\xi_2) \, \dot{\Omega}(\xi_1) \, \dot{\Omega}(\xi_2) \, \mathrm{d}\xi_1 \, \mathrm{d}\xi_2 \\ &- y^3 \int_0^t \int_0^t \int_0^t \mathcal{F}'_3(t-\xi_1, t-\xi_2, t-\xi_3) \, \dot{\Omega}(\xi_1) \, \dot{\Omega}(\xi_2) \, \dot{\Omega}(\xi_3) \, \mathrm{d}\xi_1 \, \mathrm{d}\xi_2 \, \mathrm{d}\xi_3.\end{aligned} \tag{12.22}$$

Note that in (12.22) the position of the neutral axis from which y is to be measured has not yet been determined. Inserting (12.22) into (12.20) yields after integration with respect to y

$$-\tfrac{1}{2}\{e^2(t)-[d-e(t)]^2\}\int_0^t \mathcal{F}_1'(t-\xi_1)\dot{\Omega}(\xi_1)\,d\xi_1$$

$$+\tfrac{1}{3}\{e^3(t)+[d-e(t)]^3\}\int_0^t\int_0^t \mathcal{F}_2'(t-\xi_1,\,t-\xi_2)\,\dot{\Omega}(\xi_1)\,\dot{\Omega}(\xi_2)\,d\xi_1\,d\xi_2$$

$$-\tfrac{1}{4}\{e^4(t)-[d-e(t)]^4\}\int_0^t\int_0^t\int_0^t \mathcal{F}_3'\,(t-\xi_1,\,t-\xi_2,\,t-\xi_3)$$

$$\dot{\Omega}(\xi_1)\,\dot{\Omega}(\xi_2)\,\dot{\Omega}(\xi_3)\,d\xi_1\,d\xi_2\,d\xi_3 = 0. \qquad (12.23)$$

The position of the neutral axis $e(t)$ can be determined from (12.23) provided the kernel functions, $\mathcal{F}_1'$,and $\mathcal{F}_2'$ and $\mathcal{F}_3'$ are known and the curvature $\Omega(t)$ is prescribed.

Inserting (12.22) into (12.21) yields the following results after integration with respect to y

$$-\tfrac{1}{3}\{e^3(t)+[d-e(t)]^3\}\int_0^t \mathcal{F}_1'(t-\xi_1)\,\dot{\Omega}(\xi_1)\,d\xi_1$$

$$+\tfrac{1}{4}\{e^4(t)-[d-e(t)]^4\}\int_0^t\int_0^t \mathcal{F}_2'(t-\xi_1,\,t-\xi_2)\,\dot{\Omega}(\xi_1)\,\dot{\Omega}(\xi_2)\,d\xi_1\,d\xi_2$$

$$-\tfrac{1}{5}\{e^5(t)+[d-e(t)]^5\}\int_0^t\int_0^t\int_0^t \mathcal{F}_3'(t-\xi_1,\,t-\xi_2,\,t-\xi_3)$$

$$\dot{\Omega}(\xi_1)\,\dot{\Omega}(\xi_2)\,\dot{\Omega}(\xi_3)\,d\xi_1\,d\xi_2\,d\xi_3+M(t)/b = 0. \qquad (12.24)$$

Prescribed Curvature. The external bending moment $M(t)$ induced from the prescribed curvature can be calculated from (12.24) in which $e(t)$ has been determined from (12.23).

Thus, for a nonlinear viscoelastic beam with prescribed curvature the position of the neutral axis can be determined from (12.23) and the stress, strain and bending moment can be evaluated from (12.22), (12.16) and (12.24) respectively. It is evident that both the bending moment M and position of the neutral axis e are time dependent even for constant curvature Ω.

If $\mathcal{F}_2'$ in (12.17) is zero, which implies that the tensile and compressive stress-strain relations are the same, (12.23) becomes

$$\tfrac{1}{2}\{e^2(t)-[d-e(t)]^2\}\left\{\int_0^t \mathcal{F}_1'(t-\xi_1)\dot{\Omega}(\xi_1)\,d\xi_1\right.$$
$$+\tfrac{1}{2}\{e^2(t)+[d-e(t)]^2\}\int_0^t\int_0^t\int_0^t \mathcal{F}_3'(t-\xi_1, t-\xi_2, t-\xi_3)$$
$$\left.\dot{\Omega}(\xi_1)\,\dot{\Omega}(\xi_2)\,\dot{\Omega}(\xi_3)\,d\xi_1\,d\xi_2\,d\xi_3\right\} = 0. \tag{12.25}$$

In order that this equation be satisfied for all values of $\mathcal{F}_1'$, $\mathcal{F}_3'$ and $\Omega(t)$, $\{e^2(t)-[d-e(t)]^2\}$ has to be zero. This condition implies that

$$e(t) = d/2. \tag{12.26}$$

It is obvious from (12.26) that the assumption that the tensile and compressive stress-strain relations are the same resulted in the position of the neutral axis of the beam being coincident with the axis of the centroid and independent of time. Inserting (12.26) into (12.24) with $\mathcal{F}_2' = 0$ yields the following results

$$\frac{M(t)}{I} = \int_0^t \mathcal{F}_1'(t-\xi_1)\,\dot{\Omega}(\xi_1)\,d\xi_1$$
$$+\frac{3}{20}d^2\int_0^t\int_0^t\int_0^t \mathcal{F}_3'(t-\xi_1, t-\xi_2, t-\xi_3)$$
$$\dot{\Omega}(\xi_1)\,\dot{\Omega}(\xi_2)\,\dot{\Omega}(\xi_3)\,d\xi_1\,d\xi_2\,d\xi_3, \tag{12.27}$$

where $I = bd^3/12$ is the area moment of inertia of the cross-section of a rectangular beam with respect to the axis of the centroid.

Prescribed Bending Moment. The bending problem discussed above can be called the "relaxation formulation" in which the curvature of the beam is prescribed and the bending moment is sought. The counterpart is the creep formulation in which the external bending moment is prescribed and the curvature is sought. The later problem is more difficult since in this case (12.23) and (12.24) become two coupled nonlinear integral equations for $e(t)$ and $\Omega(t)$. Even under the assumption that $\mathcal{F}'_2 = 0$ and consequently (12.26) results, to determine $\Omega(t)$ from (12.27) requires solving a nonlinear multiple integral equation.

In the above discussion of the creep formulation of the beam problem the relaxation form of constitutive equation has been used. It may seem that using the creep form of the constitutive equation may lead to a simpler solution. The difficulty though is that the stress distribution, which is time dependent and nonlinearly dependent on y, is unknown even under constant bending moment.

Thus in the following the relaxation formulation of constitutive equation (12.17) will be considered. When the bending moment $M(t)$ is prescribed (12.23) and (12.24) are two simultaneous nonlinear multiple integral equations for $e(t)$ and $\Omega(t)$. The solution of these equations is expected to be quite difficult. Once $\Omega(t)$ and $e(t)$ have been determined from (12.23) and (12.24) the strain and stress distributions can be obtained from (12.16) and (12.22). By using the appropriate displacement boundary conditions together with $\Omega(t)$ the deflection of the beam may also be determined.

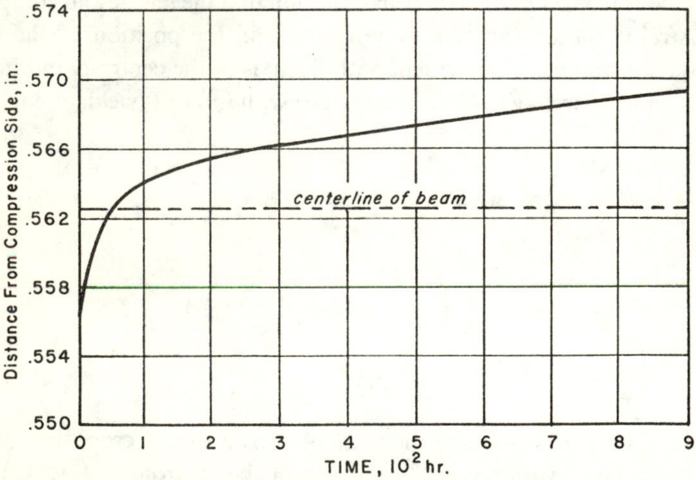

Fig. 12.3. Location of Neutral Axis During Creep in Bending when Tension and Compression Creep are Unequal. Computed for canvas laminate at 77°F under a bending moment of 679 in.-lb. on a rectangular beam of 0.478 in. width and 1.125 in. depth. From [12.2], courtesy ASME.

The solution of (12.23) and (12.24) for prescribed moment, $M(t)$, will be simplified if the multiple integral form is replaced by a single integral representation such as the modified superposition principle as described in Section 9.5 and Appendix A5 or the short-time approximation discussed in Section 7.15. In these cases (12.23) and (12.24) will reduce to nonlinear single integral equations which may be solved by a numerical iteration process such as employed by Huang and Lee [12.3]. Another approach is to first solve the problem for a linear viscoelastic material by taking $\mathcal{F}_2' = \mathcal{F}_3' = 0$. The result is then employed as a first approximation in an iteration process to improve the accuracy of solution of (12.23) and (12.24).

A somewhat different approach was used by Findley, Poczatek and Mathur [12.2] to predict creep in bending from tension and compression creep data when creep in tension and compression were unequal. In this work the memory effect was neglected and the stress dependence was described by hyperbolic sine functions of stress. For a particular material and constant bending moment the computations showed that the location of the neutral axis varied with time as shown in Fig. 12.3 and the stress distribution was as given in Fig. 12.4.

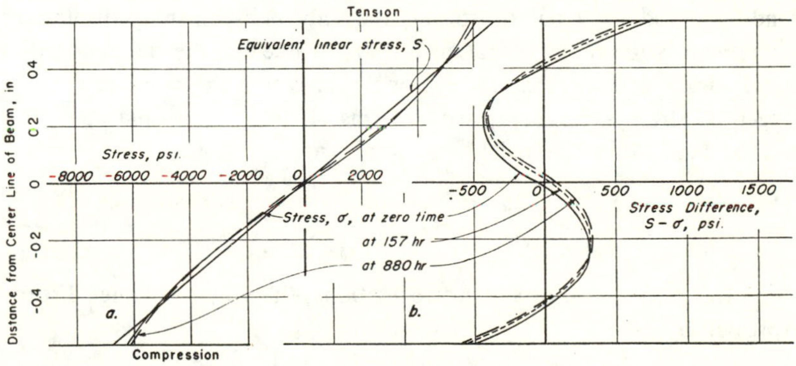

Fig. 12.4. Stress Distribution During Creep of a Rectangular Beam when Tension and Compression Creep are Unequal: (a) stress distribution during creep; (b) deviation from linear stress distribution during creep. Computed for canvas laminate at 77°F under a bending moment of 679 in.-lb. on a rectangular beam of 0.478 in. width and 1.125 in. depth. From [12.2], courtesy ASME.

12.4 Thick-walled Cylinder under Axially Symmetric Loading

A nonlinear viscoelastic thick-walled cylinder is considered, as shown in Fig. 12.5. It is subjected to axially symmetric loading and constrained axially. In order to simplify the solution of the problem the following assumptions are made: (1) small deformations; (2) linearly compressible constitutive relations. The second assumption is a closer description of material behavior than an incompressible assumption.

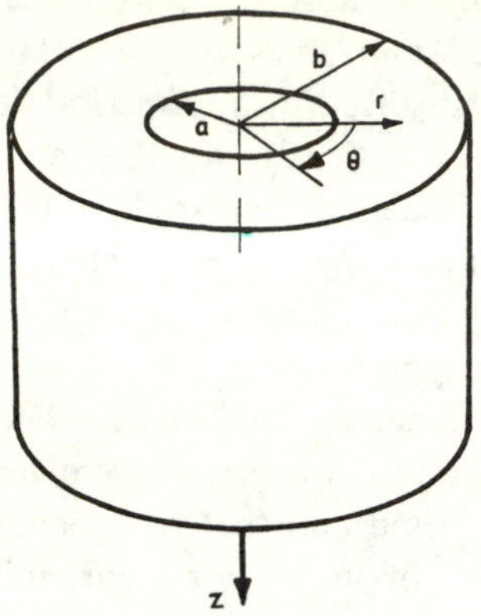

Fig. 12.5. Thick-walled Tube.

Due to the axial symmetry of the geometric configuration and loading condition, and the axial constraint, the only non-vanishing displacement component is the radial displacement $u(r)$, where r is the radial coordinate in cylindrical coordinates, r, θ, z. For this situation the radial strain ε_r and tangential strain ε_θ are expressed in terms of the radial displacement only, and the axial strain is zero. Thus the strains are as follows:

$$\varepsilon_r = \frac{\partial u}{\partial r}, \qquad \varepsilon_\theta = \frac{u}{r}, \qquad \varepsilon_z = 0. \tag{12.28}$$

Using the assumption of small deformations, the volume change V can be expressed by

$$V = \varepsilon_r + \varepsilon_\theta + \varepsilon_z = \frac{\partial u}{\partial r} + \frac{u}{r} = \frac{1}{r}\frac{\partial}{\partial r}(ru). \tag{12.29}$$

Equation (12.29) is a first order differential equation of u as a function of r with V as the undetermined parameter. The solution of (12.29) yields

$$u(r, t) = \frac{A(t)}{r} + \frac{1}{r}\int_0^r yV(y, t)\, dy, \tag{12.30}$$

where y is a dummy variable and $A(t)$ is an undetermined time function.

Substituting (12.30) into (12.28) the radial and tangential strains, ε_r, ε_θ are as follows:

$$\varepsilon_r = \frac{\partial u}{\partial r} = -\frac{A(t)}{r^2} + V(r, t) - \frac{1}{r^2}\int_0^r yV(y, t)\, dy, \tag{12.31}$$

$$\varepsilon_\theta = \frac{u}{r} = \frac{A(t)}{r^2} + \frac{1}{r^2}\int_0^r yV(y, t)\, dy. \tag{12.32}$$

If an incompressible material were considered, $V = 0$, (12.31), (12.32) would become

$$\varepsilon_r = -\frac{A(t)}{r^2}, \qquad \varepsilon_\theta = \frac{A(t)}{r^2}. \tag{12.33}$$

Thus, ε_r and ε_θ as shown in (12.31), (12.32) could be separated into an incompressible strain and a compressible strain.

Nonlinear viscoelastic constitutive equations under a linearly compressible assumption were derived in Section 7.6, and expressed in terms of the strain tensor $\boldsymbol{\epsilon}$ in Section 7.9. For a biaxial strain state, such as considered here,

$$\boldsymbol{\epsilon} = \begin{vmatrix} \varepsilon_r & 0 & 0 \\ 0 & \varepsilon_\theta & 0 \\ 0 & 0 & 0 \end{vmatrix},$$

the corresponding stress components under a linearly compressible assumption can be obtained from (7.85), (7.86) by changing notations from the creep form to relaxation, as follows:

$$\sigma_1 \to \varepsilon_r,\ \sigma_2 \to \varepsilon_\theta,\ \tau = 0,\ \varepsilon_{11} \to \sigma_r,\ \varepsilon_{22} \to \sigma_\theta.$$

Thus, from (7.85)

$$\begin{aligned}\sigma_r(t) = &\int_0^t [\psi_1(\dot\varepsilon_r + \dot\varepsilon_\theta) + \psi_2(\tfrac{2}{3}\dot\varepsilon_r - \tfrac{1}{3}\dot\varepsilon_\theta)]\, d\xi_1 \\ &+ \int_0^t\int_0^t \psi_6[\tfrac{2}{9}\dot\varepsilon_r^2 - \tfrac{1}{9}\dot\varepsilon_\theta^2 - \tfrac{2}{9}\dot\varepsilon_r\dot\varepsilon_\theta)]\, d\xi_1\, d\xi_2 \\ &+ \int_0^t\int_0^t\int_0^t \psi_{12}[\tfrac{2}{3}\dot\varepsilon_r{}^3 - \tfrac{1}{3}\dot\varepsilon_\theta{}^3 + \dot\varepsilon_r\dot\varepsilon_\theta^2 - \dot\varepsilon_r^2\dot\varepsilon_\theta]\, d\xi_1\, d\xi_2\, d\xi_3,\end{aligned} \tag{12.34}$$

from (7.86)

$$\sigma_\theta(t) = \int_0^t [\psi_1(\dot{\varepsilon}_r+\dot{\varepsilon}_\theta)+\psi_2(\tfrac{2}{3}\dot{\varepsilon}_\theta-\tfrac{1}{3}\dot{\varepsilon}_r)]\, d\xi_1$$

$$+\int_0^t\int_0^t \psi_6[\tfrac{2}{9}\dot{\varepsilon}_\theta^2-\tfrac{1}{9}\dot{\varepsilon}_r^2-\tfrac{2}{9}\dot{\varepsilon}_r\dot{\varepsilon}_\theta]\, d\xi_1\, d\xi_2$$

$$+\int_0^t\int_0^t\int_0^t \psi_{12}[\tfrac{2}{3}\dot{\varepsilon}_\theta^3-\tfrac{1}{3}\dot{\varepsilon}_r^3+\dot{\varepsilon}_r^2\dot{\varepsilon}_\theta-\dot{\varepsilon}_\theta^2\dot{\varepsilon}_r]\, d\xi_1\, d\xi_2\, d\xi_3, \tag{12.35}$$

from (7.87)

$$\sigma_z(t) = \int_0^t (\psi_1-\tfrac{1}{3}\psi_2)(\dot{\varepsilon}_r+\dot{\varepsilon}_\theta)\, d\xi_1+\int_0^t\int_0^t \psi_6(-\tfrac{1}{9}\dot{\varepsilon}_r^2-\tfrac{1}{9}\dot{\varepsilon}_\theta^2+\tfrac{4}{9}\dot{\varepsilon}_r\dot{\varepsilon}_\theta)\, d\xi_1\, d\xi_2$$

$$+\int_0^t\int_0^t\int_0^t \psi_{12}(-\tfrac{1}{3}\dot{\varepsilon}_r^3-\tfrac{1}{3}\dot{\varepsilon}_\theta^3)\, d\xi_1\, d\xi_2\, d\xi_3, \tag{12.36}$$

and from (12.34) and (12.35)

$$(\sigma_r-\sigma_\theta) = 0+\int_0^t \psi_2(\dot{\varepsilon}_r-\dot{\varepsilon}_\theta)\, d\xi_1+\int_0^t\int_0^t \tfrac{1}{3}\psi_6[\dot{\varepsilon}_r^2-\dot{\varepsilon}_\theta^2]\, d\xi_1\, d\xi_2$$

$$+\int_0^t\int_0^t\int_0^t \psi_{12}[\dot{\varepsilon}_r^3-\dot{\varepsilon}_\theta^3+2\dot{\varepsilon}_\theta^2\dot{\varepsilon}_r-2\dot{\varepsilon}_r^2\dot{\varepsilon}_\theta]\, d\xi_1\, d\xi_2\, d\xi_3. \tag{12.37}$$

Inserting (12.31), (12.32) into (12.34), (12.35) and (12.37), the dependence of stresses σ_r, σ_θ on strains ε_r, ε_θ is replaced by dependence on $A(t)$ and $V(r, t)$.

The stress components σ_r, σ_θ as given above must satisfy the equation of equilibrium which for axial symmetry is given by

$$\frac{\partial\sigma_r}{\partial r}+\frac{1}{r}(\sigma_r-\sigma_\theta) = 0. \tag{12.38}$$

$(\sigma_r-\sigma_\theta)$ and $\partial\sigma_r/\partial r$ can be expressed in terms of $A(t)$ and $V(r, t)$ straightforwardly from (12.34), (12.37) and (12.31), (12.32) as follows:

$$\sigma_r-\sigma_\theta = \int_0^t \psi_2(t-\xi_1)\,[\dot V(r,\xi_1)-\frac{2\dot A(\xi_1)}{r^2}-\frac{2}{r^2}\int_0^r y\dot V(y,\xi_1)\,dy]\,d\xi_1$$

$$+\int_0^t\int_0^t \frac{1}{3}\psi_6(t-\xi_1,t-\xi_2)\left[\dot V-2\frac{\dot A\dot V}{r^2}-\frac{2\dot V}{r^2}\int_0^r y\dot V(y,\xi_1)\,dy\right]$$

$$+\int_0^t\int_0^t\int_0^t 6\psi_{12}(t-\xi_1,t-\xi_2,t-\xi_3)\left[-\frac{\dot A^3}{r^6}+\frac{3\dot A^2\dot V}{2r^4}-\frac{3\dot A^2}{r^6}\int_0^r y\dot V(y,\xi)\,dy\right]$$

$$d\xi_1\,d\xi_2\,d\xi_3, \quad (12.39)$$

$$\sigma_r = \int_0^t\left[\psi_1\dot V+\psi_2\left(\frac{2}{3}\dot V-\frac{\dot A}{r^2}-\frac{1}{r^2}\int_0^r y\dot V dy\right)\right]d\xi_1$$

$$+\int_0^t\int_0^t\frac{1}{3}\psi_6\left[\frac{\dot A^2}{r^4}-\frac{2\dot A\dot V}{r^2}+\frac{2\dot A}{r^4}\int_0^r y\dot V dy\right]d\xi_1\,d\xi_2$$

$$+\int_0^t\int_0^t\int_0^t\psi_{12}\left[-3\frac{\dot A^3}{r^6}+\frac{5\dot A^2\dot V}{r^4}-\frac{9\dot A^2}{r^6}\int_0^r y\dot V dy\right]d\xi_1\,d\xi_2\,d\xi_3, \qquad (12.40)$$

$$\frac{\partial\sigma_r}{\partial r} = \int_0^t\left[\psi_1\frac{\partial\dot V}{\partial r}+\psi_2\left(\frac{2}{3}\frac{\partial\dot V}{\partial r}+\frac{2\dot A}{r^3}+\frac{2}{r^3}\int_0^r y\dot V dy-\frac{\dot V}{r}\right)\right]d\xi_1$$

$$+\int_0^t\int_0^t\frac{1}{3}\psi_6\left[-\frac{4\dot A^2}{r^5}+\frac{6\dot A\dot V}{r^3}-\frac{2\dot A}{r^2}\frac{\partial\dot V}{\partial r}-\frac{8\dot A}{r^5}\int_0^r y\dot V dy\right]d\xi_1\,d\xi_2$$

$$+\int_0^t\int_0^t\int_0^t\psi_{12}\left[18\frac{\dot A^3}{r^7}-\frac{29\dot A^2\dot V}{r^5}+\frac{5\dot A^2}{r^4}\frac{\partial\dot V}{\partial r}+\frac{54\dot A^2}{r^7}\int_0^r y\dot V dy\right]d\xi_1\,d\xi_2\,d\xi_3, \qquad (12.41)$$

where after (12.39) the display of time parameters ξ_1, ξ_2, ξ_3 has been omitted from the kernels. In the process of obtaining (12.38), (12.39) and (12.40) from (12.34), (12.35), and (12.31), (12.32), the terms containing nonlinear functions of V were omitted. Inserting (12.39) and (12.41) into (12.38) yields a nonlinear differential-integral equation for $V(r, t)$ as a function of r. This equation is too complicated to be solved. However, by further neglecting terms containing the product of V and A and the product of V and A^2 in (12.39), (12.40) and (12.41), the equation reduces to the following form:

$$\frac{\partial}{\partial r}\int_0^t \left[\psi_1(t-\xi_1)+\frac{2}{3}\psi_2(t-\xi_1)\right]\dot{V}(r,\xi_1)\,d\xi_1$$

$$= \frac{4}{3r^5}\int_0^t\int_0^t \psi_6(t-\xi_1, t-\xi_2)\,\dot{A}(\xi_1)\,\dot{A}(\xi_2)\,d\xi_1\,d\xi_2$$

$$-\frac{12}{r^7}\int_0^t\int_0^t\int_0^t \psi_{12}(t-\xi_1, t-\xi_2, t-\xi_3)\dot{A}(\xi_1)\dot{A}(\xi_2)\dot{A}(\xi_3)\,d\xi_1\,d\xi_2\,d\xi_3. \quad (12.42)$$

Integrating (12.42) with respect to r

$$\int_0^t \left[\psi_1(t-\xi)+\frac{2}{3}\psi_2(t-\xi)\right]\dot{V}(r,\xi)\,d\xi = Z_0(t)+\frac{Z_1(t)}{r^4}+\frac{Z_2(t)}{r^6}, \quad (12.43)$$

where $Z_0(t)$ is the integration constant (dependent on t only), and

$$Z_1(t) = -\frac{1}{3}\int_0^t\int_0^t \psi_6(t-\xi_1, t-\xi_2)\dot{A}(\xi_1)\dot{A}(\xi_2)\,d\xi_1\,d\xi_2, \quad (12.43a)$$

$$Z_2(t) = 2\int_0^t\int_0^t\int_0^t \psi_{12}(t-\xi_1, t-\xi_2, t-\xi_3)\dot{A}(\xi_1)\dot{A}(\xi_2)\dot{A}(\xi_3)\,d\xi_1\,d\xi_2\,d\xi_3. \quad (12.43b)$$

Equation (12.43) is a differential-integral equation of $V(r, t)$ which can be solved as follows. Taking the Laplace transform (see Appendix A4) of (12.43) yields

$$s\left(\hat{\psi}_1+\frac{2}{3}\hat{\psi}_2\right)\hat{V} = \hat{Z}_0+\frac{\hat{Z}_1}{r^4}+\frac{\hat{Z}_2}{r^6}. \quad (12.44)$$

Let

$$\hat{\beta}(s) = 1/s\left(\hat{\psi}_1+\frac{2}{3}\hat{\psi}_2\right). \quad (12.45)$$

Introducing (12.45) into (12.44)

$$\hat{V} = \left(\hat{Z}_0 + \frac{\hat{Z}_1}{r^4} + \frac{\hat{Z}_2}{r^6}\right)\hat{\beta}. \tag{12.46}$$

The inverse Laplace transform of (12.46) is found by using the convolution theorem, no. 29 of Table A4.1,

$$V(r, t) = \int_0^t \left[Z_0(\xi) + \frac{Z_1(\xi)}{r^4} + \frac{Z_2(\xi)}{r^6}\right] \beta(t-\xi)\, d\xi. \tag{12.47}$$

When the form of $\hat{\psi}_1(s)$ and $\hat{\psi}_2(s)$ are known $\beta(t)$ may be determined from (12.45) by taking the inverse Laplace transform. The resulting function may then be used in place of $\beta(t)$ in the following expressions.

Inserting (12.47) into (12.30), (12.31) and (12.32) yields

$$u(r, t) = \frac{A(t)}{r} + \int_0^t \left[\frac{r}{2} Z_0(\xi) - \frac{Z_1(\xi)}{2r^3} - \frac{Z_2(\xi)}{4r^5}\right] \beta(t-\xi)\, d\xi, \tag{12.48}$$

$$\varepsilon_r(r, t) = -\frac{A(t)}{r^2} + \frac{1}{2}\int_0^t \left[Z_0(\xi) + \frac{3}{r^4} Z_1(\xi) + \frac{5}{2r^6} Z_2(\xi)\right] \beta(t-\xi)\, d\xi, \tag{12.49}$$

$$\varepsilon_\theta(r, t) = \frac{A(t)}{r^2} + \frac{1}{2}\int_0^t \left[Z_0(\xi) - \frac{1}{r^4} Z_1(\xi) - \frac{1}{2r^6} Z_2(\xi)\right] \beta(t-\xi)\, d\xi. \tag{12.50}$$

Inserting (12.49), (12.50) into (12.34), (12.35), (12.36) yields the stress components σ_r, σ_θ, σ_Z, as follows:

$$\sigma_r(r, t) = \int_0^t \left[\psi_1 \dot{V} + \psi_2 \left(\frac{2}{3}\dot{V} - \frac{\dot{A}}{r^2} - \frac{1}{r^2}\int_0^r y\dot{V}\, dy\right)\right] d\xi_1$$

$$+ \int_0^t\int_0^t \frac{1}{3}\psi_6 \cdot \frac{\dot{A}^2}{r^4}\, d\xi_1\, d\xi_2 - \int_0^t\int_0^t\int_0^t 3\psi_{12}\frac{\dot{A}^3}{r^6}\, d\xi_1\, d\xi_2\, d\xi_3$$

$$= Z_0(t) + \frac{Z_1(t)}{r^4} + \frac{Z_2(t)}{r^6} - \int_0^t \psi_2(t-\xi)\left\{\left[\frac{Z_0(\xi)}{2} - \frac{Z_1(\xi)}{2r^4} - \frac{Z_2(\xi)}{4r^6}\right]\beta(0)\right.$$

$$\left. + \int_0^\xi \left[\frac{Z_0(x)}{2} - \frac{Z_1(x)}{2r^4} - \frac{Z_2(x)}{4r^6}\right]\frac{\partial\beta(\xi - x)}{\partial \xi}\, dx\right\} d\xi$$

$$-\frac{1}{r^2}\int_0^t \psi_2(t-\xi)\,\dot{A}(\xi)\mathrm{d}\xi+\frac{1}{3r^4}\int_0^t\int_0^t \psi_6(t-\xi_1,\,t-\xi_2)\,\dot{A}(\xi_1)\,\dot{A}(\xi_2)\,\mathrm{d}\xi_1\,\mathrm{d}\xi_2$$

$$-\frac{3}{r^6}\int_0^t\int_0^t\int_0^t \psi_{12}(t-\xi_1,\,t-\xi_2,\,t-\xi_3)\,\dot{A}(\xi_1)\,\dot{A}(\xi_2)\,\dot{A}(\xi_3)\,\mathrm{d}\xi_1\,\mathrm{d}\xi_2\,\mathrm{d}\xi_3, \tag{12.51}$$

$$\begin{aligned}\sigma_\theta(r,t)=&\int_0^t\left[\psi_1\dot{V}+\psi_2\left(-\frac{1}{3}\dot{V}+\frac{\dot{A}}{r^2}+\frac{1}{r^2}\int y\dot{V}\,\mathrm{d}y\right)\right]\mathrm{d}\xi\\&+\int_0^t\int_0^t\frac{1}{3}\psi_6\frac{\dot{A}^2}{r^4}\,\mathrm{d}\xi_1\,\mathrm{d}\xi_2+\int_0^t\int_0^t\int_0^t\frac{5}{3}\psi_{12}\frac{\dot{A}^3}{r^6}\,\mathrm{d}\xi_1\,\mathrm{d}\xi_2\,\mathrm{d}\xi_3,\\&=Z_0(t)+\frac{Z_1(t)}{r^4}+\frac{Z_2(t)}{r^6}\\&-\int_0^t\psi_2(t-\xi)\left\{\left[\frac{Z_0(\xi)}{2}+\frac{3}{2}\,\frac{Z_1(\xi)}{r^4}+\frac{5}{4}\,\frac{Z_2(\xi)}{r^6}\right]\beta(0)\right.\\&\left.+\int_0^\xi\left[\frac{Z_0(x)}{2}+\frac{3}{2}\,\frac{Z_1(x)}{r^4}+\frac{5}{4}\,\frac{Z_2(x)}{r^6}\right]\frac{\partial\beta(\xi-x)}{\partial\xi}\,\mathrm{d}x\right\}\mathrm{d}\xi\\&+\frac{1}{r^2}\int_0^t\psi_2(t-\xi)\,\dot{A}(\xi)\,\mathrm{d}\xi\\&+\frac{1}{3r^4}\int_0^t\int_0^t\psi_6(t-\xi_1,\,t-\xi_2)\,\dot{A}(\xi_1)\,\dot{A}(\xi_2)\,\mathrm{d}\xi_1\,\mathrm{d}\xi_2\\&+\frac{5}{3r^6}\int_0^t\int_0^t\int_0^t\psi_{12}(t-\xi_1,\,t-\xi_2,\,t-\xi_3)\,\dot{A}(\xi_1)\,\dot{A}(\xi_2)\,\dot{A}(\xi_3)\,\mathrm{d}\xi_1\,\mathrm{d}\xi_2\,\mathrm{d}\xi_3.\end{aligned} \tag{12.52}$$

$$\sigma_z(r,t)=\frac{2Z_1(t)}{r^4}+\int_0^t(\psi_1-\tfrac{1}{3}\psi_2)\dot{V}\,\mathrm{d}\xi, \tag{12.53}$$

where x is a dummy time variable.

The deformation, strains and stresses as shown in (12.48), (12.49), (12.50), (12.51), (12.52) and (12.53) depend on two undetermined time functions $A(t)$ and $Z_0(t)$ which can be evaluated from the appropriate boundary conditions.

Deformation boundary conditions. If deformation boundary conditions are specified,

$$\begin{aligned} u(r, t) &= U_1(t) \quad \text{at} \quad r = a, \\ u(r, t) &= U_2(t) \quad \text{at} \quad r = b. \end{aligned} \tag{12.54}$$

Inserting (12.54) into (12.48) yields

$$U_1(t) = \frac{A(t)}{a} + \frac{a}{2} \int_0^t \left[Z_0(\xi) - \frac{Z_1(\xi)}{a^4} - \frac{Z_2(\xi)}{2a^6} \right] \beta(t-\xi)\, d\xi,$$

$$U_2(t) = \frac{A(t)}{b} + \frac{b}{2} \int_0^t \left[Z_0(\xi) - \frac{Z_1(\xi)}{b^4} - \frac{Z_2(\xi)}{2b^6} \right] \beta(t-\xi)\, d\xi.$$

Eliminating $Z_0(t)$ from these two equations yields

$$\frac{U_1}{a} - \frac{U_2}{b} = \left(\frac{1}{a^2} - \frac{1}{b^2} \right) A(t) - \frac{1}{2} \int_0^t \left[\left(\frac{1}{a^4} - \frac{1}{b^4} \right) Z_1(\xi) \right.$$
$$\left. + \frac{1}{2} \left(\frac{1}{a^6} - \frac{1}{b^6} \right) Z_2(\xi) \right] \beta(t-\xi)\, d\xi. \tag{12.55}$$

Equation (12.55) is a nonlinear integral equation for $A(t)$. Notice that in this equation both $Z_1(t)$ and $Z_2(t)$ are functions of $A(t)$ as given in (12.43a) and (12.43b).

Stress boundary conditions. If stress boundary conditions are specified,

$$\begin{aligned} \sigma_r(r, t) &= f_1(t) \quad \text{at} \quad r = a, \\ \sigma_r(r, t) &= f_2(t) \quad \text{at} \quad r = b. \end{aligned} \tag{12.56}$$

Inserting (12.56) into (12.51) after eliminating $Z_0(t)$ yields

$$f_1(t) - f_2(t) = Z_1 \left(\frac{1}{a^4} - \frac{1}{b^4} \right) + Z_2 \left(\frac{1}{a^6} - \frac{1}{b^6} \right)$$

$$+ \int_0^t \psi_2(t-\xi) \left\{ \left[\frac{Z_1(\xi)}{2} \left(\frac{1}{a^4} - \frac{1}{b^4} \right) + \frac{Z_2(\xi)}{4} \left(\frac{1}{a^6} - \frac{1}{b^6} \right) \right] \beta(0) \right.$$

$$+\int_0^{\xi}\left[\frac{Z_1(x)}{2}\left(\frac{1}{a^4}-\frac{1}{b^4}\right)+\frac{Z_2(x)}{4}\left(\frac{1}{a^6}-\frac{1}{b^6}\right)\right]\frac{\partial(\xi-x)}{\partial\xi}\,dx\Bigg\}\,d\xi$$

$$-\left(\frac{1}{a^2}-\frac{1}{b^2}\right)\int_0^t \psi_2\dot{A}\,d\xi+\frac{1}{3}\left(\frac{1}{a^4}-\frac{1}{b^4}\right)\int_0^t\int_0^t \psi_6\dot{A}\dot{A}\,d\xi_1\,d\xi_2$$

$$-3\left(\frac{1}{a^6}-\frac{1}{b^6}\right)\int_0^t\int_0^t\int_0^t \psi_{12}\dot{A}\dot{A}\dot{A}\,d\xi_1\,d\xi_2\,d\xi_3, \tag{12.57}$$

Again, (12.57) is a nonlinear integral equation for $A(t)$.

After determining $A(t)$ and then $Z_0(t)$ from given boundary conditions using either (12.55) or (12.57) substitution of $A(t)$ and $Z_0(t)$ into (12.51), (12.52) and (12.53) yields the distribution of the required stresses σ_r, σ_θ and σ_z both with respect to space and time.

Other approaches to the solution of thick walled tubes in the nonlinear range are found in [12.4, 12.5].

CHAPTER 13

EXPERIMENTAL METHODS

13.1 Introduction

In this chapter some general principles and precautions which should be considered in setting up experimental apparatus and executing creep and stress relaxation experiments will be discussed. Specific apparatus (especially electrical and electronic) will not be discussed in detail, since there are so many alternatives and new instrumentation is developing so rapidly. No attempt will be made to discuss all methods which have been used nor to describe simplified or rough methods. Only methods which would be useful when high precision is desired will be considered. For a general discussion of mechanical testing see references [13.1], [13.2].

13.2 Loading Apparatus for Creep

While a creep test is generally considered to be a constant stress experiment, constant load tests are usually performed and usually satisfactory. Under uniaxial tension the cross-sectional area decreases during a test; hence the stress changes if the load is constant. However, the amount of creep strain that is of interest for structural applications requiring high precision is seldom more than one to three percent. In this case the cross-sectional area will change as much as one to three percent if the material is incompressible and less if it is compressible. Thus constant load tests are usually performed and the change in stress under constant load is neglected. When this is not permissible the load may be programmed to produce constant stress. The programming may be approximate by adjusting it to correspond to predetermined changes in cross-section as a function of axial strain or more exact by using measured lateral strain as input to servo control of load. Either method is cumbersome and will not be considered further. For torsion creep tests changes in cross-sectional area are negligible or zero.

For constant load creep testing dead weight loading is the simplest and generally the most satisfactory. For some situations servo control may be used because it is available or for some other reason, such as to provide a controllable and repeatable loading pattern during application of load. The simplest dead weight loading is direct application of a weight suspended from one end of a specimen. If the required weight exceeds, say, 50 pounds this method is not very satisfactory because of difficulty of applying the load without undesirable disturbances. For greater loads (or any load) dead weights applied to a lever may be employed. One advantage of dead weight loading is that the stress may be determined directly from the load and geometry of the specimen; otherwise the force on the specimen must be determined by a load cell or other force measuring device.

If the stress is to be determined from the weight (or force) applied to the end of a lever the lever must be properly designed. It must be essentially frictionless. Ball and roller bearings are seldom satisfactory. Knife edges and seats of good quality properly supported to avoid deflection are perhaps the most reliable. Flexible pivots may be used, but their stiffness may affect the load on the specimen at small loads or large deflections. For a creep machine the lever is not designed like a beam balance. It must have the axes of rotation of all three knife edges in the same plane, the center of gravity of the lever must also lie in this plane, and the lever must be sufficiently stiff that the above relationships are not disturbed by bending.

Friction must be kept to a minimum if the stress is to be controlled and measured by means of dead weights. For this purpose the knife edges and seats must be carefully made and must be properly aligned with each other. Application of Molycote to the knife edges is also helpful. The friction should always be checked as for example by first balancing the lever and then applying small weights to a scale pan at the loading end of the lever to determine how much weight is required to cause motion. Preferably this would also be done with the lever supporting a substantial weight balanced by another weight on the other end of the lever.

Other methods of load application include hydraulic or pneumatic (by means of pistons) and power screws. Pistons involving sliding seals, rings or glands in the small sizes required generally involve too much friction to either control the load or determine the stress from the fluid pressure. One exception is a rolling diaphragm type of piston. When devices for applying the load, such as pistons, which involve friction are employed the load must be measured separately by means of a load measuring device placed between the specimen and any apparatus which is subject to friction. If a lever is involved which does not have negligible friction the load cell must be placed between the specimen and the lever.

The supporting frame of the creep machine must be stiff enough that its deflection under load will not disturb the load measuring or strain measuring systems.

13.3 Load Application

In the analysis and interpretation of creep tests it is usually assumed that the load is applied instantaneously and remains constant after application. This is of course impossible to achieve because of the dynamics of the spring-mass system involved in a creep machine. Thus in practice the load is applied gradually over a short time span so as to avoid vibration or overshooting the desired load. If weights are applied by hand this must be done with care and consistently so as to apply the load each time in approximately the same time (a few seconds) and so as to avoid bouncing, vibration or swinging of the weight. Automatic systems of load application must accomplish the same ends. When servo control is employed a step change in input is not a satisfactory means of load application if overshoot and vibration result. Instead programmed loading may be employed or the load may be applied by manual control by first balancing the servo under zero load then applying the load by rotating a continuously variable control such as zero suppression until the desired load is reached.

While different rates of loading and careless load application will undoubtedly affect the resulting creep, the authors have had no difficulty achieving repeatable and consistent results by using careful hand application of loads.

13.4 Test Specimen

The most common specimens for creep testing are rods or thin-walled tubes of circular cross-section having enlarged ends with suitable means for attachment to the testing machine. In creep tests under torsion or internal pressure thin-walled tubular specimens have almost always been used to achieve uniformity in stress distribution (see Section 13.5). For all these specimens it is important that the cross-sectional area be uniform over the gage length covered by the extensometer within say 1/2 percent. For thin-walled tubes the most critical measurement is the wall thickness. This should be measured directly, not inferred from inside and outside diameters. The wall thickness should be measured at several equally spaced positions around the circumference at each interval. The average of these readings should be used in computing the area. One means of measuring the wall thickness is to insert a steel ball (as from a ball bearing) in the surface of a stiff mandrel, place the tubular specimen over the mandrel, mount the mandrel between centers, as in a lathe, and measure the thickness with a one ten-thousandths dial gage or other sensitive instrument suitably mounted directly over the steel ball.

13.5 Uniform Stressing or Straining

Uniform stressing is very important in creep testing in order to permit accurate determination of stress and strain. To achieve uniform stressing in either tension, compression or torsion it is necessary to eliminate bending stresses as far as possible. Some of the means by which this may be accomplished are as follows. Suitable low friction shackles such as crossed knife edges [13.3] may be employed in which provision is made for adjusting the position of the crossing point with respect to the specimen axis. Such crossed knife edges or their equivalent may be employed at each end of the specimen in tension or compression [13.4]. To position them properly means must be provided to observe any bending which occurs when a small test load is applied. This may be accomplished by using four strain gages on the specimen located so as to indicate bending or by using a bending detector [13.3] which is clamped to the specimen while the knife edges are adjusted and then re-

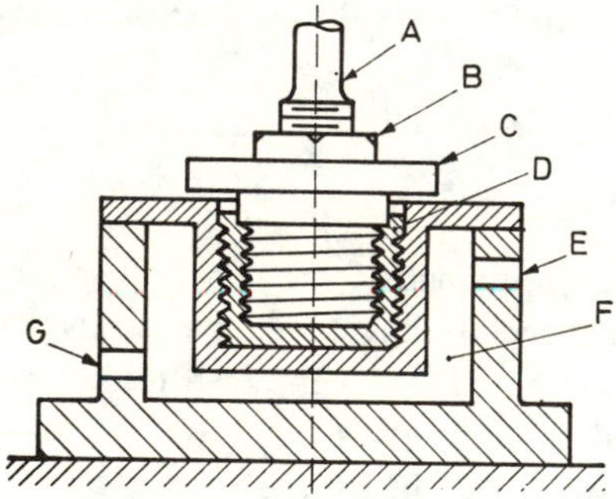

Fig. 13.1. Wood's Metal Grip.
A. Specimen
B. Lock nut
C. Socket
D. Wood's metal
E. Inlet for heating and cooling medium
F. Heat transfer chamber
G. Outlet for heating and cooling medium

moved, see Section 13.11. Another method which may be employed with a testing machine whose frame and loading mechanism are such that no bending of the specimen is induced by the machine during loading is as follows. This method employs a cup of Wood's metal which forms part of the lower specimen grip [13.5]. The Wood's metal is in a molten state while the upper end of the specimen is being fixed in the machine. Solidification of the Wood's metal causes the lower end of the specimen to be locked in the machine without introducing bending or twisting. See Fig. 13.1. Specimens which are to be tested at room temperature must be isolated from the heat of the Wood's metal grip if this grip is to be used.

Another problem in uniform stressing arises in torsion and in the circumferential stress induced by internal pressure in tubular specimens. In both cases a stress gradient and strain gradient are inherent in the method of stressing. They cannot be eliminated—only their effects minimized by making the wall thickness to diameter ratio as small as possible. This ratio cannot be made as small as might be desired because of buckling [13.6] in torsion (or compression) and because the wall thickness should be many times (20 perhaps) the grain size or cell size of the material.

In torsion the strain gradient is a linear function of the radial distance from the axis of the tube, as prescribed by geometry. The stress gradient will also be linear for elastic or linearly viscoelastic material but nonlinear for nonlinear material. These gradients should be taken into account when considering a bulk phenomenon such as creep to avoid errors of 6 percent or more even in tubes having a ratio of wall thickness to diameter of 0.06 made of linear material. One way to take these gradients into account is to average the values computed at the inner and outer surfaces of the tube assuming a linear variation through the wall thickness. This assumption is true, except for stress, when the material is nonlinear. The resulting equations for torsion in the linear range are as follows.

The average shearing stress τ_{av} due to torque T of a tubular specimen is

$$\tau_{av} = \frac{32\mathrm{T}(D-h)/2}{\pi(D^4-d^4)}, \tag{13.1}$$

and the average engineering shearing strain γ_{av} due to angle of twist θ is

$$\gamma_{av} = (\theta/l)(D-h)/2, \tag{13.2}$$

where D, d, h and l are the outside diameter, inside diameter, wall thickness and gage length, respectively.

Internal pressure in tubes generates nonlinear stress and strain distributions even in thin-walled tubes of linear material. The circumferential stress at the inner surface of an elastic tube for which $h/D = 0.06$ is about 15 percent greater than the circumferential stress at the outer surface, and the corresponding strains for a material having Poisson's ratio equal 0.4 differ by about 20 percent. In both stress and strain the deviation from a linear distribution is about one percent. For a linearly viscoelastic or elastic material the circumferential stress σ_θ at the outer surface may be computed from (6.63) by making $r = b$. The circumferential strain ε_θ at the outer surface may be computed from the observed radial displacement u by substituting b for r in (6.70). However, for nonlinear material of unknown characteristics the stress distribution will be unknown so that the accuracy of stress calculated from (6.63) will be unknown. The surface strain calculated from (6.70) is valid, however. The strain distribution for nonlinear viscoelastic material depends strongly on the material characteristics. Thus it is even more uncertain than the stress distribution. Hence, there appears to be no simple method of calculating the circumferential stress and strain for nonlinear material that is any more accurate than described above for linear material.

13.6 Strain Measurement

Resistance strain gages may be employed but have certain disadvantages. The stiffening effect of the gage and the local heating due to the current in the gage are of great importance for plastics [13.7]. Another factor of importance in combined stress experiments is that it is very difficult to align the axis of strain gages accurately enough so that different components of strain measurement are completely separated.* Errors due to such misalignment even as large as two percent are difficult to avoid. For long-time experiments the stability of resistance strain gages becomes important. Factors such as creep of the adhesive or creep of the gage backing can result in unacceptable drift of the gage output. When strain gages must be used, the longer the gage the better for resistance to drift. Also, foil gages show less drift than wire gages.

Perhaps the most reliable method is the use of an extensometer covering a gage length of at least 2 inches. The extensometer should average the strain in the specimen so that it is not affected by any bending which may occur. It should have high sensitivity, stability and accuracy. For combined stress

* For example a false indication of twist may be indicated under a purely axial loading of a circular tube.

experiments such as combined tension and torsion it should be designed so that strain in one mode (say tension) does not produce false indications of strain in the other mode (say torsion). It should also be free of frictional effects. Frictional effects should be verified by (for example) placing small loads on the specimen in the elastic range and removing them to verify that the strain indicated by the extensometer returns to the original value. By reducing the size of the test loads the sensitivity of the extensometer may be verified. For this purpose a fairly stiff specimen should be used; otherwise an apparent lack of sensitivity may be due to friction in the loading system rather than the extensometer. An extensometer which the authors have found satisfactory for tension, compression and combined tension and torsion is described in [13.3].

For some situations it is convenient to employ a strip chart recorder or even digital recorder. The latter may be preferred since the data must usually be reduced to digital form. When manual readings are employed the frequency of observation must be considered. Because of the decreasing rate of creep in at least the early stages it is important that observations be made frequently for say the first hour after which the interval may be increased.

Measurement of circumferential strains in plastic tubes under internal pressure may be accomplished by means of an instrument consisting of four direct current differential transformers mounted on an invar ring so as to sense the change in diameter at two orientations 90° apart [13.8].

13.7 **Temperature Control**

Good temperature control is necessary not only because temperature affects creep rate but because temperature fluctuations cause fluctuations in thermal expansion. Thermal expansion is difficult to isolate from creep strains, except for shear strain (torsion) which is insensitive to thermal expansion.

Various means of control of temperature have been employed. At elevated temperatures the most difficult problem is to achieve uniform temperature over the gage length of the specimen. The stack effect of rising air currents causes great difficulty. When a fluid heat-transfer medium may be used the temperature distribution is easier to control, but the creep of many plastics and other materials is affected by absorption of fluids.

The following may be used to improve the temperature distribution: rapid air circulation, introducing heat at more than one position along the gage length, surrounding the specimen surface with high conductivity metal, such as copper, placed so as to allow only a small air space between the specimen and enclosure, and a forced flow of heated air downward around the specimen to counteract the stack effect.

Temperature sensing instrumentation may employ resistance elements (thermistors) or voltage elements (thermocouples). Usually the latter are more suitable than the former because their small size makes possible better thermal contact with the specimen. Closed loop servotype automatic controllers are to be preferred over on-off or proportional type controllers for controlling the heat input. When heat is introduced by separate elements at different positions along the specimen it may be advantageous to employ a controller for each heater, however, the controllers may interfere with each other.

13.8 Humidity and Temperature Controlled Room

Both temperature and relative humidity must be controlled during creep or relaxation experiments on materials which absorb moisture, such as many plastics, concrete and wood. For experiments at room temperature the temperature and humidity of the room itself may be controlled.

There are many ways of approaching the design of a temperature and humidity controlled room. Factors to consider in the room itself include the following. Where possible a room in the interior of a building which has temperature control of the building is desirable. If this is not possible exterior walls and ceilings should be heavily insulated and exterior windows avoided. An entry vestibule is desirable to minimize disturbances due to opening the door. To permit good control of humidity throughout the room all surfaces of the enclosure should have a moisture barrier. Where this cannot be installed as part of the construction the application of a low permeability enamel is beneficial.

For room temperature control heating and cooling are generally required, though in some situations heating only may be sufficient. With modern thermostats and proper air circulation the temperature may be controlled within $\pm\frac{1}{2}$°F ($\pm\frac{1}{4}$°C). Electrical heating and mechanical refrigeration are usually the easiest to install and the most reliable.

Control of relative humidity is the feature which introduces the most problems. All systems involve circulation of air through some type of duct work which may also serve to distribute the air uniformly throughout the room. Means of control of humidity include the following:

(a) All air may be passed through a chilling coil whose surfaces are covered with water (from a water spray). Air emerges from this coil at a lowered temperature and saturated (100 percent relative humidity). This air is next passed over a heating coil where it is reheated to the desired room temperature. In so doing the relative humidity is reduced a predictable amount. Since the amount of moisture saturated air will hold is a function of its temperature, the relative humidity of the reheated air can be controlled by controlling the temperature of the saturated air from the chilling coil.

(b) Another method is to pass all outside make-up air plus a fixed portion of the return air from the room through a dehumidifier and then through a water spray [13.9]. In this method the dehumidifier and water spray are controlled by a humidistat so that only the function needed at the time is operating. The dehumidifier consists of a small compressor pumping refrigerant from an expansion (cooling) coil and compressing it into a condensing (heating) coil. Both coils are placed in series in the air duct so that the air to be dehumidified is first chilled and its moisture removed; then it is reheated by the same heat which was removed from the cooling coil and pumped into the heating coil. Thus the net change in temperature of the air stream is small. With a sensitive humidistat and suitable switching the relative humidity can be controlled by this system within ± 1 percent. In this system a separate larger compressor is used to cool the air when needed. This compressor and electric heaters are controlled by a thermostat so that either heating or cooling, as required, is supplied to the circulating return air from the room.

13.9 Internal Pressure

Experiments using internal pressure in thin-walled tubes to achieve biaxial stress conditions encounter special problems. The pressurizing medium may affect the creep behavior. This is especially true of liquids and is even true of some gases [13.10]. When gases are used as the pressurizing medium heat is generated when the pressure is applied to the gas. The resulting temperature rise is more than usually can be tolerated. Of course stressing solid specimens also causes a temperature change, but the magnitude of the change is usually negligible. The temperature change resulting from use of gases as a pressurizing medium may be made negligible by inserting a high-thermal-conductivity core bar inside the specimen so as to take up almost all the space [13.10]. Thus the volume of gas to be compressed is small and the heat liberated is quickly absorbed by the core bar.

13.10 **Strain Control and Stress Measurement for Relaxation**

There are at least three approaches to strain control for relaxation experiments. When the stiffness of the loading mechanism including everything outside the gage length of the specimen—grips, ends of specimen, dynamometer, etc.—is very much greater than the stiffness of the gage length portion of the specimen no control apparatus is needed if thermal expansion is not involved. For example, the strain of a long sample of rubber-like material when stretched considerably over a stiff frame will remain essentially constant. An Instron testing machine is often employed in this manner.

Another approach is to conduct a relaxation test as a succession of steps of constant load creep such that whenever the creep strain exceeds a specified deviation from the desired relaxational strain the load is reduced to return the strain to the desired value. This may be accomplished automatically by a beam balance or manually by dead weights.

The third approach makes use of modern automatic control and is the most desirable means. This method requires a servo controlled loading system (closed loop) in which the strain measuring device is the output sensor for the feed-back loop [13.11]. This system is not an automated version of the second approach since with proper equipment the strain may be maintained constant within the ability of the strain sensor to detect changes. Such a system usually requires a strain sensor whose output is an electrical signal (such as a DC voltage) which is proportional to the strain. Thus both high sensitivity and great stability are required of the electrical system. An example of this method is described in [13.12].

Measurement of stress in a relaxation apparatus may be accomplished in several ways. A load cell having adequate sensitivity, stability, low hysteresis and high stiffness may be employed in the first and third methods listed above. In the second method the stress may be determined from the load indicated by the beam balance or dead weight.

Since strain is held constant in a relaxation test, variations in temperature will cause changes in stress to accommodate the thermal expansion which would otherwise occur in tension or compression experiments. This change in stress is not due to relaxation of course. Hence it is necessary to maintain constant temperature.

13.11 A Machine for Combined Tension and Torsion

As an example the machine [13.3] shown in Fig. 13.2 will be described briefly. This machine was designed to permit creep or stress relaxation tests in tension and/or torsion and/or internal pressure. It also permitted simultaneous creep in one loading mode and relaxation in another mode, or tests under varying load or strain. It could accommodate both low and high strength materials and small or large strains.

The loading was accomplished by a lever, A Fig. 13.2, for tension and a lever B, pulleys C, and steel bands D for torsion. Either dead weights applied at the ends of the levers or servo control of loads were used [13.12]. In the latter a hydraulic cylinder and load cell replaced the dead weights at the end of the lever. The pivots of the levers and pulleys were provided with knife edges. Beams, E Fig. 13.2 (also provided with knife edges), joined the tension and torsion levers. These beams could be loaded at any point in their spans, thus producing proportional loading in tension and torsion, or unloading in tension and loading in torsion.

Internal pressure in the specimen was controlled by a servomechanism operating on a hydraulic cylinder which controlled the pressure of a suitable gas [13.10]. The pressure was sensed by means of a manganin cell.

Stresses were computed from the known loads applied to the specimen as determined from the dead weights or output of load cells, if used.

Axial alignment of the specimen was achieved by adjusting the crossing point of two sets of crossed knife-edge shackles, F Fig. 13.2, in accordance with bending indicated by the bending detector shown in Fig. 13.3. The bending was checked by applying a small test load to the specimen. The bending detector consisted of eight differential transformers, A Fig. 13.3, arranged as shown so that the difference in output of opposite pairs indicated bending in the plane of the pair. The instrument shown was used to indicate the bending in either of two gage lengths 2 inches long. Adjustment of the knife-edge shackles so as to minimize bending in both gage lengths tended to eliminate bending over the entire 4-inch gage length.

A mechanism described in [13.3] was employed to permit large axial strains without disturbing the position of the torque bands. This would otherwise cause unknown axial forces to be produced by the torque mechanism. An air bearing was used to support the weight of the upper torque drum, G Fig. 13.2, and a special low-friction hydraulic thrust bearing, H Fig. 13.2, was incorporated to permit a known torque to be applied to the specimen while the specimen was also supporting a large tensile load.

Fig. 13.2. Machine for Combined Tension and Torsion.
A. Tension lever
B. Torsion lever
C. Pulleys for torsion
D. Bands for torsion
E. Beam for proportional loading
F. Crossed knife-edge shackles
G. Upper torque drum
H. Hydraulic thrust bearing
I. Differential transformer for axial strain measurement

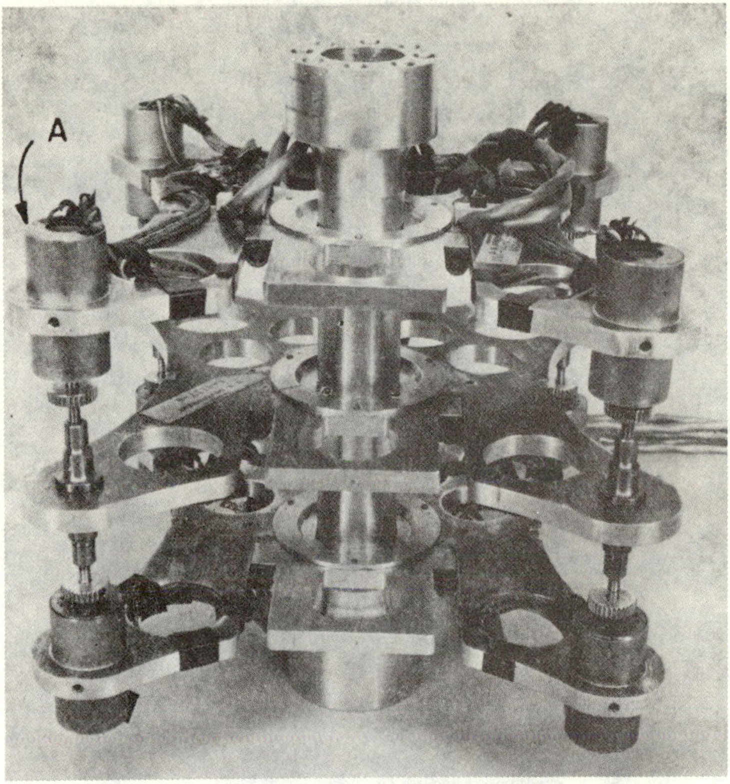

Fig. 13.3. Bending Detector. From [13.3], courtesy ASTM.
A. Differential transformers

Axial and torsional strains were determined by use of the instrument shown in Fig. 13.4. It consisted of two pairs of invar rods which transferred the motion of the upper and lower gage sections of the specimen to location I Fig. 13.2, below the lower pull-rod and above the lower knife-edge shackle, at which the motion was compared. The axial relative movement was determined by a differential transformer of one inch travel, A Fig. 13.4b. The angle of twist was determined by measuring with a measuring microscope the relative positions of reference lines on the drum B and pointer C Fig. 13.4b. The drum and core of the differential transformer were attached to the lower end of a pair of rods attached to the upper gage section of the specimen A Fig. 13.4a by means of spring loaded points on a pair of rings arranged so as to act as a gimbal, B Fig. 13.4a. The pointer and coil of the differential transformer were attached in a similar manner to the lower gage section of

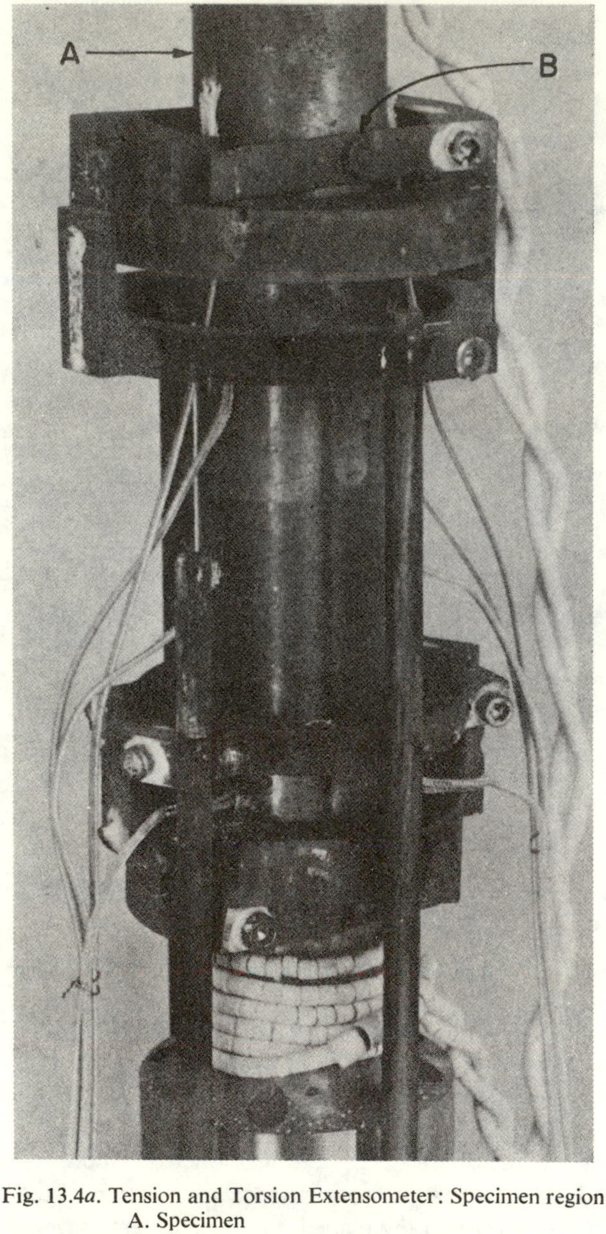

Fig. 13.4*a*. Tension and Torsion Extensometer: Specimen region.
A. Specimen
B. Gimbaled rings and gage points

Fig. 13.4*b*. Tension and Torsion Extensometer: Lower region.
A. Differential transformer
B. Drum for twist measurement
C. Pointer for twist measurement

the specimen. The weight of each section of this instrument was balanced by supporting it by a pair of very flexible coil springs. This flexibility insured that straining the specimen during a test would not significantly affect the balance. At room temperature circumferential strain was measured by four DCDT's mounted on an invar ring [13.8].

Elevated temperature was maintained by means of servo type temperature controllers sensing the emf of thermocouples and controlling the heat input either externally by means of an oven or internally by means of a quartz-tube heating lamp or resistance heaters in a copper core bar. Generally three separately controlled heat sources were required, one within the gage length and one outside each end of the gage length.

APPENDIX A 1

LIST OF SYMBOLS

Symbol	*Meaning*	*Page of first appearance for each item, respectively*
A	Constant, parameter, area	13, 62, 110
A_i	Parameter in a sum	77
A_1 to A_6	Kernel functions	315
$A(t)$	Time function	280
$\{A\}$, $\{A'\}$, $\{\mathsf{A}\}$, $\{\mathsf{A}'\}$	Functions of K_α, σ_{ij}	180, 181, 183, 188
a	Number of terms, inside radius of tube, constant	65, 122, 170
a_i	A vector, coefficient in a series	23, 333
a_0, a_1, a_2	Material constant	88, 256
a_σ, a_ε	Shift factors	172, 173
a_T	Temperature shift factor	105
a_1, a_2, a_3	Functions of temperature	256
B	Constant, parameter	13, 125
B_1 to B_4	Kernel functions	315
B'_1, B'_2, B'_3, B_1, B_2, B_3	Functions of K_α and principal stresses	187, 179
$\{B\}$, $\{B'\}$, $\{\mathsf{B}'\}$, $\{\mathsf{B}'\}$	Functions of K_α, σ_{ij}	180, 181, 183, 188
b	Number of terms, width of beam, outside radius of tube, constant	71, 111, 122, 170
b_j	Coefficient in a series	333
b_0, b_1	Material constant	88, 256
b_1, b_2, b_3	Functions of temperature	256
C	Parameter	127
C_i	Stress rate	247
C_1, C_2	Kernel functions, constants	315, 334

Symbol	*Meaning*	*Page of first appearance*
C_{ijkl}	A 4th order tensor describing the elastic moduli	46
$C_{ijkl}(t)$	Creep functions for anisotropic material	87
$\{C\}, \{C'\}, \{\mathsf{C}\}, \{\mathsf{C}'\}$	Functions of K_α, σ_{ij}	180, 181, 183, 188
c	Constant, number of terms	5, 77
D, D	Constant, differential operator, parameter, outside diameter, denominator	9, 65, 204, 293, 333
D_1	Kernel function	315
d	Depth of beam, diameter, inside diameter	111, 268, 293
d_{ij}, $\mathbf{d}$	Deviatoric strain tensor	38
E, E_e	Elastic modulus	14, 114
E_1, E_2	Storage modulus and loss modulus	92
E_{kl}	Green's strain tensor	36
E_v	Equivalent modulus for viscoelastic material (relaxation modulus)	79
E^*	Complex relaxation modulus	92
$E(t)$	Relaxation modulus	81
e	Base of natural logarithm	8
$\boldsymbol{e}_{kl}$	Cauchy's strain tensor	36
$e(t)$	Position of neutral axis	111
$F(\,)$, F	Function of ( ), force	18, 90
$F_{ij}[\,]$	Functional of [ ]	18
F_0	Amplitude of force	90
F_1 to F_7	Combination of kernel functions for creep	158
F_i	Force components; components of body force per unit volume	22, 41
F_θ	Tangential force	125
$\mathsf{F}^+(\sigma)$	Function of stress	257
$\mathcal{F}_1$ to $\mathcal{F}_7$	Combination of kernel functions for relaxation	163
$\mathcal{F}_1'$ to $\mathcal{F}_3'$	Lumped kernel functions for conventional relaxation test	274
$ff,(\,)$	Function of ( )	9

Symbol	*Meaning*	*Page of first appearance*
f, f_m	Frequency	90
f_1, f_2	Boundary stresses	287
G	One of two Lamé elastic constants (the shearing modulus)	46
G^*	Function of time and extension ratio	171
G_1 to G_4	Combination of kernel functions for creep	158
G_{pq}	Function of Green's strain tensor	313
$G_{ij}[\quad]$	Functional of [ ]	18
$G(t)$	Stress relaxation modulus in shear	87
$\mathsf{G}(t)$	A separable time function	257
$\mathcal{G}_1$ to $\mathcal{G}_4$	Combination of kernel functions for relaxation	163
$\mathcal{G}_1^*$, $\mathcal{G}_2^*$	Lumped kernel functions	207
g, $g(\quad)$	Function of ( )	9
g_0, g_1, g_2	Functions of stress	172
g_i, g_j	Special kernel functions	336
$H(t)$	Heaviside function	57
h	Thickness of sheath, wall thickness	124, 293
$h_0, h_1 \ldots, h_\infty$	Functions of strain	173
h	Planck constant	253
I	Moment of inertia of area	112
I_1, I_2, I_3	Invariants of the stress tensor	25
$\mathbf{I} = \delta_{ij}$	The Kronecker delta or unit tensor	24
i, i	Dummy subscript, $\sqrt{-1}$	11, 91
J_1, J_2, J_3	Invariants of the stress deviator tensor	12, 27
$\mathsf{J}(t)$, $J(t)$	Creep compliance	17, 81
J_1, J_2	Storage compliance, loss compliance	93
J^*	Complex creep compliance	92
$\mathcal{J}$	Polar moment of inertia of area	271
j	Dummy subscript	11
K	Bulk modulus of elasticity	48
$K_\alpha(\quad)$, $K'_\alpha(\quad)$	Kernel function for creep	139, 154
$K(t)$	Bulk stress relaxation modulus	87
$K_v(t)$	Viscoelastic resilience modulus	117
K	Boltzmann constant	253

Symbol	*Meaning*	*Page of first appearance*
k	Constant, dummy subscript	8, 25
ksi	1000 pounds per square inch	254
k_1, k_2	Material constants	106
k_e	Resiliance modulus of foundation material	116
k_v	Spring constant of Kelvin foundation material	118
L	Constant	14
$L(\ )$	Kernel function	221
L_{000}, L_{001}, etc.	Kernel function of the time parameters	222
$\mathcal{L}\{\ \}$	Laplace transformation	101
l	Length, dummy subscript, limit	8, 36, 327
M	Bending moment, maximum stress, constant	110, 168, 326
M	Parameter	123
m	Coefficient of time dependent term, index	69, 116
N	Number of terms, numerator	64, 333
N	Parameter	123
n	Time exponent, integer	9, 318
n_j, $\mathbf{n}$	Vector denoting normal to a plane	23, 24
P	Linear differential operator for stress	15
P, P'	Points in a body	35
p	Stress exponent, pressure, circular frequency of natural vibration	8, 48, 90
psi	Pounds per square inch	10
p_a	Material constants	62
p, p'	Displaced positions of points in a body	35
Q	Differential operator for strain	15
$Q_i(\ )$	Kernel function	137
q	Stress exponent, time exponent	14, 89
q_b	Material constants	62
R	Spring constant in mechanical models	52
R_2, R_3, R_4	Remainder	167, 169
$R(\ )$	Kernel function	221
$R(t_R)$	Relaxation spectrum	69
r	Dummy index, number of terms,	71, 83, 122

Symbol	*Meaning*	*Page of first appearance*
	radial coordinate	
r_1, r_2	Roots of an equation	62
$S_i(\ )$	Kernel function	137
$S_{kl}(\ )$	Function of ( )	313
$\mathbf{S}$, $\mathbf{S}^{(a)}$	a-tuple integral	313
s	Laplace transform variable, dummy time variable	59, 172
$s_i = \mathbf{s}$	Tensor of the stress deviator	11
s, s_0	Line segments	36
T	Period of a cycle, temperature	91, 105
T_i	Surface tractions	23
T	Torque	268
t	Time	4
t_c	Retardation time	56
t_R	Relaxation time	54
U_i	Surface displacements	48
u	Displacement in the x or x_1 direction	37
u_i	Displacement vector	37
V	Volume change	280
v	Displacement in the y or x_2 direction	37
W	Energy	95
w	Displacement in the z or x_3 direction, deflection of beam	37, 113
X_i	Fixed Cartesian coordinates	35
X_i	Function of x_j	337
x	Coordinate, reciprocal of retardation time	42, 69
x_i	Coordinate axes, coordinates of deformed position	22, 35
x	Dummy variable	285
Y_i	Function of y_j	337
y	Coordinate,	42
y	Dummy variable	280
$Z_1(t)$, $Z_2(t)$, $Z_3(t)$	Functions of time	284
z	Coordinate	42

Symbol	*Meaning*	*Page of first appearance*
z	Dummy variable	338
α	Angle, parameter, angle of twist per unit length, constant	60, 124, 270, 326
α_i	Functions of p_i	77
$\alpha(T)$	Thermal expansion	250
$[\alpha_i]$	Functions of K_α and σ_{ij}	178
β	Constant, angle	8, 60
β_i	Functions of q_i	77
$\beta_i(t), \beta(t)$	Time functions	170, 384
$\beta_1, \beta_2, \beta_3$	Functions of ψ_i and principal strains	183
$\Gamma, \Gamma_a, \Gamma_T, \Gamma_u$	Surface and portions of surface of a body	48
$\Gamma(\ )$	Gamma function	70
γ, γ_{xy}	Conventional engineering shearing strain	38
γ_i	Components of conventional shear strain	178
$\Delta_1, \Delta_2, \Delta_3$	First, second, third order integral respectively	166
Δf^*	Free energy of barrier	253
δ	Phase angle, loss angle	91
$\delta_{ij} = \mathbf{I}$	The Kronecker delta or unit tensor	23
$\delta(t)$	Dirac delta function	53
$\varepsilon, \varepsilon^e, \varepsilon^c, \varepsilon^r$	Strain, elastic strain, creep strain, recovery strain, respectively	4, 230
$\varepsilon^V, \varepsilon^{VE}$	Irrecoverable and recoverable time dependent strain, respectively	20
ε_v	Volumetric strain	38
$\varepsilon_{\text{I}}, \varepsilon_{\text{II}}, \varepsilon_{\text{III}}$	Principal strains	183
$\dot{\varepsilon}_{\text{I}}$	Principal strain rate	11
$\varepsilon_{ij} = \boldsymbol{\epsilon}$	Strain tensor	18
$\dot{\varepsilon}_{ij} = \dot{\boldsymbol{\epsilon}}$	Strain rate tensor	11
$\varepsilon^\circ, \varepsilon^+$	Time-independent strain, coefficient of time-dependent term, respectively	13, 14
$\varepsilon_{xx}, \varepsilon$	Linear (or normal) strain, shear strain	37
ε^*	Complex strain amplitude	91
$\hat{\varepsilon}$	Non-stress induced volumetric strain	145
$\boldsymbol{\epsilon}^t$	Strain due to stress and time	250

Symbol	*Meaning*	*Page of first appearance*
$\boldsymbol{\epsilon}^T$	Tensor of thermal expansion	250
$\bar{\boldsymbol{\epsilon}} = \mathrm{tr}\varepsilon_{ij}$	Trace of strain tensor	148
$\overline{\boldsymbol{\epsilon\epsilon}} = \mathrm{tr}(\boldsymbol{\epsilon\epsilon})$	Trace of product $\boldsymbol{\epsilon\epsilon}$	148
$\zeta, \zeta_\sigma, \zeta_\varepsilon$	Reduced times	105, 172, 173
η	Coefficient of viscosity in mechanical models	52
θ	Angle, coordinate in cylindrical coordinates	104, 122
$\varkappa$	Unprescribed strain component	164
λ	One of two Lamé elastic constants, distance between equilibrium positions	46, 255
μ_1, μ_2, μ_3	Constants	257
ν	Poisson's ratio, frequency of jump of energy barrier	46, 253
ν_v	Equivalent Poisson's ratio for visco-elastic material	80
ξ	Arbitrary time	17
Π	Product of N-terms	66
ϱ	Mass per unit volume, infinitesimal	42, 320
$\varrho_1, \varrho_2, \varrho_3$	Functions of temperature	257
σ	Stress, normal stress	5.28
σ_e	Effective stress	11
σ_v	Mean normal stress	26
σ^+, σ^0	Material constants	9,14
$\sigma_{\mathrm{I}}, \sigma_{\mathrm{II}}, \sigma_{\mathrm{III}}$	Principal stresses	11
σ^*	Complex stress amplitude	92
$\hat{\sigma}$	Non-strain induced negative pressure	148
$\sigma_{11}, \sigma_{22}, \sigma_{33}$	Normal stresses	11
$\sigma_{12}, \sigma_{23}, \sigma_{31}$	Shearing stresses	11
$\sigma_{ij} = \boldsymbol{\sigma}$	Stress tensor	18
σ_x	Normal stress	24
$\bar{\boldsymbol{\sigma}} = \mathrm{tr}\sigma_{ij}$	Trace of the tensor σ_{ij}	30
$\overline{\boldsymbol{\sigma\sigma}} = \mathrm{tr}(\boldsymbol{\sigma\sigma})$	Trace of the product of two tensors	31
$\overline{\boldsymbol{\sigma\sigma\sigma}} = \mathrm{tr}(\boldsymbol{\sigma\sigma\sigma})$	Trace of the product of three tensors	32
τ_{xy}, τ	Shear stress	24, 28
$\tau_{\mathrm{I}}, \tau_{\mathrm{II}}, \tau_{\mathrm{III}}$	Principal shear stresses	24

Symbol	*Meaning*	*Page of first appearance*
Φ^0, Φ^+	Functions of stress and temperature	256
$\varphi_\alpha, \varphi_{1111}, \varphi_{1112}$, etc.	Kernel functions	132, 244
$\varphi(t)$	Compliance or reciprocal modulus, time-dependent creep function	18, 84
$\varphi(t_c)$	Retardation spectrum	68
Ψ	Angle of twist	268
ψ_a, ψ_b, ψ_c	Kernel functions	206
$\psi_\alpha(\ ), \psi'_\alpha(\ )$	Kernel functions for relaxation	140, 155
$\psi(t)$	Time-dependent stress relaxation function	85
$\psi^{(a)}(\ )$	Kernel functions for relaxation	314
Ω	Curvature, extension ratio	110, 170
Ω_{ij}	Matrix of a triple product of a tensor	31
ω	Angular frequency	90
$(\bar{\ })$	Trace of ( )	30
$(\hat{\ })$	The caret over a symbol indicates Laplace transformation	59
$(\dot{\ }), \dot{\varepsilon}, \dot{\sigma}$	Strain rate, stress rate	4, 54
$(\ddot{\ }), \ddot{\varepsilon}, \ddot{\sigma}$	Second derivative with respect to time	62
$(\ )_i, (\ )_{ij}, (\ )_{kl}$	Dummy indices	11
$(\ )_{\mathrm{I}}, (\ )_{\mathrm{II}}, (\ )_{\mathrm{III}}$	Principal values of ( )	24

APPENDIX A2

MATHEMATICAL DESCRIPTION OF NONLINEAR VISCO-ELASTIC CONSTITUTIVE RELATION

A2.1 Introduction

In Chapter 7 a nonlinear viscoelastic constitutive equation was derived using the concept of the linear superposition principle but extending it to the nonlinear range. In this appendix a formal and more rigorous derivation of a nonlinear viscoelastic constitutive relation will be presented using the principles of continuum mechanics. The basic framework of the following derivation is based on the original work by Green and Rivlin [A2.1, A2.2], Green, Rivlin and Spencer [A2.3] and Noll [A2.4] and Pipkin [A2.5].

A2.2 Material Properties

The materials considered are the class of materials in which the stress components $\sigma_{ij}(X_l, t)$ at a given position X_l and time t are functions of the deformation* gradient history $\partial x_k(\xi)/\partial X_l$ for time ξ from $-\infty$ up to time t. Therefore

$$\sigma_{ij}(t) = F_{ij}\{[\partial x_k(\xi)/\partial X_l], t\}, \tag{A2.1}$$

where F_{ij} is a tensorial function of the parameter shown in the bracket: $\xi = 0 \rightarrow \infty$.

If it is assumed further that there is no aging process taking place in the material during the period of investigation, the material behavior is unaltered by an arbitrary shift of the time scale. Hence (A2.1) also can be expressed as

$$\sigma_{ij}(t) = F_{ij}[\partial x_k(t-\xi)/\partial X_l], \quad \xi = 0 \rightarrow \infty. \tag{A2.2}$$

From the continuum mechanics viewpoint not every form of the right-hand side of (A2.2) is physically permissible. For example, the relation between stress and deformation must be independent of the orientation of the body relative to any external reference coordinates. This is sometimes called the principle of objectivity.

* See Chapter 3 for a discussion of deformation and kinematics.

It is shown in references [A2.1] and [A2.4] that in order to satisfy the principle of objectivity (A2.2) must have the following form:

$$\sigma_{ij}(t) = \frac{\partial x_i(t)}{\partial X_k}\frac{\partial x_j(t)}{\partial X_l} S_{kl}[G_{pq}(t-\xi)], \qquad \text{(A2.3)}$$

$$\xi = 0 \rightarrow \infty,$$

where

$$G_{pq}(t-\xi) = \frac{\partial x_m(t-\xi)}{\partial X_p}\frac{\partial x_m(t-\xi)}{\partial X_q}.$$

Since $G_{pq}(t-\xi)$ can be expressed as a function of Green's strain tensor $E_{pq}(t-\xi)$ (3.61), (A2.3) can be expressed as a function of E_{pq} as follows,

$$\sigma_{ij}(t) = \frac{\partial x_i(t)}{\partial X_k}\frac{\partial x_j(t)}{\partial X_l} S_{kl}(t), \qquad \text{(A2.4)}$$

where

$$S_{kl}(t) = S_{kl}[E_{pq}(t-\xi)], \qquad \text{(A2.4a)}$$

$$\xi = 0 \rightarrow \infty.$$

Notice in (A2.4) that σ_{ij} is referred to the deformed state while S_{kl} and E_{pq} are referred to the undeformed state. The deformed and undeformed states are not comparable. The term $\frac{\partial x_i(t)}{\partial X_k}\frac{\partial x_j(t)}{\partial X_l}$ in (A2.4) can be interpreted as the transformation from the undeformed coordinates to the deformed coordinates.

A2.3 Multiple Integral Representation of Initially Isotropic Materials (Relaxation Form)

By assuming continuity, small changes in strain history cause only small changes in the stress. Thus using the Weierstrass theorem, which is an approximation of continuous functions by polynomials, Green and Rivlin [A2.1] derived an approximation of the function S_{kl} of (A2.4) in the form of a sum of multiple integrals. This sum may be expressed in the following form using the notation employed by Pipkin [A2.5].

$$\mathbf{S} = \sum_{a=0}^{N} \mathbf{S}^{(a)} \qquad \text{(A2.5)}$$

where $\mathbf{S}^{(a)}$ is an a-tuple integral.

$\mathbf{S}^{(0)}$ and $\mathbf{S}^{(1)}$ have the forms,

$$\mathbf{S}^{(0)} = \psi^{(0)}I,$$

$$\mathbf{S}^{(1)} = \int_0^\infty \psi^{(1)}(\xi_1)\dot{\mathbf{E}}(t-\xi_1)\,\mathrm{d}\xi_1.$$

Let $\xi_1 = t-y_1$,

$\mathrm{d}\xi_1 = -\mathrm{d}y_1.$

Then $\mathbf{S}^{(1)}$ becomes

$$\mathbf{S}^{(1)} = \int_t^{-\infty} \psi^{(1)}(t-y_1)\dot{\mathbf{E}}(y_1)\,(-\mathrm{d}y_1)$$

$$= -\int_t^{-\infty} \psi^{(1)}(t-y_1)\dot{\mathbf{E}}(y_1)\,\mathrm{d}y_1 = \int_{-\infty}^{t} \psi^{(1)}(t-y_1)\dot{\mathbf{E}}(y_1)\,\mathrm{d}y_1,$$

or changing notation, ξ for y,

$$\mathbf{S}^{(1)} = \int_{-\infty}^{t} \psi^{(1)}(t-\xi_1)\dot{\mathbf{E}}(\xi_1)\,\mathrm{d}\xi_1.$$

Similarly $\mathbf{S}^{(2)}$, $\mathbf{S}^{(3)}$ and $\mathbf{S}^{(a)}$ may be written as follows

$$\mathbf{S}^{(2)} = \int_{-\infty}^{t}\int_{-\infty}^{t} \psi^{(2)}(t-\xi_1, t-\xi_2)\,\dot{\mathbf{E}}(\xi_1)\dot{\mathbf{E}}(\xi_2)\,\mathrm{d}\xi_1\,\mathrm{d}\xi_2,$$

$$\mathbf{S}^{(3)} = \int_{-\infty}^{t}\int_{-\infty}^{t}\int_{-\infty}^{t} \psi^{(3)}(t-\xi_1, t-\xi_2, t-\xi_3)\,\dot{\mathbf{E}}(\xi_1)\,\dot{\mathbf{E}}(\xi_2)\,\dot{\mathbf{E}}(\xi_3)\,\mathrm{d}\xi_1\,\mathrm{d}\xi_2\,\mathrm{d}\xi_3,$$

$$\mathbf{S}^{(a)} = \int_{-\infty}^{t}\cdots\int_{-\infty}^{t} \psi^{(a)}(t-\xi_1, t-\xi_2, \ldots t-\xi_a)\dot{\mathbf{E}}(\xi_1)\dot{\mathbf{E}}(\xi_2)\ldots\dot{\mathbf{E}}(\xi_a)\,\mathrm{d}\xi_1\ldots\mathrm{d}\xi_a, \quad \text{(A2.6)}$$

where $\mathrm{E}(\xi)$ stands for $E_{kl}(\xi)$, $\dot{\mathbf{E}}(\xi) = \partial\mathrm{E}(\xi)/\partial\xi$, $\xi_1, \xi_2, \ldots, \xi_a$ are arbitrary time variables and $\psi^{(a)}$ are kernel functions.

For initially isotropic material, the kernel functions $\psi^{(0)}, \psi^{(1)}, \psi^{(2)}, \psi^{(3)}, \ldots$ are functions of time, the invariants of the strain tensor and material constants under isothermal conditions, as follows:

$$
\begin{aligned}
\psi^{(0)} = & \int_0^t A_1(t-\xi_1)\,\bar{\dot{\mathbf{E}}}(\xi_1)\mathrm{d}\xi_1 \\
& + \int_0^t\int_0^t A_2(t-\xi_1, t-\xi_2)\,\bar{\dot{\mathbf{E}}}(\xi_1)\bar{\dot{\mathbf{E}}}(\xi_2)\,\mathrm{d}\xi_1\,\mathrm{d}\xi_2 \\
& + \int_0^t\int_0^t A_3(t-\xi_1, t-\xi_2)\,\overline{\dot{\mathbf{E}}(\xi_1)\,\dot{\mathbf{E}}}(\xi_2)\,\mathrm{d}\xi_1\,\mathrm{d}\xi_2 \\
& + \int_0^t\int_0^t\int_0^t A_4(t-\xi_1, t-\xi_2, t-\xi_3)\,\overline{\dot{\mathbf{E}}(\xi_1)\,\dot{\mathbf{E}}(\xi_2)\,\dot{\mathbf{E}}}(\xi_3)\,\mathrm{d}\xi_1\,\mathrm{d}\xi_2\,\mathrm{d}\xi_3 \\
& + \int_0^t\int_0^t\int_0^t A_5(t-\xi_1, t-\xi_2, t-\xi_3)\,\bar{\dot{\mathbf{E}}}(\xi_1)\,\overline{\dot{\mathbf{E}}(\xi_2)\,\dot{\mathbf{E}}}(\xi_3)\,\mathrm{d}\xi_1\,\mathrm{d}\xi_2\,\mathrm{d}\xi_3 \\
& + \int_0^t\int_0^t\int_0^t A_6(t-\xi_1, t-\xi_2, t-\xi_3)\,\bar{\dot{\mathbf{E}}}(\xi_1)\,\bar{\dot{\mathbf{E}}}(\xi_2)\,\bar{\dot{\mathbf{E}}}(\xi_3)\,\mathrm{d}\xi_1\,\mathrm{d}\xi_2\,\mathrm{d}\xi_3 + \ldots,
\end{aligned}
\tag{A2.7a}
$$

$$
\begin{aligned}
\psi^{(1)}(t-\xi_1) = & \; B_1(t-\xi_1) \\
& + \int_0^t B_2(t-\xi_1, t-\xi_2)\,\bar{\dot{\mathbf{E}}}(\xi_2)\,\mathrm{d}\xi_2 \\
& + \int_0^t\int_0^t B_3(t-\xi_1, t-\xi_2, t-\xi_3)\,\bar{\dot{\mathbf{E}}}(\xi_2)\,\bar{\dot{\mathbf{E}}}(\xi_3)\,\mathrm{d}\xi_2\,\mathrm{d}\xi_3 \\
& + \int_0^t\int_0^t B_4(t-\xi_1, t-\xi_2, t-\xi_3)\,\overline{\dot{\mathbf{E}}(\xi_2)\,\dot{\mathbf{E}}}(\xi_3)\,\mathrm{d}\xi_2\,\mathrm{d}\xi_3 + \ldots,
\end{aligned}
\tag{A2.7b}
$$

$$
\begin{aligned}
\psi^{(2)}(t-\xi_1, t-\xi_2) = & \; C_1(t-\xi_1, t-\xi_2) \\
& + \int_0^t C_2(t-\xi_1, t-\xi_2, t-\xi_3)\,\bar{\dot{\mathbf{E}}}(\xi_3)\,\mathrm{d}\xi_3 + \ldots,
\end{aligned}
\tag{A2.7c}
$$

$$
\psi^{(3)}(t-\xi_1, t-\xi_2, t-\xi_3) = D_1(t-\xi_1, t-\xi_2, t-\xi_3) + \ldots. \tag{A2.7d}
$$

However, the integral over $A_6\bar{\dot{\mathbf{E}}}(\xi_1)\bar{\dot{\mathbf{E}}}(\xi_2)\bar{\dot{\mathbf{E}}}(\xi_3)$ in (A2.7a) is redundant because it can be expressed through the Hamilton-Cayley equation (3.39) as a combination of other terms. $A_1, \ldots, A_5, B_1, \ldots, B_4, C_1, \ldots, D_1, \ldots$ in (A2.7) are functions of time and material constants, $\bar{\dot{\mathbf{E}}}$ represents the trace of the strain rate tensor $\dot{\mathbf{E}}$, $\overline{\dot{\mathbf{E}}\dot{\mathbf{E}}}$ represents the trace of the tensor $\dot{\mathbf{E}}\dot{\mathbf{E}}$ etc. The lower limits of integration in (A2.7a), (A2.7b), (A2.7c) are usually taken as zero instead of $-\infty$, since the material is assumed to be undisturbed prior to the time $t = 0$. All the kernel functions above were assumed to be symmetric with respect to all the arguments.

For infinitesimal (or small) deformation the Green's strain tensor E_{kl} as defined by (3.61) reduces to the usual infinitesimal strain tensor ε_{kl}. Inserting (A2.7) into (A2.6) and eliminating A_6, the open form (A2.5) can be obtained. Then inserting this equation into (A2.4) the resulting equation for infinitesimal strain is found to be the same as (7.12).

A2.4 The Inverse Relation (Creep Form)

The above described the derivation of stress in terms of strain history (relaxation). The inverse relation, the strain in terms of stress history (creep) can be obtained formally as follows. If (A2.4a) has an inverse, it can be expressed in the form.

$$E_{kl}(t) = E_{kl}[S_{pq}(t-\xi)], \qquad \xi = 0 \rightarrow \infty, \tag{A2.8}$$

where $S_{pq}(t)$ is related to the stress tensor in the deformed state σ_{ij} by (A2.4). Therefore, the stress history (A2.8) involves not only the instantaneous stress components, σ_{ij}, but also the deformation gradients. This complicates the problem. However, for small deformations and rotations, the dependence of strain on stress history may be nonlinear even for small deformations. Thus $\dfrac{\partial x_i}{\partial X_j} \cong \delta_{ij}$ and (A2.8) reduces to the following for small deformations and rotations.

$$\varepsilon_{kl}(t) = \varepsilon_{kl}[\sigma_{pq}(t-\xi)], \qquad \xi = 0 \rightarrow \infty. \tag{A2.9}$$

Again, the method discussed in Section A2.3 can be used to describe the creep strain tensor as a sum of multiple integrals of stress history which lead to the same expressions as given in (7.9) for the third order theory.

UNIT STEP FUNCTION AND UNIT IMPULSE FUNCTION*

A3.1 Unit Step Function or Heaviside Unit Function

The unit step function is defined as follows:

$$H(t-a) = \begin{cases} 1 & \text{if } t > a, \\ \frac{1}{2} & \text{if } t = a, \\ 0 & \text{if } t < a, \end{cases} \tag{A3.1}$$

where a is an arbitrary number and t is time. Function (A3.1) is shown graphically in Fig. A3.1.

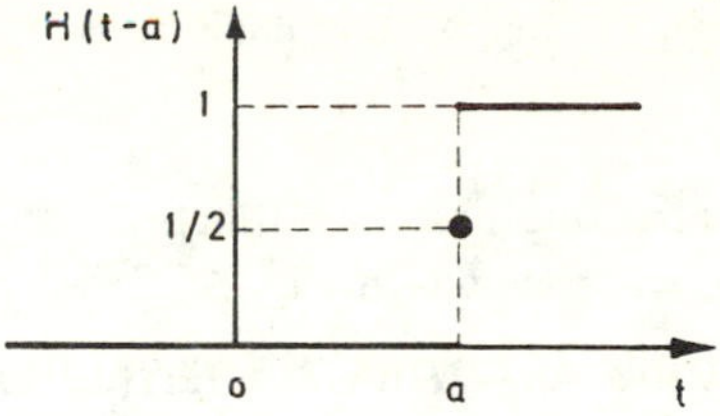

Fig. A3.1. Unit Step.

Any function $f(t)$ when multiplied by $H(t-a)$ will have a value of zero for $t < a$ and a value of $f(t)$ in the region $t > a$. Thus a step change in stress from $\sigma = 0$ to $\sigma = \sigma_0$ at $t = a$ may be represented by

$$\sigma(t) = \sigma_0 H(t-a) \tag{A3.2}$$

The Laplace transform of the unit step function is given in Table A4.1 of Appendix A4.

Typical applications of the unit step function are shown in the following examples:

Example 1.

The step function $f(t)$ shown in Fig. A3.2 can be expressed as follows by adding the appropriate value of (A3.2) at each step.

$$f(t) = H(t-a)+(1/2)H(t-2a)+(3/2)H(t-3a), \tag{A3.3}$$

* For additional information on this subject consult references [A3.1, A3.2].

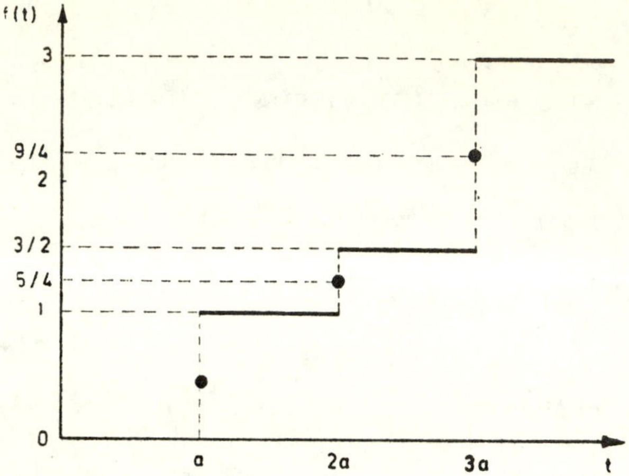

Fig. A3.2. Multiple Steps.

Example 2.

By adding and subtracting unit step functions at values of t equal to a, $2a$, $3a$, $4a$ the square wave shown in Fig. A3.3 can be obtained as follows:

$$f(t) = H(t) - 2H(t-a) + 2H(t-2a) - 2H(t-3a) + 2H(t-4a). \qquad \text{(A3.4)}$$

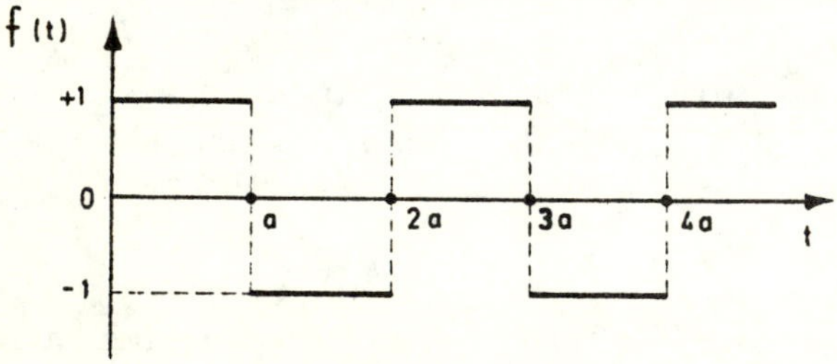

Fig. A3.3. Square Wave.

Example 3.

The rectified square wave shown in Fig. 3.4 can be represented by a function defined as follows:

$$f(t) = H(\sin t), \qquad \text{(A3.5)}$$

where

$$H(\sin t) = \begin{cases} 1 & \text{if} \quad 2n\pi < t < (2n+1)\pi, \\ 0 & \text{if} \quad (2n+1)\pi < t < 2(n+1)\pi. \end{cases}$$

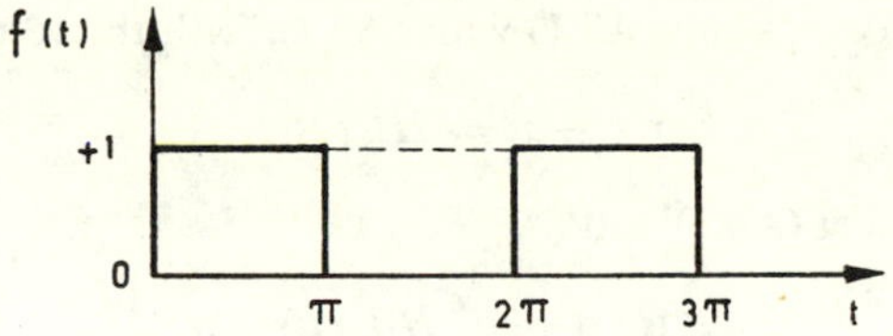

Fig. A3.4. Rectified Square Wave.

Thus for $n = 0$	$0 < t < \pi$	$f(t) = 1,$
	$\pi < t < 2\pi$	$f(t) = 0,$
and for $n = 1$	$2\pi < t < 3\pi$	$f(t) = 1,$
	$3\pi < t < 4\pi$	$f(t) = 0.$

Example 4.

Using the definition of $H(\sin t)$ given in Example 3, a half-wave rectified sinusoidal function as shown in Fig. A3.5, may be represented by the following equation:

$$f(t) = \sin t\, H(\sin t). \tag{A3.6}$$

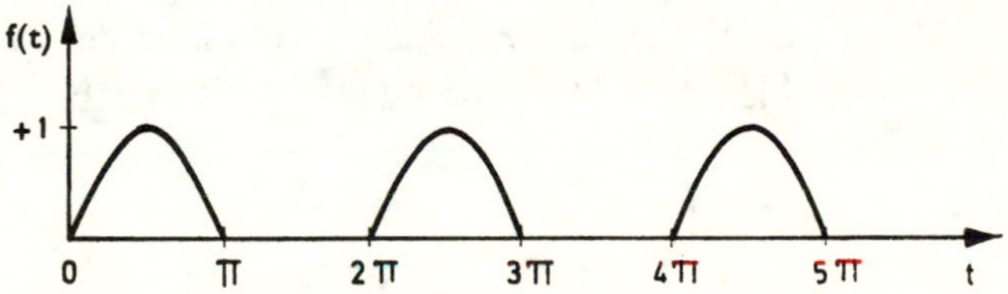

Fig. A3.5. Half-wave Rectified Sine Wave.

According to (A3.5) when $2n\pi < t < (2n+1)\pi$ the function $H(\sin t) = 1$. Thus $\sin t\, H(\sin t) = \sin t$ which is positive in these intervals. Also, when $(2n+1)\pi < t < 2(n+1)\pi$, according to (A3.5), $H(\sin t) = 0$. Thus $\sin t$ $H(\sin t) = 0$ in these intervals as shown in Fig. A3.5.

A3.2 Signum Function

The signum function sgn (t) is defined as follows:

$$\operatorname{sgn}(t-a) = \begin{cases} 1 & \text{if } t > a, \\ 0 & \text{if } t = a, \\ -1 & \text{if } t < a. \end{cases} \tag{A3.7}$$

The relation between the signum function and the unit step function can be obtained by comparing (A3.1) with (A3.7), with the result,

$$\operatorname{sgn}(t-a) = 2H(t-a)-1. \tag{A3.8}$$

For example, using (A3.1) and (A3.8),

$$\text{for } t > a,\ H(t-a) = 1$$
$$\operatorname{sgn}(t-a) = 2H(t-a)-1 = 2-1 = 1$$

which agrees with (A3.7).

Similarly, for $t = a$, $H(t-a) = 1/2$,

$$\operatorname{sgn}(t-a) = 2H(t-a)-1 = 1-1 = 0$$

and for $t < a$, $H(t-a) = 0$,

$$\operatorname{sgn}(t-a) = 2H(t-a)-1 = 0-1 = -1.$$

A3.3 Unit Impulse or Dirac Delta Function

In many physical phenomena a quantity changes by a very large magnitude over a very short time in response to a certain excitation, yet the total magnitude of the change resulting from the excitation has a fixed finite value. The function shown in Fig. A3.6 has such properties. It is zero everywhere

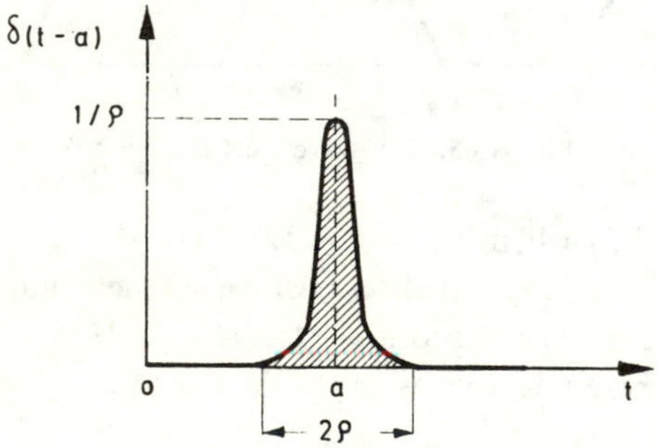

Fig. A3.6. Dirac Delta Function.

except in the infinitesimal region $a-\varrho < t < a+\varrho$, where it has the maximum value $1/\varrho$. As $\varrho \to 0$, the maximum value of the function approaches infinity, but its value integrated over time indicated by the hatched area in Fig. A3.6 is always unity. In the limiting case, $\varrho \to 0$, the shape of the function is immaterial. Any function possessing the above properties is called the unit impulse function, or the Dirac delta function and is defined by the symbolic function $\delta(t-a)$ where

$$\delta(t-a) = 0 \quad \text{for } t \neq a \tag{A3.9}$$

and is positive definite for $t = a$ such that

$$\int_0^\infty \delta(t-a)\,\mathrm{d}t = 1, \tag{A3.10}$$

where $a > 0$. Similarly, $\delta(t)$ is defined such that

$$\delta(t) = 0 \quad \text{for } t \neq 0$$

and

$$\int_{-\infty}^{\infty} \delta(t)\,\mathrm{d}t = \int_{0-}^{0+} \delta(t)\,\mathrm{d}t = 1. \tag{A3.10a}$$

If the delta function $\delta(t)$ is multiplied by any function $f(t)$ the product is zero everywhere except at $t = a$. Thus the integral

$$\int_0^\infty f(t)\,\delta(t-a)\mathrm{d}t \tag{A3.11}$$

is zero except when $t = a$. For this value of $t = a, f(a)$ is independent of t and may be removed from under the integral sign. According to (A3.10) the remaining integral is unity. Thus

$$\int_0^\infty f(t)\,\delta(t-a)\,\mathrm{d}t = f(a) \tag{A3.12}$$

and

$$\int_{-\infty}^{\infty} f(t-a)\,\delta(t)\,\mathrm{d}t = f(-a) \tag{A3.13}$$

Since $H(t)$ is zero for $t < a$ it follows that

$$\int_{-\infty}^{\infty} \delta(t)\,H(t)\,\mathrm{d}t = \int_0^\infty \delta(t)\,\mathrm{d}t = 1/2 \tag{A3.14}$$

by making use of (A3.10a). Hence from (A3.13) and (A3.14)

$$\int_{-\infty}^{\infty} f(t-a)\,\delta(t)\,H(t)\,\mathrm{d}t = (1/2)f(-a). \tag{A3.15}$$

The Laplace transform of the Dirac delta function is given in Table A4.1 of Appendix A4.

A3.4 Relation Between Unit Step Function and Unit Impulse Function

Since the derivative of the unit step function is zero everywhere except at $t = 0$, where it approaches infinity, and since the impulse function is also zero everywhere except at $t = 0$ where it approaches infinity, it can be considered that the delta function is the derivation of the unit function

$$\delta(t-a) = \frac{\mathrm{d}}{\mathrm{d}t} H(t-a) = \dot{H}(t-a). \tag{A3.16}$$

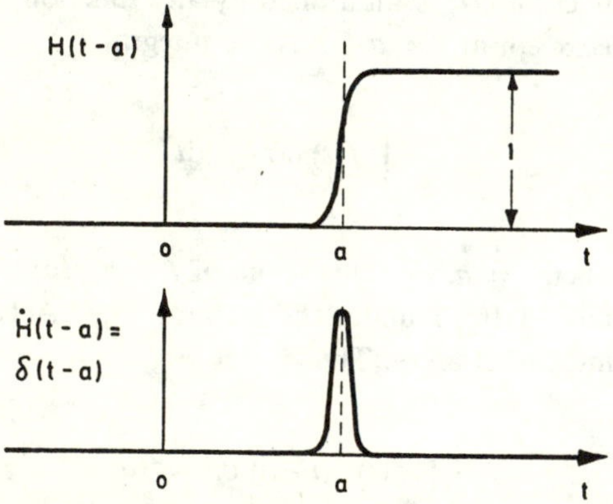

Fig. A3.7. Relation between Heaviside and Dirac Delta Functions.

Thus the relationship between the unit step function and unit impulse function may be illustrated as in Fig. A3.7.

Due to their unique characteristics, integration of the unit function and delta function needs to be handled carefully as follows:

$$\int_{0}^{\infty} H(t-a)\,\delta(t-b)\,\mathrm{d}t = \begin{cases} 0 & \text{if } a > b, \\ 1/2 & \text{if } a = b,\ a \geqslant 0, \\ 0 & \text{if } a = b,\ a < 0, \\ 1 & \text{if } a < b,\ b \geqslant 0, \\ 0 & \text{if } a < b,\ b < 0. \end{cases} \tag{A3.17}$$

The following result relating the integration of the unit function and impulse function can be obtained readily from (A3.17)

$$\int_0^\infty H^N(t-a)\,\delta(t-a)\,\mathrm{d}t = \begin{cases} \dfrac{1}{N+1} & \text{if } a > 0 \\ 0 & \text{if } a < 0 \end{cases} \tag{A3.18}$$

where N is any positive integer or zero.

A3.5 Dirac Delta Function or Heaviside Function in Evaluation of Integrals

Example 1: Evaluate the integral

$$\int_{-\infty}^{t} \varphi(t-\xi)\,\dot{\sigma}(\xi)\,\mathrm{d}\xi \tag{A3.19}$$

for a step input in $\sigma(t)$ from 0 for $t < a$ to σ_0 for $t > a$. This input can be expressed as

$$\sigma(t) = \sigma_0 H(t-a),$$

or

$$\dot{\sigma}(t) = \sigma_0 \dot{H}(t-a) = \sigma_0 \delta(t-a) \tag{A3.20}$$

Substituting (A3.20) into (A3.19) yields

$$\sigma_0 \int_{-\infty}^{t} \varphi(t-\xi)\,\delta(\xi-a)\,\mathrm{d}\xi. \tag{A3.21}$$

By definition (A3.9) $\delta(\xi-a) = 0$ except when $\xi = a$. Thus (A3.21) goes to zero except when $\xi = a$. For $\xi = a$, $\varphi(t-a)$ is independent of ξ and may be removed from the integral, and the remaining integral is unity. Thus the integral (A3.21) becomes

$$\sigma_0 \varphi(t-a), \quad t > a. \tag{A3.22}$$

Example 2. Evaluate the integral

$$\int_{-\infty}^{t}\int_{-\infty}^{t} \varphi_2(t-\xi_1)\varphi_2(t-\xi_2)\dot{\sigma}(\xi_1)\dot{\sigma}(\xi_2)\,\mathrm{d}\xi_1\,\mathrm{d}\xi_2 \tag{A3.23}$$

following two step changes in $\sigma(t)$ described by

$$\sigma(t) = \sigma_1 H(t)+(\sigma_2-\sigma_1)H(t-t_1), \quad t > t_1,$$

or

$$\dot{\sigma}(t) = \sigma_1\delta(t)+(\sigma_2-\sigma_1)\delta(t-t_1), \quad t > t_1. \tag{A3.24}$$

Equation (A3.24) indicates that a constant stress of σ_1 was applied at $t = 0$ and changed to a constant stress of σ_2 at $t = t_1$.

Rearrange (A3.23) as follows:

$$\left[\int_{-\infty}^{t} \varphi_2(t-\xi_1)\dot{\sigma}(\xi_1)\,\mathrm{d}\xi_1\right]\left[\int_{-\infty}^{t} \varphi_2(t-\xi_2)\dot{\sigma}(\xi_2)\,\mathrm{d}\xi_2\right]. \tag{A3.23a}$$

Substitute (A3.24) for $\dot{\sigma}(\xi_1)$ and $\dot{\sigma}(\xi_2)$ in (A3.23a) and consider only one integral since they are similar

$$\int_{-\infty}^{t} \varphi_2(t-\xi_1)\,[\sigma_1\delta(\xi_1)+(\sigma_2-\sigma_1)\,\delta(\xi_1-t_1)]\,\mathrm{d}\xi_1. \tag{A3.25}$$

Using (A3.13) on the first and (A3.12) on the second integral of (A3.25), yields

$$\varphi_2(t)\sigma_1+\varphi_2(t-t_1)\,(\sigma_2-\sigma_1). \tag{A3.26}$$

Thus (A3.23a) equals (A3.26) squared or

$$[\varphi_2(t)\,\sigma_1]^2+2\varphi_2(t)\,\varphi_2(t-t_1)\,\sigma_1(\sigma_2-\sigma_1)+[\varphi_2(t-t_1)\,(\sigma_2-\sigma_1)]^2. \tag{A3.27}$$

Example 3. Integrate the following

$$\int_{-\infty}^{t}\int_{-\infty}^{t} \varphi_2(t-\xi_1, t-\xi_2)\dot{\sigma}(\xi_1)\dot{\sigma}(\xi_2)\,\mathrm{d}\xi_1\,\mathrm{d}\xi_2 \tag{A3.28}$$

for two step changes in $\sigma(t)$ as given in (A3.24). Rearrange (A3.28);

$$\int_{-\infty}^{t}\left[\int_{-\infty}^{t} \varphi_2(t-\xi_1, t-\xi_2)\dot{\sigma}(\xi_1)\,\mathrm{d}\xi_1\right]\dot{\sigma}(\xi_2)\,\mathrm{d}\xi_2. \tag{A3.28a}$$

Introducing (A3.24) for $\dot{\sigma}(\xi_1)$ using ξ_1 instead of t, in the inner integral of

(A3.28a) and making use of (A3.13), (A3.12) or (A3.26) yields

$$\int_{-\infty}^{t} [\varphi_2(t, t-\xi_2)\,\sigma_1 + \varphi_2(t-t_1, t-\xi_2)\,(\sigma_2-\sigma_1)]\,\dot{\sigma}(\xi_2)\,d\xi_2. \tag{A3.29}$$

Then substituting (A3.24) for $\dot{\sigma}(\xi_2)$ using ξ_2 instead of t, and integrating as before, results in

$$\begin{aligned}&\varphi_2(t, t)\,\sigma_1^2 + \varphi_2(t, t-t_1)\,\sigma_1(\sigma_2-\sigma_1) + \varphi_2(t-t_1, t)\,(\sigma_2-\sigma_1)\sigma_1\\ &\quad + \varphi_2(t-t_1, t-t_1)(\sigma_2-\sigma_1)^2.\end{aligned} \tag{A3.30}$$

LAPLACE TRANSFORMATION*

A4.1 Definition of the Laplace Transformation

Let $f(t)$ be a function of t, for $t > 0$. Its Laplace transform, denoted by $\mathcal{L}\{f(t)\}$ or $\hat{f}(s)$, is defined by

$$\mathcal{L}\{f(t)\} = \hat{f}(s) = \int_0^\infty e^{-st} f(t)\, dt. \tag{A4.1}$$

Thus the above operation transforms the function $f(t)$ of t into a new function $\hat{f}(s)$ of the transformed variable s, which may be real or complex.

Example 1. $f(t) = e^{at}$, then its Laplace transform is

$$\hat{f}(s) = \int_0^\infty e^{at} e^{-st}\, dt = \int_0^\infty e^{-(s-a)t}\, dt = \frac{1}{s-a}.$$

A4.2 Sufficient Conditions for Existence of Laplace Transforms

If the integral (A4.1) is to converge, certain limitations must be placed on the function $f(t)$ and the variable s. The sufficient conditions for the existance of $\hat{f}(s)$ (that is that $f(t)$ be transformable) are:

(a) $f(t)$ is *piecewise continuous* in an interval $\alpha \leq t \leq \beta$;
(b) $|f(t)| \leq Me^{at}$ for some choice of the constants M and a.

If conditions (a) and (b) are satisfied then $\hat{f}(s)$ exists for all $s > a$. Piecewise continuous implies that the interval can be subdivided into a *finite number of intervals* in each of which the function is *continuous* and has *finite right*

---

* For a more extensive treatment of this subject consult references [A4.1], [A4.2] and [A4.3].

and left limits. Figure A4.1 shows a discontinuous function which is piecewise continuous between α and β.

The proof of the existence of $\hat{f}(s)$ by using (b) is as follows:

$$\left| \int_0^l f(t) e^{-st} \, dt \right| \leq \int_0^l |f(t)| e^{-st} \, dt \leq \int_0^l M e^{at-st} \, dt$$

$$\leq \int_0^\infty M e^{-(s-a)t} \, dt = \frac{M}{s-a}. \tag{A4.2}$$

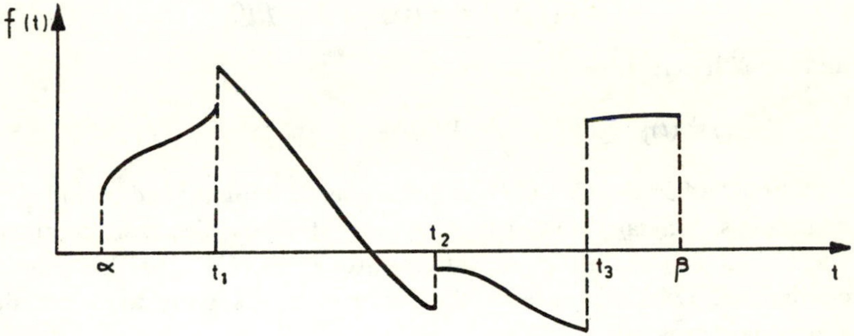

Fig. A4.1. A Piecewise Continuous Function which is Discontinuous at t_1, t_2 and t_3.

Since $\dfrac{M}{s-a}$ has a finite value for $s > a$, the integral on the left remains bounded (finite) as $l \to \infty$.

A4.3 Some Important Properties of Laplace Transforms

(a) Linearity

$$\mathcal{L}\{\alpha_1 f_1(t) + \alpha_2 f_2(t) + \ldots\} = \alpha_1 \hat{f}_1(s) + \alpha_2 \hat{f}_2(s) + \ldots, \tag{A4.3}$$

where $\alpha_1, \alpha_2, \ldots,$ are independent of t. Equation (A4.3) can be proven by applying (A4.1) to each term.

(b) Laplace Transform of Derivatives

$$\mathcal{L}\{f'(t)\} = s\hat{f}(s) - f(0), \text{ where } f'(t) = \frac{d}{dt} f(t). \tag{A4.4}$$

To prove this theorem $\mathcal{L}\{f'(t)\}$ is integrated by parts to obtain

$$\mathcal{L}\{f'(t)\} = \int_0^\infty e^{-st} f'(t)\, dt = e^{-st} f(t) \Big|_0^\infty + s \int_0^\infty e^{-st} f(t) dt$$

$$= -f(0) + s\hat{f}(s).$$

Thus, the Laplace Transformation reduces the operation of differentiation to a simple algebraic operation of functions of the transform variable s. A similar procedure can be applied to obtain the Laplace Transform of the second derivative

$$\mathcal{L}\{f''(t)\} = s^2\hat{f}(s) - sf(0) - f'(0) \tag{A4.5}$$

and the Nth derivative

$$\mathcal{L}\{f^{(N)}(t)\} = s^N\hat{f}(s) - s^{(N-1)}f(0) - \ldots - sf^{(N-2)}(0) - f^{(N-1)}(0). \tag{A4.6}$$

Equation (A4.5) is valid if f'' is piecewise continuous and f and f' are continuous. Similarly (A4.6) is valid only if f^N is piecewise continuous and only if $f(t)$ and all its derivatives through the $(N-1)$st derivative are continuous. Discontinuities in the function or its derivatives introduce additional terms.

(c) Laplace Transform of Integrals

$$\mathcal{L}\left\{\int_0^t f(\xi) d\xi\right\} = \hat{f}(s)/s. \tag{A4.7}$$

In proving this theorem the following substitutions are employed in (A4.4). Let

$$\int_0^t f(\xi) d\xi = g(t), \text{ then } g'(t) = f(t), \text{ and } g(0) = 0.$$

From (A4.4)

$$\mathcal{L}\{g'(t)\} = \mathcal{L}\{f(t)\} = \hat{f}(s) = s\hat{g}(s),$$

since $g(0) = 0$.

Therefore $\hat{g}(s) = \mathcal{L}\left\{\int_0^t f(\xi)\, d\xi\right\} = \dfrac{\hat{f}(s)}{s}.$

Equation (A4.7) can be extended readily to higher order integrals as follows "for N integrals"

$$\mathscr{L}\left\{\int_0^t \cdots \int_0^t f(\xi)\, \mathrm{d}\xi\, \mathrm{d}\xi \ldots \mathrm{d}\xi\right\} = \frac{\hat{f}(s)}{s^N}, \quad N = 1, 2, \ldots. \tag{A4.8}$$

(d) Convolution Theorem

Let $\hat{f}(s)$ and $\hat{g}(s)$ be the transforms of $f(t)$ and $g(t)$. Then the product of the transforms is

$$\hat{f}(s)\hat{g}(s) = \left[\int_{\xi'=0}^{\infty} \mathrm{e}^{-s\xi'} f(\xi')\, \mathrm{d}\xi'\right]\left[\int_{\xi=0}^{\infty} \mathrm{e}^{-s\xi} g(\xi)\, \mathrm{d}\xi\right], \tag{A4.9}$$

where ξ and ξ' are dummy variables representing time. Rewrite this product as a double integral:

$$\hat{f}(s)\hat{g}(s) = \int_{\xi=0}^{\infty} \int_{\xi'=0}^{\infty} \mathrm{e}^{-s(\xi+\xi')} f(\xi')g(\xi)\, \mathrm{d}\xi'\, \mathrm{d}\xi. \tag{A4.10}$$

Changing variables in the inner, ξ', integral (ξ fixed) let $\xi'+\xi = t$. Then $\xi' = t-\xi$ and $\mathrm{d}\xi' = \mathrm{d}t$. The range of integration over t is from $t = \xi$ (corresponding to $\xi' = 0$) to $t = \infty$ (corresponding to $\xi' = \infty$). Thus,

$$\hat{f}(s)\hat{g}(s) = \int_{\xi=0}^{\infty} \int_{t=\xi}^{\infty} \mathrm{e}^{-st} f(t-\xi)g(\xi)\, \mathrm{d}t\, \mathrm{d}\xi. \tag{A4.11}$$

The double integral (A4.11) represents the shaded area under the diagonal line $\xi = t$ in the ξ vs. t plane shown in Fig. A4.2.

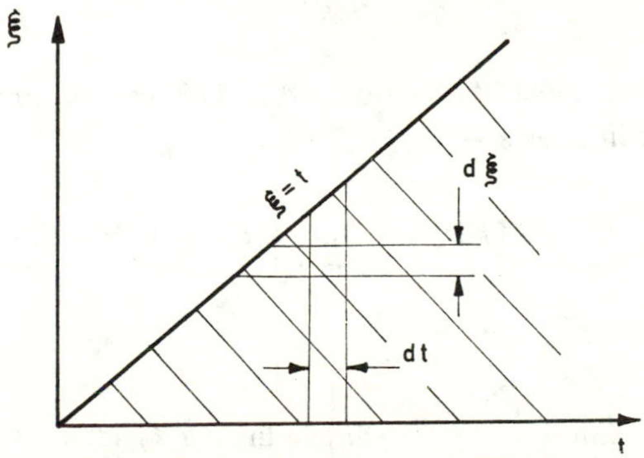

Fig. A4.2. Double Integral (A4.11)

The inner integral (the t-variable) ranges from the line $t = \xi$ to $t = \infty$ as shown by the horizontal strip of width $d\xi$ in Fig. A4.2. Then the outer integral (the ξ variable) sums the horizontal strips over the range $\xi = 0$ to $\xi = \infty$. To interchange the order of integration involves making the ξ the inner variable. This requires that the variable ξ range from 0 to the line $\xi = t$ as shown by the vertical strip of width dt. Then the outer integral sums the vertical strips over the interval from $t = 0$ to $t = \infty$. Hence (A4.11) may be rewritten as follows by interchanging the order of integration

$$\hat{f}(s)\hat{g}(s) = \int_{t=0}^{\infty} \int_{\xi=0}^{\xi=t} e^{-st} f(t-\xi) g(\xi)\, d\xi\, dt,$$

$$\hat{f}(s)\hat{g}(s) = \int_{0}^{\infty} e^{-st} \left[\int_{0}^{t} f(t-\xi) g(\xi)\, d\xi \right] dt,$$

$$\hat{f}(s)\hat{g}(s) = \mathcal{L}\left[\int_{0}^{t} f(t-\xi) g(\xi)\, d\xi \right] = \mathcal{L}\{f(t) * g(t)\}. \tag{A4.12}$$

Equation (A4.12) is the convolution theorem. The permutation between $f(t)$ and $g(t)$, namely,

$$\mathcal{L}\{f(t)*g(t)\} = \mathcal{L}\{g(t)*f(t)\} \tag{A4.13}$$

can be proven by using a change of variable, $u = t-\xi$.

(e) The initial Value Theorem is

$$\lim_{t\to 0} f(t) = \lim_{s\to\infty} s\hat{f}(s), \tag{A4.14}$$

when $f(t)$ and $f'(t)$ satisfy Section A4.2. This can be proven from (A4.4) by taking the limit as $s \to \infty$,

$$\lim_{s\to\infty} [\mathcal{L}\{f'(t)\}] = \lim_{s\to\infty} \int_{0}^{\infty} e^{-st} f'(t)\, dt = \lim_{s\to\infty} [s\hat{f}(s) - f(0)], \tag{A4.14a}$$

but

$$\lim_{s\to\infty} \left[\int_{0}^{\infty} e^{-st} f'(t)\, dt \right] = \lim_{s\to\infty} [\mathcal{L}\{f'(t)\}] = 0.$$

Table A4.1
Laplace Transform Pairs

	Function of Time, $f(t)$ for $t > 0$	Laplace Transform, $\hat{f}(s) = \mathcal{L}[f(t)]$
1.	$f(t)$	$\int_0^\infty f(t)e^{-st}\,dt = \hat{f}(s)$
2.	1	$1/s$
3.	a	a/s
4.	$H(t)$	$1/s$
5.	$H(t-a)$	$(e^{-as})/s$
6.	$\delta(t) = \dot{H}(t) = \frac{d}{dt}H(t)$	1
7.	$\delta(t-a)$	e^{-as}
8.	$\dot{\delta}(t) = \frac{d}{dt}\delta(t)$	s
9.	t	$1/s^2$
10.	t^n for $n > -1$, $n = 0, 1, 2, 3$	$\frac{n!}{s^{n+1}}$
11.	t^k for $k > -1$	$\frac{\Gamma(k+1)}{s^{k+1}}$
12.	e^{-at}	$\frac{1}{s+a}$
13.	$t^n e^{-at}$	$\frac{n!}{(s+a)^{n+1}}$
14.	$\sin at$	$\frac{a}{s^2+a^2}$
15.	$\cos at$	$\frac{s}{s^2+a^2}$
16.	$e^{-bt}\sin at$	$\frac{a}{(s+b)^2+a^2}$
17.	$e^{-bt}\cos at$	$\frac{s+b}{(s+b)^2+a^2}$
18.	$\sinh at$	$\frac{a}{s^2-a^2}$
19.	$\cosh at$	$\frac{s}{s^2-a^2}$
20.	$e^{-at}-e^{-bt}$	$\frac{b-a}{(s+a)(s+b)}$
21.	$ae^{-at}-be^{-bt}$	$\frac{(a-b)s}{(s+a)(s+b)}$

Table A4.1
Laplace Transform Pairs, Continued

	Function of Time, $f(t)$ for $t > 0$	Laplace Transform, $\hat{f}(s) = \mathcal{L}[f(t)]$
22.	$1-e^{-at}$	$\frac{a}{s(s+a)}$
23.	$\frac{t}{a}-\frac{1}{a^2}(1-e^{-at})$	$\frac{1}{s^2(s+a)}$
24.	$af(t)$	$a\hat{f}(s)$
25.	$f(t)+g(t)$	$\hat{f}(s)+\hat{g}(s)$
26.	$\frac{df(t)}{dt}$	$s\hat{f}(s)-f(0)$
27.	$\frac{d^2f(t)}{dt^2}$	$s^2\hat{f}(s)-sf(0)-\frac{df(0)}{dt}$
28.	$\int_0^t f(\xi)d\xi$	$\frac{1}{s}\hat{f}(s)$
29.	$\int_0^t f(t-\xi)g(\xi)\,d\xi = \int_0^t f(\xi)g(t-\xi)d\xi = f(t)*g(t)$ (convolution theorem)	$\hat{f}(s)\hat{g}(s)$

Hence,

$$0 = \lim_{s\to\infty} s\hat{f}(s)-f(0)$$

or,

$$\lim_{s\to\infty} s\hat{f}(s) = f(0) = \lim_{t\to 0} f(t).$$

(f) The final value theorem is

$$\lim_{t\to\infty} f(t) = \lim_{s\to 0} s\hat{f}(s), \tag{A4.15}$$

when $f(t)$ and $f'(t)$ satisfy Section A4.2. This situation can also be proven from (A4.4) by taking the limit of (A4.4) as $s \to 0$,

$$\lim_{s\to 0} [\mathcal{L}\{f'(t)\}] = \lim_{s\to 0} \int_0^\infty e^{-st}f'(t)\,dt = \lim_{s\to 0} [s\hat{f}(s)-f(0)]. \tag{A4.15a}$$

The integral on the left-hand side may be evaluated as follows:

$$\lim_{s \to 0} \int_0^{\infty} e^{-st} f'(t) dt = \int_0^{\infty} f'(t) dt = \lim_{t \to \infty} \int_0^{t} f'(t)\, dt = \lim_{t \to \infty} [f(t) - f(0)]. \tag{A4.15b}$$

Substituting (A4.15b) for the left-hand side of (A4.15a) yields (A4.15), since $f(0)$ is not a function of either s or t.

Other transform pairs and important properties of Laplace transforms are given in Table A4.1.

A4.4 The Inverse Laplace Transform

If the Laplace transform of a function $f(t)$ is $\hat{f}(s)$, then $f(t)$ is the inverse Laplace transform of $\hat{f}(s)$, which is usually written in the following form

$$f(t) = \mathcal{L}^{-1}\{\hat{f}(s)\}. \tag{A4.16}$$

If $f(t)$ satisfies the conditions stated in Section A4.2, then the inverse Laplace transform is unique.

Although an integral inversion formula can be used to obtain the inverse Laplace transform, in most cases it proves to be too complicated. Instead the Laplace transform table is used by looking up the function $f(t)$ corresponding to the transform $\hat{f}(s)$ in the Laplace transform table. For more complicated functions which are not available in the table, various methods can be used to obtain the approximate inversion, see for example [A4.4].

A4.5 Partial Fraction Expansion

The function $\hat{f}(s)$ for which an inverse transform $f(t)$ is needed is often a ratio of two polynomials of s such as

$$\hat{f}(s) = \frac{a_i s^i + a_{i-1} s^{i-1} + \ldots a_1 s + a_0}{b_j s^j + b_{j-1} s^{j-1} + \ldots b_1 s + b_0}. \tag{A4.17}$$

The procedure of obtaining the inverse transform of such functions is simplified by using the partial fraction expansion. To form the partial fraction expansion factor the denominator $D(s)$ so that $\hat{f}(s)$ has the form

$$\hat{f}(s) = \frac{N(s)}{D(s)} = \frac{N(s)}{(s+r_1)(s+r_2)\ldots(s+r_j)}, \tag{A4.18}$$

where r_1, r_2, etc. are the roots of the denominator $D(s)$, i.e., roots of $D(s) = 0$. Only the case when all roots are different will be considered here. The partial fraction expansion of (A4.18) is then

$$\hat{f}(s) = \frac{C_1}{s+r_1} + \frac{C_2}{s+r_2} + \ldots \frac{C_j}{s+r_j} \tag{A4.19}$$

and the inverse transform $f(t)$ is obtained by using item 12 of Table A4.1 as follows

$$f(t) = C_1 e^{-r_1 t} + C_2 e^{-r_2 t} + \ldots C_j e^{-r_j t}, \tag{A4.20}$$

and the constants C_j are determined from the expression

$$C_j = \lim_{s \to -r_j} \left\{ \frac{N(s)}{D(s)} (s+r_j) \right\}. \tag{A4.21}$$

For example if

$$\hat{f}(s) = \frac{s+2}{s(s+4)(s+1)} = \frac{C_1}{(s+0)} + \frac{C_2}{(s+4)} + \frac{C_3}{(s+1)} \tag{A4.22}$$

then C_2, for example, is found from (A4.21) to be

$$C_2 = \lim_{s \to -4} \left\{ \frac{(s+2)(s+4)}{s(s+4)(s+1)} \right\} = \left. \frac{s+2}{s(s+1)} \right|_{s=-4} = -\frac{1}{6}. \tag{A4.23}$$

A4.6 Some Uses of the Laplace Transform

Example 2. Find the solution of the differential equation

$$\frac{d^2y}{dt^2} + k^2 y = F(t), \tag{A4.24}$$

$$y(0) = 1, \quad y'(0) = -2.$$

The Laplace transform of (A4.24) can be written

$$s^2\hat{y}(s) - sy(0) - y'(0) + k^2\hat{y}(s) = \hat{F}(s). \tag{A4.25}$$

Inserting the initial conditions and rearranging terms yields the following equation

$$(s^2+k^2)\,\hat{y}(s) = \hat{F}(s) + s - 2, \tag{A4.26}$$

or

$$\hat{y}(s) = \frac{s-2}{s^2+k^2} + \frac{\hat{F}(s)}{s^2+k^2}. \tag{A4.26a}$$

The inverse Laplace transformation of (A4.26a) yields

$$y(t) = \mathscr{L}^{-1}\left\{\frac{s-2}{s^2+k^2}\right\} + \mathscr{L}^{-1}\left\{\frac{\hat{F}(s)}{s^2+k^2}\right\}$$

$$= \cos kt - \frac{2}{k}\sin kt + \frac{1}{k}\int_0^t F(t-\xi)\sin(k\xi)\,d\xi. \tag{A4.27}$$

In obtaining the inverse Laplace transform (A4.27), items 14, 15 and 29 of Table A4.1 were employed.

Example 3. Find the solution of the integral equation

$$y(t) = t^2 + \int_0^t y(\xi)\sin(t-\xi)\,d\xi. \tag{A4.28}$$

Taking the Laplace transform yields

$$\hat{y}(s) = \frac{2}{s^3} + \hat{y}(s)\left(\frac{1}{s^2+1}\right) \tag{A4.29}$$

where the convolution theorem (item 29 of Table A4.1) was employed for the last term. Rearranging terms of (A4.29) yields

$$\hat{y}(s)\left(1 - \frac{1}{s^2+1}\right) = \frac{2}{s^3} \tag{A4.30}$$

or

$$\hat{y}(s) = \frac{2(s^2+1)}{s^5} = \frac{2}{s^3} + \frac{2}{s^5}. \tag{A4.30a}$$

The inversion of (A4.30a) yields

$$y(t) = t^2 + \frac{1}{12}t^4. \tag{A4.31}$$

APPENDIX A5

DERIVATION OF THE MODIFIED SUPERPOSITION PRINCIPLE FROM THE MULTIPLE INTEGRAL REPRESENTATION

A5.1 Second Order Term

It was shown by Nolte and Findley [A5.1] that the modified superposition principle could be derived from the multiple integral representation by redefining the kernel functions as functions of their smaller time arguments only (or the more recent event in the case of step loadings). This relation may be derived as follows for a second order term.

Consider a second order term of the multiple integral representation

$$\int_0^t \int_0^t F(t-\xi_1, t-\xi_2) \frac{d\mathsf{x}(\xi_1)}{d\xi_1} \frac{d\mathsf{y}(\xi_2)}{d\xi_2} d\xi_1 \, d\xi_2 \tag{A5.1}$$

and apply it to the multiple step history defined by

$$\mathsf{x}(t) = \sum_{i=1}^{N} \mathsf{x}_i H(t-t_i),$$

and

$$\mathsf{y}(t) = \sum_{j=1}^{N} \mathsf{y}_j H(t-t_j),$$

where x, y represent either stresses or strains and x_i or y_j is the change in $\mathsf{x}(t)$ or $\mathsf{y}(t)$ at $t = t_i$ or $t = t_j$; $H(t-t_i)$ is the Heaviside function which equals zero for $t < t_i$, and equals one for $t > t_i$. For the multiple step history (A5.1) becomes

$$\sum_{i=1}^{N} \sum_{j=1}^{N} F(t-t_i, t-t_j) \mathsf{x}_i \mathsf{y}_j. \tag{A5.2}$$

To restrict $F(t-t_i, t-t_j)$ to a function of its smaller argument only, define

$$F(t-t_i, t-t_j) = \begin{cases} g_i \text{ for } i \geqslant j, \; t_i \geqslant t_j \\ g_j \text{ for } i < j, \; t_i < t_j, \end{cases} \tag{A5.3}$$

or

$$F(t-t_i, t-t_j) = F(t-t_i, t-t_i)\,H(t_i-t_j)+F(t-t_j, t-t_j)\,H(t_j-t_i).$$

Using this definition, (A5.2) becomes

$$\sum_{i=1}^{N}\sum_{j=1}^{N} F(t-t_i, t-t_j)\mathsf{x}_i\mathsf{y}_j \rightarrow$$

$$\begin{aligned}&\mathsf{x}_1(\mathsf{y}_1g_1+\mathsf{y}_2g_2+\mathsf{y}_3g_3+\ldots\mathsf{y}_Ng_N)\\ &+\mathsf{x}_2(\mathsf{y}_1g_2+\mathsf{y}_2g_2+\mathsf{y}_3g_3+\ldots\mathsf{y}_Ng_N)\\ &+\mathsf{x}_3(\mathsf{y}_1g_3+\mathsf{y}_2g_3+\mathsf{y}_3g_3+\ldots\mathsf{y}_Ng_N)\\ &\quad\vdots\qquad\vdots\\ &+\mathsf{x}_N(\mathsf{y}_1g_N+\mathsf{y}_2g_N+\mathsf{y}_3g_N+\ldots\mathsf{y}_Ng_N)\\ &=\sum_{i=1}^{N}\mathsf{x}_i\left[\sum_{j=1}^{i}\mathsf{y}_jg_i+\sum_{j=i}^{N}\mathsf{y}_jg_j-\mathsf{y}_ig_i\right].\end{aligned} \tag{A5.4}$$

If the identities $\sum_{i=1}^{N}(g_i\mathsf{x}_i\mathsf{y}_i-g_i\mathsf{x}_i\mathsf{y}_i)=0$ and $\sum_{i=1}^{N}\left(\sum_{j=1}^{i}\mathsf{x}_j\mathsf{y}_ig_i-\sum_{j=1}^{N}\mathsf{x}_i\mathsf{y}_jg_j\right)=0$

are added to (A5.4), the g_j terms vanish and (A5.4) becomes

$$\sum_{i=1}^{N}\left[\mathsf{x}_i\sum_{j=1}^{i}\mathsf{y}_jg_i+\sum_{j=1}^{i}\mathsf{x}_j\mathsf{y}_ig_i-\mathsf{x}_i\mathsf{y}_ig_i\right]=\sum_{i=1}^{N}g_i\left[\mathsf{x}_i\sum_{j=1}^{i-1}\mathsf{y}_j+\mathsf{y}_i\sum_{j=1}^{i-1}\mathsf{x}_j+\mathsf{x}_i\mathsf{y}_i\right]. \tag{A5.5}$$

By defining $\mathsf{X}_i=\sum_{j=1}^{i-1}\mathsf{x}_j$, $\mathsf{Y}_i=\sum_{j=1}^{i-1}\mathsf{y}_j$, (A5.5) can be written as

$$\sum_{i=1}^{N}g_i[\mathsf{Y}_i\mathsf{x}_i+\mathsf{X}_i\mathsf{y}_i+\mathsf{x}_i\mathsf{y}_i]$$

or

$$\sum_{i=1}^{N}g_i[(\mathsf{X}_i+\mathsf{x}_i)(\mathsf{Y}_i+\mathsf{y}_i)-\mathsf{X}_i\mathsf{Y}_i]=\sum_{i=1}^{N}g_i\Delta_i(\mathsf{XY}), \tag{A5.6}$$

where $\Delta_i(\mathsf{XY})=[(\mathsf{X}_i+\mathsf{x}_i)(\mathsf{Y}_i+\mathsf{y}_i)\ -\mathsf{X}_i\mathsf{Y}_i]$ is the step change in the product $x(t)y(t)$ of (A5.1) at $t = t_i$. Equation (A5.6) derived here is merely a restatement of a second order term for (9.20).

If the step history given in (A5.1) is used to approximate a finite time interval of a varying stress history and if N in (A5.1) approaches infinity,

(A5.6) becomes

$$\int_0^t F(t-\xi)\frac{d}{d\xi}[x(\xi)\, y(\xi)]\, d\xi, \tag{A5.7}$$

where $F(t-\xi) = F(t-\xi,\ t-\xi)$. Thus restricting the kernel of (A5.1) to its smallest argument reduced the double integral (A5.1) to a single integral (A5.7).

A5.2 **Third Order Term**

A third order term of the multiple integral representation having the form

$$\int_0^t\int_0^t\int_0^t K(t-\xi_1, t-\xi_2, t-\xi_3)\frac{dx(\xi_1)}{d\xi_1}\frac{dy(\xi_2)}{d\xi_2}\frac{dz(\xi_3)}{d\xi_3}\, d\xi_1\, d\xi_2\, d\xi_3 \tag{A5.8}$$

and subjected to the following multiple step history

$$x(t) = x_i H(t-t_i)$$
$$y(t) = y_i H(t-t_i)$$
$$z(t) = z_i H(t-t_i)$$

has the following result after integration with respect to ξ_1, ξ_2, and ξ_3

$$\sum_{i=1}^N\sum_{j=1}^N\sum_{k=1}^N K(t-t_i, t-t_j, t-t_k)x_i y_j z_k \tag{A5.9}$$

The kernel function K in (A5.9) is defined in the following such that the kernel function is restricted to its smallest argument only,

$$\begin{aligned} K(t-t_i, t-t_j, t-t_k) &= K_{iii}H(t_i-t_j)\, H(t_i-t_k) \\ &\quad + K_{jjj}H(t_j-t_i)\, H(t_j-t_k) \\ &\quad + K_{kkk}H(t_k-t_i)\, H(t_k-t_j) \\ &\quad \text{etc.} \end{aligned}$$

where $H(t_i-t_j)$ etc. is the Heaviside Function.

By applying to (A5.9) the same procedures as used in the derivation of (A5.6) and (A5.7) from (A5.3), the triple integral (A5.8) is reduced to the following single integral

$$\int_0^t K(t-\xi)\,\frac{d}{d\xi}\,[x(\xi)\,y(\xi)\,z(\xi)]\,d\xi, \tag{A5.10}$$

where $K(t-\xi) = K(t-\xi,\ t-\xi,\ t-\xi)$. This result also can be demonstrated by using (A5.7) as follows. Using (A5.7) for the inner two integrals of (A5.8) yields

$$\int_0^t \left\{ \int_0^t K(t-\xi',\ t-\xi',\ t-\xi_3)\,\frac{d}{d\xi'}\,[x(\xi')\,y(\xi')]\,d\xi' \right\} \frac{dz\,(\xi_3)}{d\xi_3}\,d\xi_3.$$

Using (A5.7) again to the above equation yields (A5.10).

A5.3 Application to Third Order Multiple Integrals for Creep

Applying (A5.7) and (A5.10) to the three term multiple integral representation of creep yields the corresponding "single" integral representation, the "modified superposition principle". For example, inserting (A5.7) and (A5.10) into (7.9) of the multiple integral representation for creep yields the following results:

$$\begin{aligned}
\boldsymbol{\epsilon}(t) = \mathbf{I} \int_0^t \Big[& K_1 \frac{d}{d\xi}\,\bar{\boldsymbol{\sigma}} + K_3 \frac{d}{d\xi}\,(\bar{\boldsymbol{\sigma}}\,\bar{\boldsymbol{\sigma}}) + K_4 \frac{d}{d\xi}\,(\overline{\boldsymbol{\sigma}\boldsymbol{\sigma}}) \\
& + K_7 \frac{d}{d\xi}\,(\overline{\boldsymbol{\sigma}\boldsymbol{\sigma}\boldsymbol{\sigma}}) + K_8 \frac{d}{d\xi}\,(\overline{\boldsymbol{\sigma}\boldsymbol{\sigma}}\,\bar{\boldsymbol{\sigma}}) \Big]\,d\xi \\
+ \int_0^t \Big[& K_2 \frac{d}{d\xi}\,\boldsymbol{\sigma} + K_5 \frac{d}{d\xi}\,(\bar{\boldsymbol{\sigma}}\boldsymbol{\sigma}) + K_6 \frac{d}{d\xi}\,(\boldsymbol{\sigma}\boldsymbol{\sigma}) \\
& + K_9 \frac{d}{d\xi}\,(\bar{\boldsymbol{\sigma}}\,\bar{\boldsymbol{\sigma}}\boldsymbol{\sigma}) + K_{10} \frac{d}{d\xi}\,(\overline{\boldsymbol{\sigma}\boldsymbol{\sigma}}\boldsymbol{\sigma}) \\
& + K_{11} \frac{d}{d\xi}\,(\bar{\boldsymbol{\sigma}}\boldsymbol{\sigma}\boldsymbol{\sigma}) + K_{12} \frac{d}{d\xi}\,(\boldsymbol{\sigma}\boldsymbol{\sigma}\boldsymbol{\sigma}) \Big]\,d\xi,
\end{aligned} \tag{A5.11}$$

where K_1 to K_{12} are functions of the time variable $(t-\xi)$ only. These kernel functions may be determined completely from constant stress creep experiments. Equation (A5.11) illustrates how to obtain the constitutive relation under the modified superposition assumption from the multiple integral representation. Other constitutive relations under the modified superposition principle at different stress states (for creep) or strain states (for stress relaxation) can be obtained in the same manner. For example, the constitutive relation for the axial strain and shear strain under combined time-dependent axial stress and torsion can be obtained from (8.43), (8.44), (8.45) and (8.46) as follows:

$$\varepsilon_{11}(t) = \int_0^t \left[F_1 \frac{d\sigma(\xi)}{d\xi} + F_2 \frac{d\sigma^2(\xi)}{d\xi} + F_3 \frac{d\sigma^3(\xi)}{d\xi} + F_4 \frac{d\sigma(\xi)\,\tau^2(\xi)}{d\xi} + F_5 \frac{d}{d\xi}\tau^2(\xi) \right] d\xi, \tag{A5.12}$$

$$\varepsilon_{22}(t) = \ldots$$

$$\varepsilon_{33}(t) = \ldots$$

$$\varepsilon_{12}(t) = \int_0^t \left[G_1 \frac{d\tau(\xi)}{d\xi} + G_2 \frac{d\,\tau^3(\xi)}{d\xi} + G_3 \frac{d\sigma(\xi)\tau(\xi)}{d\xi} + G_4 \frac{d\sigma^2(\xi)\,\tau(\xi)}{d\xi} \right] d\xi, \tag{A5.13}$$

where F_1, F_2, F_3, F_4, F_5 and G_1, G_2, G_3, G_4 are time functions which can be determined completely from constant stress creep experiments as discussed in Section 8.12.

Note that some of the terms in (A5.12), (A5.13) involve derivatives of products. Thus (A5.13) for example may be written as follows after expanding indicated derivatives and using the dot to indicate differentiation.

$$\varepsilon_{12}(t) = \int_0^t \{G_1\,\dot{\tau}(\xi) + 3G_2\,\tau^2(\xi)\,\dot{\tau}(\xi) + G_3[\sigma(\xi)\,\dot{\tau}(\xi) + \tau(\xi)\,\dot{\sigma}(\xi)] + G_4[\sigma^2(\xi)\,\dot{\tau}(\xi) + 2\tau(\xi)\sigma(\xi)\,\dot{\sigma}(\xi)]\}\,d\xi. \tag{A5.14}$$

Conversion Table †

Conversions: *English-cgs-SI Units*

Quantity	English to:	cgs*;	SI*	cgs to:	English*;	SI*
Length	1 inch (in)	2.54 centimeter (cm)	0.0254 meter (m)	1 centimeter (cm)	0.39370 in	10^{-2} m
	1 foot (ft)	30.48 cm	0.3048 m	1 meter (m)	39.370 in	1 m
Area	1 in^2	6.4516 cm^2	6.4516×10^{-4} m^2	1 cm^2	0.15500 in^2	10^{-4} m^2
	1 ft^2	929.03 cm^2	0.092903 m^2	1 m^2	15,500 in^2	1 m^2
Volume	1 in^3	16.387 cm^3	0.16387×10^{-6} m^3	1 cm^3	0.061023 in^3	10^{-6} m^3
	1 ft^3	28,317 cm^3	0.028317 m^3	1 liter	61.023 in^3	0.001 m^3
Mass	1 pound mass (lbm) (avoirdupois)	453.59 gram (gm)	0.45359 kilogram (kg)	1 gm	0.0022046 lbm	0.001 kg
				1 kg	2.2046 lbm	1 kg
Density	1 lbm/in^3	27.680 gm/cm^3	2.7680×10^4 kg/m^3	1 gm/cm^3	0.03613 lbm/in^3	1000 kg/m^3
	1 lbm/ft^3	0.016018 gm/cm^3	16.018 kg/m^3	1 kg/cm^3	36.13 lbm/in^3	10^6 kg/m^3
Velocity	1 in/second (in/s)	2.540 cm/s	0.0254 m/s	1 cm/s	0.3937 in/s	10^{-2} m/s
	1 ft/s	30.48 cm/s	0.3048 m/s	1 m/s	39.37 in/s	1 m/s
Force	1 pound force (lbf)	444.82×10^3 dynes (d)	4.4482 newton (N)	1 d	2.248×10^{-6} lbf	1×10^{-5} N
Stress or pressure (see Fig. A6.1)	1 lbf/in^2 (psi)	68,948 d/cm^2	6,894.8 N/m^2	1 d/cm^2	1.4504×10^{-5} psi	10^{-1} N/m^2
	1 ksi = 10^3 psi	70.31 kg/cm^2	6.8948 N/m^2	1 kg/cm^2	14.223 psi	1.020×10^{-5}N/m^2

* All values rounded off to 5 significant figures where applicable.

† For additional information consult references [A6.1, A6.2, A6.3].

Conversion Table (cont'd.)

Conversions: English-cgs-SI Units

Quantity	English to:	cgs*;	SI*	cgs to:	English*;	SI*
Work or energy	1 ft-lbf	0.13826 kg-m	1.3558 joule (J)	1 kg-m = 2.3427 g-cal (mean)	7.2330 ft-lbf 0.0092972 BTU (mean)	9.8067 J
	1 ft-lbf	0.32389 g-cal (mean)		1 g-cal (mean)	0.0039685 BTU (mean)	4.1900 J
	1 BTU (mean) = 777.97 ft-lbf	251.98 g-cal (mean)	1,054.4 J			
Strain (dimensionless)	1 in/in	1 cm/cm	1 m/m	1 cm/cm	1 in/in	1 m/m
Viscosity	1 poise = 100 centipoise	1 d-s/cm²	10^{-1} N-s/m²	1 d-s/cm²	1 poise	10^{-1} N-s/m²
Temperature (see Fig. A6.2)	Degrees Fahrenheit (F) = (9/5)(C)+32			Degrees Celsius (C) = (5/9)(*F*−32)		

* All values rounded off to 5 significant figures where applicable.

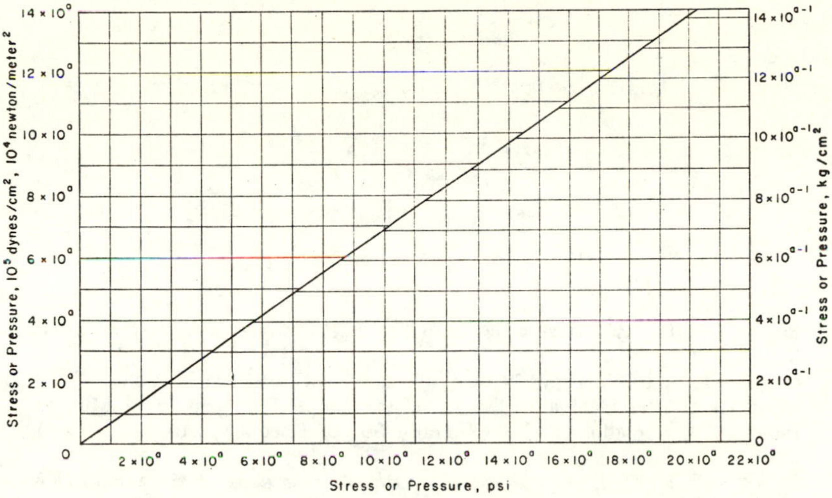

Fig. A6.1. Stress or Pressure Conversion: English-Metric.

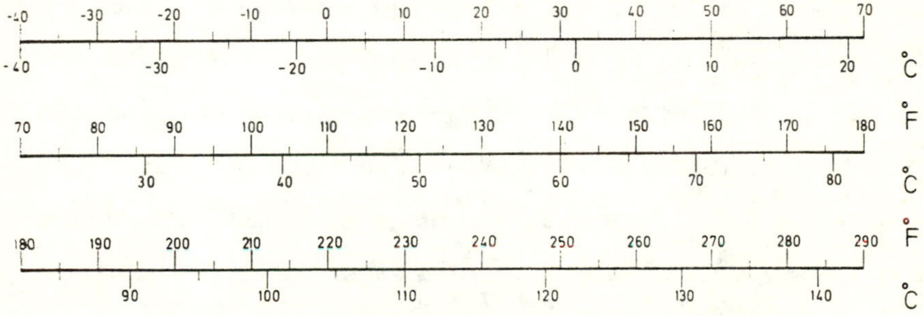

Fig. A6.2. Temperature Conversion: Fahrenheit-Centigrade.

BIBLIOGRAPHY

[1.1] Boltzmann, L., "Zur Theorie der elastischen Nachwirkung," Pogg. Ann. Physik, Vol. 7 (1876), p. 624.

[1.2] Onaran, K. and Findley, W. N. "Combined Stress Creep Experiments on a Nonlinear Viscoelastic Material to Determine the Kernel Functions for a Multiple Integral Representation of Creep," Trans. Soc. of Rheology, Vol. 9, II (1965), p. 299.

[2.1] Norton, F. H., *The Creep of Steel at High Temperature*. McGraw Hill, New York (1929).

[2.2] Tapsell, H. J., *Creep of Metals*, Oxford University Press (1931).

[2.3] Hoff, N. J. (Editor), *High Temperature Effects in Aircraft Structures*, Pergamon, London (1958).

[2.4] Finnie, I. and Heller, W. R., *Creep of Engineering Materials*, McGraw Hill, New York (1959).

[2.5] Kachanov, L. M., *Theory of Creep* (in Russian), Gos. Izdat. Fis.-Mat. Lit., Moscow (1960).

[2.6] Lubahn, J. D. and Felgar, R. P., *Plasticity and Creep of Metals*, John Wiley, New York (1961).

[2.7] Dorn, J. E. (Editor), *Mechanical Behavior of Materials at Elevated Temperature*, McGraw Hill, New York (1961).

[2.8] Odqvist, F. K. G. and Hult, J., *Creep Strength of Metallic Materials*, Springer, Berlin (1962).

[2.9] Kennedy, A. J., *Processes of Creep and Fatigue in Metals*, Wiley, New York (1963).

[2.10] Odqvist, F. K. G., *Mathematical Theory of Creep and Creep Rupture*, Oxford Mathematics Monograms (1966).

[2.11] Hult, J., *Creep in Engineering Structures*, Blaisdell Pub. Co., Waltham, Mass. (1966).

[2.12] Rabotnov, Y. N., *Creep Problems in Structural Members*, North-Holland Pub. Co., Amsterdam (1969).

[2.13] Smith, A. I. and Nicolson, A. M. (Editors), *Advances in Creep Design*, A. E. Johnson Memorial Volume, Applied Science Publishers, London (1971).

[2.14] Vicat, L. T., "Note Sur l'Allongement Progressif du Fil de Fer Soumis à Diverses Tensions", Annales, Ponts et Chaussées, Mémoires et Docum., Vol. 7 (1834).

[2.15] Thurston, R. H., *Materials of Construction*, John Wiley, New York (1895).

[2.16] Andrade, E. N., "The Viscous Flow in Metals and Allied Phenomena", Proceedings of Royal Society, Series A, Vol. 84 (1910), p. 1.

[2.17] Bailey, R. W., "Creep of Steel Under Simple and Compound Stresses and The Use of High Initial Temperature in Steam Power Plants", Transactions, Tokyo Sectional Meeting of the World Power Conference, Tokyo (1929), p. 1089.

[2.18] Ludwik, P., *Elemente der Technologischen Mechanik*, Springer, Berlin (1909).

[2.19] SODERBERG, C. R., "The Interpretation of Creep Tests for Machine Design", Transaction of ASME, Vol. 58 (1936), p. 733.

[2.20] NADAI, A., "On the Creep of Solids at Elevated Temperatures", Journal of Applied Physics, Vol. 8 (1937), p. 418.

[2.21] PRANDTL, L., "Ein Gedankenmodell zur kinetischen Theorie der festen Körper", Zeitschrift für Angewandte Mathematik und Mechanik, Vol. 8 (1928), p. 85.

[2.22] SCOTT BLAIR, G. W. and VEINOGLOU, B. C., "A Study of the Firmness of Soft Materials Based on Nuttings Equation", Journal of Science and Instruments, Vol. 21 (1944), p. 149.

[2.23] HOFF, N. J., "Mechanics Applied to Creep Testing", Society for Experimental Stress Analysis, Proceedings, Vol. XVII (1958), p. 1.

[2.24] FINDLEY, W. N. and MARK, R., "Multiaxial Creep Behavior of 304 Stainless Steel", Annual Report No. 1 to Oak Ridge National Laboratory, Brown University, Division of Engineering, Oct. 1972.

[2.25] JOHNSON, A. E., "Complex-Stress Creep of Metals", Metallurgical Reviews, Vol. 5 (1960), p. 447.

[2.26] MISES, R. von, "Mechanik der festen Körper im Plastisch Deformablen Zustand", Göttinger Nachrichten, Math.-Phys. Kl. 1913 (1913), p. 582.

[2.27] ODQVIST, F. K. G., "Theory of Creep Under Combined Stresses With Application to High Temperature Machinery", Royal Swedish Academy of Eng. Research, Proceedings, Vol. 141 (1936), p. 31.

[2.28] BAILEY, R. W., "The Utilization of Creep Test Data in Engineering Design", Proceedings of Inst. of Mechanical Engineers, London, Vol. 131 (1935), p. 131.

[2.29] PRAGER, W., "Strain Hardening Under Combined Stresses", Journal of Applied Physics, Vol. 16 (1945), p. 837.

[2.30] JOHNSON, A. E., "Creep Under Complex Stress Systems at Elevated Temperature", Proceedings, Inst. of Mechanical Engineers, London, Vol. 164 (1962).

[2.31] MARIN, J., "Design of Members Subjected to Creep at High Temperatures", Journal of Applied Mechanics, Vol. 4 (1937), p. 21.

[2.32] KANTER, J. J., "The Interpretation and Use of Creep Tests", Transactions of American Society for Metals, Vol. 24 (1936), p. 870.

[2.33] TAPSELL, H. J. and JOHNSON, A. E., "Creep Under Combined Tension and Torsion", Engineering, Vol. 150 (1940), p. 24.

[2.34] MARIN, J., *Mechanical Properties of Materials and Design*, McGraw Hill, New York (1942).

[2.35] RABOTNOV, Y. N., "Some Problems on the Theory of Creep", Vestnik MGU, No. 10 (1948), NACA, Technical Memo., No. 1353 (1953).

[2.36] ZHUKOV, A. M., RABOTNOV, Y. N. and CHURIKOV, F. S., "Experimental Verification of Certain Creep Theories", Inzhen. Sbornik. Vol. 17, Izd-vo AN SSSR, Moscow (1953), p. 163.

[2.37] DRUCKER, D. C., "On the Macroscopic Theory of Inelastic Stress-Strain-Time-Temperature Behavior", Advances in Materials Research in the NATO Nations, Pergamon, London (1963), p. 193.

[2.38] WANG, T. T. and ONAT, E. T., "Non-linear Mechanical Behavior of 1100 Aluminum at 300°F", Acta Mechanica, Vol. 5, No. 1 (1968), p. 54.

[2.39] BLASS, J. J. and FINDLEY, W. N., "Short-Time Biaxial Creep of an Aluminum Alloy With Abrupt Changes of Temperature and State of Stress", Trans. ASME, Journal Applied Mechanics, Vol. 38, Series E (1971), p. 489.

[2.40] FINDLEY, W. N. and KHOSLA, G., "Application of the Superposition Principle and Theories of Mechanical Equation of State, Strain and Time Hardening to Creep of Plastics under Changing Loads", Journal Applied Physics, Vol. 26, No. 7 (1955), p. 821.

[2.41] LEADERMAN, H., *Elastic and Creep Properties of Filamentous Materials*, Textile Foundation, Washington, D. C. (1943).

[2.42] PHILLIPS, P., "The Slow Stretch in India Rubber, Glass and Metal Wires When Subjected to a Constant Pull", Phil. Mag., Vol. 9 (1905), p. 513.

[2.43] FINDLEY, W. N., "Creep Characteristics of Plastics", Symposium on Plastics, ASTM (1944), p. 118.

[2.44] FINDLEY, W. N. and KHOSLA, G., "An Equation for Tension Creep of Three Unfilled Thermoplastics", SPE Journal, Vol. 12 (1956), p. 20.

[2.45] FINDLEY, W. N. and PETERSON, D. B., "Prediction of Long-Time Creep with Ten-Year Creep Data on Four Plastic Laminates", Proccedings of ASTM, Vol. 58 (1958), p. 841.

[2.46] LAI, J. S. Y. and FINDLEY, W. N., "Elevated Temperature Creep of Polyurethane under Nonlinear Torsional Stress with Step Changes in Torque", Transactions of the Society of Rheology, Vol. 17, Issue 1 (1973), p. 129.

[2.47] FINDLEY, W. N. and TRACY, J. F., "16-Year Creep of Polyethylene and PVC" Polymer Engr. and Science. Vol. 14 (1974), p. 577.

[2.48] FINDLEY, W. N., ADAMS, C. H. and WORLEY, W. J., "The Effect of Temperature on the Creep of Two Laminated Plastics as Interpreted by the Hyperbolic-sine Law and Activation Energy Theory", Proc. ASTM, Vol. 48 (1948), p. 1217.

[2.49] EYRING, H., "Viscositiy, Plasticity and Diffusion as Examples of Absolute Reaction Rates", Journal of Chemistry and Physics, Vol. 4 (1936), p. 263.

[2.50] KAUZMANN, W., "Flow of Solid Metals from the Standpoint of the Chemical Rate Theory", Transactions of AIMME, Inst. of Metals Division, Vol. 143 (1941), p. 57.

[2.51] NUTTING, P. G., "A Study of Elastic Viscous Deformation", Proceedings of ASTM, Vol. 21 (1921), p. 1162.

[2.52] SCOTT BLAIR, G. W., VEINOGLOU, B. C. and CAFFYN, J. E., "Limitation of the Newtonian Time Scale in Relation to Non-equilibrium Rheological States and a Theory of Quasi-properties", Proceedings of Royal Society, Series A 189 (1947), p. 69.

[2.53] SCOTT BLAIR, G. W. and CAFFYN, J. E., "An Application of the Theory of Quasi-Properties to the Treatment of Anomalous Stress-Strain Relations", Philosophical Magazine, J. Science, Vol. 40 (1949), p. 80.

[2.54] ALFREY, T. and DOTY, P. M., "Methods of Specifying the Properties of Viscoelastic Materials", J. Appl. Phys., Vol. 16 (1945), p. 700.

[2.55] ALFREY, T., Jr., *Mechanical Behavior of High Polymers*, Interscience Publishers, New York (1948).

[2.56] LEE, E. H., "Viscoelasticity", in *Handbook of Engineering Mechanics*, W. Flügge (Ed.), McGraw-Hill, New York (1962).

[2.57] HILTON, H. H., "Viscoelastic Analysis", in *Engineering Design for Plastics*, E. Baer (Ed.), Reinhold, New York (1964).

[2.58] LEE, E. H., "Viscoelastic Stress Analysis", Proceedings of First Symposium on Naval Structures and Mechanics, Pergamon Press (1960), p. 456.

[2.59] BLAND, D. R., *The Theory of Linear Viscoelasticity*, Pergamon, London (1960).

[2.60] LEE, E. H. and MORISSON, J. A., "A Comparison of the Propagation of Longitudinal Waves in Rods of Viscoelastic Materials", Journal of Polymer Science; Vol. 19 (1956), p. 93.

[2.61] VOLTERRA, V., "Sulle Equazioni integro-differenziali della Teoria dell' Elasticita" Atti della Reale Academia dei Lencei, Vol. 18 (1909), p. 295.

[2.62] GROSS, B., *Mathematical Structures of the Theories of Viscoelasticity*, Paris, Hermann and Co. (1953).

[2.63] WILLIAMS, M. L., Jr., "Structural Analysis of Viscoelastic Materials", AIAA Jour., Vol. 2 (1964), p. 785.

[2.64] FLÜGGE, W., *Viscoelasticity*, Blaisdell Pub. Co., Waltham, Mass. (1967).

[2.65] ONARAN, K. and FINDLEY, W. N., "Combined Stress Creep Experiments on Viscoelastic Material with Abrupt Changes in State of Stress", Proc., Joint International Confrence on Creep, I. Mech. E., London (1963), p. 2.85.

[2.66] GREEN, A. E. and RIVLIN, R. S., "The Mechanics of Nonlinear Materials with Memory, Part I", Archive for Rational Mechanics and Analysis, Vol. 1 (1957), p. 1.

[2.67] NOLL, W., "A Mathematical Theory of the Mechanical Behavior of Continuous Media", Archive for Rational Mechanics and Analysis, Vol. 2 (1958), p. 197.

[2.68] GREEN, A. E., RIVLIN, R. S. and SPENCER, A. J. M., "The Mechanics of Nonlinear Materials With Memory, Part II", Archive for Rational Mechanics and Analysis, Vol. 3 (1959), p. 82.

[2.69] GREEN, A. E. and RIVLIN, R. S., "The Mechanics of Nonlinear Materials With Memory, Part III", Archive for Rational Mechanics and Analysis, Vol. 4 (1960), p. 387.

[2.70] COLEMAN, B. D. and NOLL, W., "Foundations of Linear Viscoelasticity", Reviews of Modern Physics, Vol. 33 (1961), p. 239.

[2.71] ERINGEN, A. C., *Nonlinear Mechanics of Continua*, McGraw Hill, New York (1962).

[2.72] ONARAN, K. and FINDLEY, W. N., "Combined Stress Creep Experiments on a Nonlinear Viscoelastic Material to Determine the Kernel Functions for a Multiple Integral Representation of Creep", Transactions of the Society of Rheology, Vol. 9 (1965), p. 299.

[2.73] PIPKIN, A. C., "Small Finite Deformations of Viscoelastic Solids", Rev. of Modern Physics, Vol. 36 (1964), p. 1034.

[2.74] BERNSTEIN, B., KEARSLEY, E. A. and ZAPAS, L. J., "A Study of Stress Relaxation with Finite Strain", Trans. Soc. Rheol., Vol. 7 (1963), p. 391.

[2.75] SCHAPERY, R. A., "A Theory of Nonlinear Thermoviscoelasticity Based on Irreversible Thermodynamics", Proc. 5th U. S. Nat. Cong. Appl. Mech. ASME (1966).

[2.76] SCHAPERY, R. A., "On a Thermodynamic Constitutive Theory and Its Application to Various Nonlinear Materials", Proc. IUTAM Symp. on Thermoinelasticity, Springer-Verlag (1969).

[2.77] SCHAPERY, R. A., "On the Characterization of Nonlinear Viscoelastic Materials", Polymer Engineering and Science, Vol. 9, No. 4 (1969), p. 295.

[2.78] LIANIS, G., "Studies on Constitutive Equations of Viscoelastic Materials Under Large Deformation", Purdue University, School of Aero and Engineering Science, Report No. ES 63–3, May (1963) and Report No. ES 63–11. Dec. (1963).

[2.79] KO, W. L. and BLATZ, P. J., "Application of Finite Viscoelastic Theory to the Deformation of Rubber like Materials", Firestone Flight Science Laboratory, C.I.T. Rep. CALCIT, SM 64–4 (1964).

[2.80] NAKADA, O. "Theory of Nonlinear Responses", Journal of Physical Society of Japan, Vol. 15 (1960). p, 2280.

[2.81] LEADERMAN, H., MCCRACKIN, F. and NAKADA, O., "Large Longitudinal Retarded Elastic Deformation of Rubber-like Network Polymers, II. Application of a General Formulation of Nonlinear Response", Transactions of Society of Rheology, Vol. 7 (1963), p. 111.

[2.82] WARD, I. M. and ONAT, E. T., "Nonlinear Mechanical Behavior of Oriented Polypropylene", Journal of Mechanics and Physics of Solids, Vol. 11 (1963), p. 217.

[2.83] ONARAN, K. and FINDLEY, W. N., "Experimental Determination of Some Kernel Functions in the Multiple Integral Method for Nonlinear Creep of Polyvinyl Chloride", Transactions, American Society of Mechanical Engineers, Journal of Applied Mechanics, Vol. 38, Series E (1971), p. 30.

[2.84] NEIS, V. V. and SACKMAN, J. L., "An Experimental Study of a Nonlinear Material with Memory", Transactions of the Society of Rheology, Vol. 11, Issue 3 (1967), p. 307.

[2.85] LOCKETT, F. J., "Creep and Stress Relaxation Experiments for Nonlinear Materials" International Journal of Engineering Sciences, Vol. 3 (1965), p. 59.

[2.86] LIFSHITZ, J. M. and KOLSKY, H., "Nonlinear Viscoelastic Behavior of Polyethylene", International Journal of Solids and Structures, Vol. 3 (1967), p. 383.

[2.87] PIPKIN, A. C. and ROGERS, T. G., "A Nonlinear Integral Representation for Viscoelastic Behavior", Journal of Mechanics and Physics of Solids, Vol. 16 (1968), p. 59.

[2.88] FINDLEY, W. N. and LAI, J. S. Y., "A Modified Superposition Principle Applied to Creep of Nonlinear Viscoelastic Material Under Abrupt Changes in State of Combined Stress", Transactions of the Society of Rheology, Vol. 11, Part 3 (1967), p. 361.

[2.89] NOLTE, K. G. and FINDLEY, W. N., "A Linear Compressibility Assumption for the Multiple Integral Representation of Nonlinear Creep of Polyurethane", Trans. ASME, Journal of Applied Mechanics, Vol. 37E (1970), p. 441.

[2.90] NOLTE, K. G. and FINDLEY, W. N., "Multiple Step, Nonlinear Creep of Polyurethane Predicted From Constant Stress Creep by Three Integral Representations", Transactions of the Society of Rheology, Vol. 15, Issue 1 (1971), p. 111.

[2.91] GOTTENBERG, N. G., BIRD, J. O. and AGRAWAL, G. L., "An Experimental Study of a Nonlinear Viscoelastic Solid in Uniaxial Tension", Trans. ASME, J. Appl. Mech., Vol. 36 E (1969), p. 558.

[2.92] STAFFORD, R. O., "On Mathematical Forms for the Material Functions in Nonlinear Viscoelasticity", J. Mech. Phys Solids, Vol. 17 (1969), p. 339.

[2.93] SMART, J. and WILLIAMS, J. G., "A Comparison of Single-Integral Nonlinear Viscoelasticity Theories", J. Mech. Phys. of Solids, Vol. 20 (1972), p. 313.

[2.94] HUANG, N. C. and LEE, E. H., "Nonlinear Viscoelasticity for Short Time Ranges", Trans. ASME, J. App. Mech., Vol. 33E (1966), p. 313.

[2.95] LAI, J. S. and ANDERSON, D., "Irrecoverable and Recoverable Nonlinear Viscoelastic Properties of Asphalt Concrete", Highway Research Record 468 (1974).

[2.96] ERMAN, B. and ONARAN, K., "Time Dependent Behavior of Nonlinear Orthotropic Materials", Istanbul Technical University, Civil Engineering Dept. (1974); in Turkish with English Summary.

*Additional Literature**

[2.97] MAZILU, P., "On the Constitutive Law of Boltzmann-Volterra," Revue Roumaine de Mathématiques Pures et Appliquées, Vol. 18 (1973), p. 1067.

[2.98] RABOTNOV, YU, N., PAPERNIK, L. Kh., and STEPANYCHEY, E. I., "Application of the Nonlinear Theory of Heredity to the Description of Time Effects in Polymeric Materials," Polymer Mechanics, Vol. 7 (1973), p. 63.

[3.1] PRAGER, W., *Introduction to Mechanics of Continua*, Ginn and Co., Boston (1961).

[3.2] TRUESDELL, C. and NOLL, W., "The Nonlinear Field Theories of Mechanics", *Handbuch der Physik*, Ed. by Flügge S., Vol. III/3, Springer, Berlin (1965), p. 26.

[3.3] FUNG, Y. C., *Foundations of Solid Mechanics*, Prentice-Hall, Englewood Cliffs, New Jersey (1965).

[4.1] KOITER, W. T., "Couple-Stresses in the Theory of Elasticity", Proc. Kon. Ned. Akad. v. Wetensch., Amsterdam, Ser. B, Vol. 67 (1964), p. 17.

[4.2] MINDLIN, R. D. and TIERSTEN, H. F., "Effects of Couple-Stresses in Linear Elasticity", Arch. Rational Mechanics and Analysis, Vol. 11 (1962), p. 415.

*Further additional literature pertaining to Chapter 2 is to be found beginning on page 360.

[4.3] SOKOLNIKOFF, I. S., *Mathematical Theory of Elasticity*, 2nd Ed., McGraw-Hill, New York (1956).

[4.4] ERINGEN, A. C., *Nonlinear Theory of Continuous Media*, McGraw-Hill, New York (1962).

[4.5] RIVLIN, R. S., "On the Principles of Equipresence and Unification", Quart. App. Math., Vol. 30 (1972), p. 227.

[4.6] LOVE, A. E. H., *A Treatise on the Mathematical Theory of Elasticity*, reprinted by Dover Publications, New York (1963).

[4.7] ZENER, C., *Elasticity and Anelasticity of Metals*, University of Chicago Press, Chicago (1948).

[4.8] PRAGER, W., *Introduction to Mechanics of Continua*, Ginn and Co., Boston (1961).

[4.9] FUNG, Y. C., *Foundations of Solid Mechanics*, Prentice-Hall, Inc., Englewood Cliffs, N. J. (1965).

[5.1] LEE, E. H., "Viscoelastic Stress Analysis", Proceedings of First Symposium on Naval Structural Mechanics, Pergamon Press, London (1960), p. 456.

[5.2] KOLSKY, H., *Stress Waves in Solids*, Clarendon Press, Oxford (1953).

[5.3] GROSS, B., *Mathematical Structure of the Theories of Viscoelasticity*, Hermann and Company, Paris (1953).

[5.4] BLAND, D. R., *The Theory of Linear Viscoelasticity*, Pergamon Press, London (1960).

[5.5] LAI, J. S. Y. and FINDLEY, W. N., "Combined Tension-Torsion Creep Experiments on Polycarbonate in the Nonlinear Range", Polymer Engineering and Science, Vol. 9 (1969), p. 378.

[5.6] KOLSKY, H., "Experimental Studies of the Mechanical Behavior of Linear Viscoelastic Solids", Proceedings of the 4th Symposium on Naval Structural Mechanics, Pergamon Press, London (1965), p. 381.

[5.7] KOLSKY, H. and SHI, Y. Y., "The Validity of Model Representation for Linear Viscoelastic Behavior", Brown University Report NONR 562(14)/5 (1958).

[5.8] ALFREY, T., Jr., *Mechanical Behavior of High Polymers*, Interscience Publishers, New York (1948).

[5.9] FERRY, J. D., *Viscoelastic Properties of Polymers*, 2nd Ed., John Wiley, New York (1961).

[5.10] WILLIAMS, M. Lr., "Structural Analysis of Viscoelastic Materials", AIAA Journal, Vol. 2 (1964), p. 785.

[5.11] TIMOSHENKO, S. and GOODIER, J. N., *Theory of Elasticity*, 3rd Ed., McGraw-Hill, New York (1970).

[5.12] FINDLEY, W. N., REED, R. M. and STERN, P., "Hydrostatic Creep of Solid Plastics", Trans. ASME, J. Applied Mech., Vol. 34E (1967), p. 895.

[5.13] NOLTE, K. G. and FINDLEY, W. N., "A Linear Compressibility Assumption for the Multiple Integral Representation of Nonlinear Creep of Polyurethane", Trans. ASME, J. Applied Mech., Vol. 37E (1970), p. 441.

[5.14] LAI, J. S. Y. and FITZGERALD, J. E., "Thermorheological Properties of Asphalt Mixtures", Highway Research Record No. 313 (1970), p. 18.

[5.15] HOPKINS, I. L. and HAMMING, R. W., "On Creep and Stress Relaxation", J. of Appl. Physics, Vol. 28 (1957), p. 906.

[5.16] SECOR, K. E. and MONISMITH, C. L., "Analysis and Interrelation of Stress-Strain-Time Data for Asphalt Concrete", Trans. Soc. Rheology, Vol. 8 (1964), p. 19.

[5.17] SHU, L. S. and ONAT, E. T., "On Anisotropic Linear Viscoelastic Solids", Proceedings of the Fourth Symposium on Naval Structural Mechanics, Pergamon Press, London (1967), p. 203.

[5.18] WILLIAMS, M. L., BLATZ, P. J. and SCHAPERY, R. A., "Fundamental Studies Relating to System Analysis of Solid Propellants", Graduate Aeronautical Laboratory, Cal. Tech. Report SM 61–5 (1961).

[5.19] ONARAN, K. and FINDLEY, W. N., "Combined Stress Creep Experiments on Viscoelastic Materials with Abrupt Changes in State of Stress", Proceedings of Joint Int. Conference on Creep, I. Mech. E., London (1963), p. 2.85.

[5.20] FLÜGGE, W., *Viscoelasticity*, Blaisdell, Waltham, Mass. (1967).

[5.21] TOBOLSKY, A. V., *Structure and Properties of Polymers*, John Wiley, New York (1960).

[5.22] SCHWARZL, F. and STAVERMAN, A. J., "Time-Temperature Dependence of Linear Viscoelastic Behavior", J. of Appl. Physics, Vol. 23 (1952), p. 838.

[5.23] WILLIAMS, M. L., LANDEL, R. F. and FERRY, J. D., "The Temperature Dependence of Relaxation Mechanism in Amorphous Polymers and Other Glass-Liquids", J. of Am. Chem. Soc., Vol. 77 (1955), p. 370.

[5.24] MORLAND, L. W. and LEE, E. H., "Stress Analysis for Linear Viscoelastic Materials with Temperature Variation", Trans. Soc. Rheology, Vol. 4 (1960), p. 223.

[5.25] JONES, J, W. in *ICRPG Solid Propellant Mechanical Behavior Manual*, published by Chemical Propellant Information Agency, The Johns Hopkins University, Silver Springs, Md. (1963).

Additional Literature

[5.26] CHRISTENSEN, R. M., *Theory of Viscoelasticity—an Introduction*, Academic Press, New York (1971).

[5.27] WARD, I. M., *Mechanical Properties of Solid Polymers*, Wiley-Interscience, London (1971).

[5.28] MESHKOV, S. I., "The Integral Representation of Functionally Exponential Functions and their Application to Dynamic Problems of Linear Viscoelasticity," J. App. Mech. and Tech. Phys., Vol. 11 (1972), p. 100.

[5.29] PIPKIN, A. C., *Lectures on Viscoelasticity Theory*, Springer-Verlag, New York (1972).

[5.30] STOUFFER, D. C., "A Thermal-Hereditary Theory for Linear Viscoelasticity," Zeitschrift für Angewandte Mathematik und Physik, Vol. 23 (1972), p. 845.

[6.1] TIMOSHENKO, S., *Strength of Materials, Parts I and II*, Third Ed., Van Nostrand, Princeton, N.J. (1956).

[6.2] FREUDENTHAL, A. M. and LORSCH, H. G., "The Infinite Elastic Beam on a Linear Viscoelastic Foundation", J. of Eng. Mech. Division, Proceedings of ASCE, Vol. 83 (1957), p. 1158.

[6.3] WOODWARD, W. B. and RADOK, J. R. M., "Stress Distribution in a Reinforced Hollow Viscoelastic Cylinder Subjected to Time Dependent Internal Pressure", Brown University Technical Report PA–TR/14 (1955).

[6.4] LEE, E. H., "Stress Analysis in Viscoelastic Bodies", Quarterly of Applied Mathematics, Vol. 13 (1955), p. 183.

[6.5] LEE, E. H. and RADOK, J. R. M., "Stress Analysis in Linearly Viscoelastic Materials", Proceedings 9th Int. Cong. Appl. Mech., Brussels (1956), p. 321.

[6.6] LEE, E. H., "Viscoelastic Stress Analysis", Proceedings of the First Symposium on Naval Structural Mechanics, Pergamon Press, London (1960), p. 456.

[6.7] HUNTER, S. C., "The Solution of Boundary Value Problems in Linear Viscoelasticity", Proceedings of the Fourth Symposium on Naval Structural Mechanics, Pergamon Press, London (1965) p. 257.

[6.8] TING, T., "Contact Problems in the Linear Theory of Viscoelasticity", Trans. ASME Journal of Applied Mechanics, Vol. 35E (1968), p. 248.

[6.9] MORLAND, L. W. and LEE, E. H., "Stress Analysis for Linear Viscoelastic Materials with Temperature Variation", Trans. Soc. Rheology, Vol. 4 (1960), p. 233.

[6.10] MUKI, R. and STERNBERG, E., "On Transient Thermal Stresses in Viscoelastic Materials with Temperature-Dependent Properties", Trans. ASME, Journal of Applied Mechanics, Vol. 28 (1961), p. 193.

[6.11] SHAPERY, R. A., "Approximate Methods of Transform Inversion for Viscoelastic Stress Analysis", Proceedings of the Fourth U. S. Nat. Congress of App. Mech., Vol. 2, ASME (1962), p. 1075.

[6.12] WILLIAMS, M. L., Jr., "Structural Analysis of Viscoelastic Materials", J. AIAA, Vol. 2 (1964), p. 785.

Additional Literature

[6.13] ADEY, R. A., and BREBBIU, C. A., "Efficient Method for Solution of Viscoelastic Problems," J. of Eng. Mech. Div., Proc. ASCE, Vol. 99 (1973), p. 1119.

[6.14] GRAHAM, G. A. C., and SABIN, G. C. W., "Correspondence Principle of Linear Viscoelasticity for Problems that Involve Time Dependent Regions," Int. J. Eng. Sci., Vol. 11 (1973), p. 123.

[7.1] WARD, I. M. and ONAT, E. T., "Non-linear Mechanical Behavior of Oriented Polypropylene", J. Mech. and Physics of Solids, Vol. 11 (1963), p. 217.

[7.2] ONARAN, K. and FINDLEY, W. N., "Combined Stress Creep Experiments on a Non-linear Viscoelastic Material to Determine the Kernel Functions for a Multiple Integral Representation of Creep", Trans. Soc. of Rheology, Vol. 9, issue 2 (1965), p. 299.

[7.3] LIFSHITZ, J. M. and KOLSKY, H., "Non-linear Viscoelastic Behavior of Polyethylene", Int. J. of Solids and Structures, Vol. 3 (1967), p. 383.

[7.4] SPENCER, A. J. M. and RIVLIN, R. S., "The Theory of Matrix Polynomials and its Application to the Mechanics of Isotropic Continua", Arch. Rational Mech. Anal., Vol. 2 (1959), p. 21.

[7.5] FINDLEY, W. N., REED, R. M. and STERN, P., "Hydrostatic Creep of Solid Plastics", Trans. ASME, J. of App. Mech., Vol. 34-E (1967), p. 895.

[7.6] NOLTE, K. G. and FINDLEY, W. N., "A Linear Compressibility Assumption for the Multiple Integral Representation of Nonlinear Creep of Polyurethane", Trans. ASME, J. of App. Mech., Vol. 37-E (1970), p. 441.

[7.7] FINDLEY, W. N. and ONARAN, K., "Incompressible and Linearly Compressible Viscoelastic Creep and Relaxation", Trans. ASME, J. Applied Mechanics, Vol. 41-E (1974), p. 243.

[7.8] LOCKETT, F. J. and STAFFORD, R. O., "On Special Constitutive Relations in Non-linear Viscoelasticity", Int. J. Eng'g. Science, Vol. 7 (1969), p. 917.

[7.9] PIPKIN, A. C., "Small Finite Deformations of Viscoelastic Solids", Reviews of Modern Physics, Vol. 36 (1964), p. 1034.

[7.10] RIVLIN, R. S., "Further Remarks on the Stress-Deformation Relations for Isotropic Materials", J. of Rational Mechanics and Analysis, Vol. 4 (1955), p. 681.

[7.11] HUANG, N. C. and LEE, E. H., "Nonlinear Viscoelasticity for Short Time Ranges", Trans. ASME, J. of Appl. Mech., Vol. 33-E (1966), p. 313.

[7.12] LUBLINER, J., "Short-Time Approximation in Nonlinear Viscoelasticity", Int. J. of Solids and Structures, Vol. 3 (1967), p. 513.

[7.13] PIPKIN, A. C. and RIVLIN, R. S., "Small Deformations Superposed on Large Deformations in Materials with Fading Memory", Arch. Rational Mech. Anal., Vol. 8 (1961), p. 297.

[7.14] APPLEBY, E. J. and LEE, E. H., "Superposed Deformations of Viscoelastic Solids", Tech. Report No. 149, Division of Engr. Mech., Stanford Univ. (1964).

[7.15] GOLDBERG, W. and LIANIS, G., "Behavior of Viscoelastic Media under Small Sinusoidal Oscillations Superposed on Finite Strain", Trans. ASME, J. of Appl. Mech., Vol. 36-E (1969), p. 558.

[7.16] LIANIS, G., "Constitutive Equations of Viscoelastic Solids under Finite Deformation", Purdue Univ. Report A & ES 63–11 (1963).

[7.17] DEHOFF, P. H., LIANIS, G. and GOLDBERG, G., "An Experimental Program for Finite Linear Viscoelasticity", Trans. Soc. Rheol., Vol. 10, issue 1 (1966), p. 385.

[7.18] GOLDBERG, W., BERNSTEIN, B. and LIANIS, G., "The Exponential Extension Ratio History, Comparison of Theory with Experiment", Int. J. Non-Linear Mechanics, Vol. 4 (1969), p. 277.

[7.19] MCGUIRT, C. W. and LIANIS, G., "Constitutive Equations for Viscoelastic Solids under Finite Uniaxial and Biaxial Deformations", Trans. Soc. Rheol., Vol. 14 II (1970), p. 117.

[7.20] FLOWERS, R. J. and LIANIS, G., "Viscoelastic Tests under Finite Strain and Variable Strain Rate", Trans. Soc. Rheol., Vol. 14 issue 4 (1970), p. 441.

[7.21] MCGUIRT, C. W. and LIANIS, G., "Experimental Investigation of Non-Linear, Non-Isothermal Viscoelasticity", Int. J. Engg. Sci., Vol. 7 (1969), p. 579.

[7.22] COLEMAN, B. D. and NOLL, W., "Foundations of Linear Viscoelasticity", Rev. Mod. Phys., Vol. 33 (1961), p. 239.

[7.23] BERNSTEIN, B., KEARSLEY, E. A. and ZAPAS, L. J., "A Study of Stress Relaxation with Finite Strain", Trans. Soc. Rheol., Vol. 7 (1963), p. 391.

[7.24] ZAPAS, L. J. and CRAFT, T., "Correlation of Large Longitudinal Deformations with Different Strain Histories", J. Res. Natn. Bur. Stand. 69A, No. 6 (1965), p. 541.

[7.25] STAFFORD, R. O., "On Mathematical Forms for the Material Functions in Nonlinear Viscoelasticity", J. Mech. Phys. Solids, Vol. 17 (1969), p. 339.

[7.26] SMART, J. and WILLIAMS, J. G., "A Comparison of Single-Integral Nonlinear Viscoelasticity Theories", J. Mech. Phys. Solids, Vol. 20 (1972), p. 313.

[7.27] SCHAPERY, R. A., "A Theory of Nonlinear Thermoviscoelasticity Based on Irreversible Thermodynamics", Proc. Fifth U. S. Nat. Congress of Appl. Mech., ASME (1966), p. 511.

[7.28] SCHAPERY, R. A., "On a Thermodynamic Constitutive Theory and Its Application to Various Nonlinear Materials", Proc. IUTAM Sym., East Kilbride (1968).

[7.29] SCHAPERY, R. A., "On the Characterization of Nonlinear Viscoelastic Materials", Polymer Engg. Sci., Vol. 9 (1969) p. 295.

[7.30] SCHAPERY, R. A., "Further Development of a Thermodynamic Constitutive Theory: Stress Formulation", Purdue University Report A & ES 69–2 (1969).

[7.31] BIOT, M. A., "Theory of Stress-Strain Relations in Anisotropic Viscoelasticity and Relaxation Phenomena", J. Applied Physics, Vol. 25 (1954), p. 1385.

[7.32] COLEMAN, B. D., "Thermodynamics of Materials with Memory", Arch. for Rational Mech. and Analysis, Vol. 17 (1964), p. 1.

[7.33] COLEMAN, B. D. and GURTIN, M. E., "Thermodynamics with Internal State Variables", J. Chem. Phys., Vol. 47 (1967), p. 597.

Additional Literature

[7.34] CLARK, R. L., FOX, W. R., and WELCH, G. B., "Representation of Mechanical Properties of Nonlinear Viscoelastic Materials by Constitutive Equations," Trans., American Society of Agricultural Engineers, Vol. 14 (1971), p. 511.

[7.35] TING, E. C., "Approximations in Nonlinear Viscoelasticity," Int. J. Engineering Science, Vol. 9 (1971), p. 995.

[7.36]* YANNAS, I. V., and HASKELL, V. C., "Utility of the Green-Rivlin Theory in Polymer Mechanics," J. Applied Physics, Vol. 42 (1971), p. 610.

[7.37] DISTEFANO, N., "Some Numerical Aspects in the Identification of a Class of Nonlinear Viscoelastic Materials," Zeitschrift für Angewandte Mathematik und Physik, Vol. 52 (1972), p. 389.

[7.38] KOVALENKO, A. D., and KARNAUKHOV, V. G., "Linearized Theory of Thermoviscoelasticity," (in Russian), Dopovidi Akademiyi Nauk Ukrainskoyi RSR, Vol. 1972 (1972), p. 69.

[7.39] LOCKETT, F. J., *Nonlinear Viscoelastic Solids*, Academic Press, London (1972).

[7.40] SMART, J., and WILLIAMS, J. G., "Power-Law Model for Multiple-Integral Theory of Nonlinear Viscoelasticity," J. Mech. Phys., Vol. 20 (1972), p. 325.

[7.41] STERNSTEIN, S. S., and HO, T. C., "Biaxial Stress Relaxation in Glassy Polymers: Polymethylmethacrylate," J. Appl. Phys., Vol. 43 (1972), p. 4370.

[7.42] YANNAS, I. V., "Transition from Linear to Nonlinear Viscoelastic Behavior. 3. Linearity Below and Above TG," J. Macr. S. Ph., Part B, Vol. 6 (1972), p. 91.

[7.43] YUAN, H., and LIANIS, G., "Experimental Investigation of Nonlinear Viscoelasticity in Combined Finite Torsion-Tension," Trans. Soc. of Rheology, Vol. 16 (1972), p. 615.

[7.44] DEHOFF, P. H., and CHAKRABA, M. K., "Nonlinear Viscoelastic Behavior of Plasticized Polyvinyl Chloride," J. of Applied Polymer Science, Vol. 17 (1973), p. 3485.

[7.45] HAUPT, P., "Zur Viskoelastizität Inkompressibler Isotroper Stoffe," Zeitschrift für Angewandte Mathematik und Mechanik, Vol. 53 (1973), p. T71.

[7.46] BRERETON, M. G., CROLL, S. G., DUCKETT, R. A., and WARD, I. M., "Nonlinear Viscoelastic Behavior of Polymers: an Implicit Equation Approach," J. Mech. Phys. Solids, Vol. 22 (1974), p. 97.

[7.47] SADD, M. H., and BERNSTEIN, B., "Small Deformations on Large Deformations in BKZ Viscoelastic Liquids," Int. J. Eng. S., Vol. 12 (1974), p. 205.

[8.1] ONARAN, K. and FINDLEY, W. N., "Combined Stress Creep Experiments on a Non-Linear Viscoelastic Material to Determine the Kernel Functions for a Multiple Integral Representation of Creep", Transactions of the Society of Rheology, Vol. 9, issue 2 (1965), p. 299.

[8.2] LENOE, E. M., HELLER, R. A. and FREUDENTHAL, A. M., "Viscoelastic Behavior of a Filled Elastomer in the Linear and Nonlinear Range", Transactions of the Society of Rheology, Vol. 9, issue 2 (1965), p. 77.

[8.3] FINDLEY, W. N. and STANLEY, C. A., "Combined Stress Creep Experiments on Rigid Polyurethane Foam in the Nonlinear Range with Application to Multiple Integral and Modified Superposition Theory", ASTM Journal of Materials, Vol. 4 (1968), p. 916.

[8.4] O'CONNOR, D. G. and FINDLEY, W. N., "Influence of Normal Stress on Creep in Tension and Compression of Polyethylene and Rigid Polyvinyl Chloride Copolymer", Trans. ASME, J. of Engineering for Industry, Vol. 84, Series B (1962), p. 237.

*While the the citations under "Additional Literature" are presented here without evaluation it seems necessary to avoid confusing the reader to make an exception regarding [7.36]. It is doubtful that the major finding of this paper (that the first and second order terms of a multiple integral series are adequate to describe nonlinear behavior) is valid for any material in a general stress state. A pure shearing state of stress or strain cannot contain a second order term because of symmetry. Thus a multiple integral representation of a nonlinear material must contain at least a third or higher odd-order term.

[8.5] Nolte, K. G. and Findley, W. N., "Relationship Between the Creep of Solid and Foam Polyurethane Resulting from Combined Stresses", Trans. ASME, J. of Basic Engineering, Series D, Vol. 92, No. 1 (1970), p. 105.

[8.6] Freudenthal, A. M., "One-dimensional Response and Coefficient of Thermal Expansion in Time-sensitive Materials", Acta Technica, Vol. 41 (1962), p. 415.

[8.7] Findley, W. N., "Creep Characteristics of Plastics", Symposium on Plastics, ASTM (1944), p. 118.

[8.8] Findley, W. N. and Peterson, D. E., "Prediction of Long Time Creep with Ten-Year Creep Data on Four Plastic Laminates", Proceedings, ASTM, Vol. 58 (1958), p. 841.

[8.9] Findley, W. N. and Tracy, J. F., "16-year Creep of Polyethylene and PVC", Polymer Engineering and Science, Vol. 14 (1974), p. 577.

[8.10] Huang, N. C. and Lee, E. H., "Non-Linear Viscoelasticity for Short Time Ranges", Trans. ASME, J. of Applied Mechanics, Vol. 33E (1966), p. 313.

[8.11] Lai, J. S. Y. and Findley, W. N., "Stress Relaxation of Non-Linear Viscoelastic Material under Uniaxial Strain", Transactions of the Society of Rheology, Vol. 12, issue 2 (1968), p. 259.

[8.12] Findley, W. N., "Effect of Crystallinity and Crazing, Aging and Residual Stress on Creep of Monochlorotrifluoroethylene, Canvas Laminate, and Polyvinyl Chloride, Respectively", Proceedings, ASTM, Vol. 54 (1954), p. 1307.

[8.13] Ward, I. M. and Onat, E. T., "Non-Linear Mechanical Behavior of Oriented Polypropylene", J. of Mechanics and Physics of Solids, Vol. 11 (1963), p. 217.

[8.14] Leaderman, H., McCrackin, F. and Nakada, O., "Large Longitudinal Retarded Elastic Deformation of Rubberlike Network Polymers II. Application of a General Formulation of Non-Linear Response", Transactions of the Society of Rheology, Vol. 7 (1963), p. 111.

[8.15] Hadley, D. W. and Ward, I. M., "Non-Linear Creep and Recovery Behavior of Polypropylene Fibres", J. Mech. Phys. Solids, Vol. 13 (1965), p. 397.

[8.16] Ward, I. M. and Wolfe, J. M., "The Non-Linear Mechanical Behavior of Polypropylene Fibres under Complex Loading Programs", J. Mech. Phys. Solids, Vol. 14 (1966), p. 131.

[8.17] Lifshitz, J. M. and Kolsky, H., "Non-Linear Viscoelastic Behavior of Polyethylene", International Journal of Solids and Structures, Vol. 3 (1967), p. 383.

[8.18] Drescher, A. and Kwaszczynska, K., "An Approximate Description of Non-Linear Viscoelastic Materials", Int. J. Non-Linear Mechanics, Vol. 5 (1970), p. 11.

[8.19] Gottenberg, W. G., Bird, J. O. and Agrawal, G. L., "An Experimental Study of a Non-Linear Viscoelastic Solid in Uniaxial Tension", Trans. ASME, J. of Applied Mechanics, Vol. 36E (1969), p. 558.

[8.20] Adeyeri, J. B., Krizek, R. J. and Achenbach, J. D., "Multiple Integral Description of the Non-Linear Viscoelastic Behavior of a Clay Soil", Transactions of the Society of Rheology, Vol. 14, issue 3 (1970), p. 375.

[8.21] Youngs, R. L., "Mechanical Properties of Red Oak Related to Drying", Forest Products Journal, Vol. 10 (1957), p. 6.

[8.22] Davis, E. A., "Creep and Relaxation of Oxygen-Free Copper", J. App. Mech., Vol. 10 (1943), p. A101.

[8.23] Blass, J. J. and Findley, W. N., "Short-Time Biaxial Creep of an Aluminum Alloy with Abrupt Changes of Temperature and State of Stress", Trans. ASME, J. App. Mech., Vol. 38E (1971), p. 489.

[8.24] Lai, J. S. Y., "Nonlinear Viscoelastic-Plastic Properties of Asphalt Mixtures", University of Utah, UTEC CE 71–011, Jan. 1971.

[8.25] Findley, W. N. and Khosla, G., "An Equation for Tension Creep of Three Unfilled Thermoplastics", SPE Journal, Vol. 12 (1956), p. 20.

[8.26] LAI, J. S. Y. and FINDLEY, W. N., "Combined Tension-Torsion Creep Experiments on Polycarbonate in the Nonlinear Range", J. Polymer Eng. and Science, Vol. 9 (1969), p. 378.

[8.27] BOLLER, K. H., in discussion appended to reference [8.8].

[8.28] PARKER, E. R., "Modern Concepts of Flow and Fracture", Trans., ASM, Vol. 50 (1958), p. 52.

[8.29] FINDLEY, W. N. and MARK, R., "Multiaxial Creep Behavior of 304 Stainless Steel", Annual Report No. 1 to Oak Ridge National Laboratory, Brown University, Division of Engineering, October 1972.

[8.30] FINDLEY, W. N., REED, R. M. and STERN, P., "Hydrostatic Creep of Solid Plastics", Trans. ASME, J. App. Mech., Vol. 34E (1967), p. 895.

[8.31] HOJO, H. and FINDLEY, W. N., "Effect of Gas Diffusion on Creep Behavior of Polycarbonate", Polymer Engr. and Sci., Vol. 13 (1973), p. 255.

Additional Literature

[8.32] BENHAM, P. P., and MCCAMMOND, D., "Studies of Creep and Contraction Ratio in Thermoplastics," Plast. Polym., Vol. 39 (1971), p. 130.

[8.33] JAMPPANEN, P., "On Nonlinear Viscoelasticity Analysis with Application to Experiments," Acta Polytechnica Scandinavica, Physics including Nucleonics Series, Vol. 83 (1971), p. 49.

[8.34] LOU, Y. C., and SCHAPERY, R. A., "Viscoelastic Characteristics of a Nonlinear Fiber-Reinforced Plastic," J. Composite Materials, Vol. 5 (1971), p. 208.

[8.35] FOOT, C. J., and WARD, I. M., "Nonlinear Viscoelastic Behavior of Isotropic Polyethylene Terephthalate," J. Mech. Phys. Solids, Vol. 20 (1972), p. 165.

[8.36] KREGERS, A. F., "Investigation of the Integral Creep Equation (Cubic Nonlinearity) for Different Loading Paths," Polymer Mechanics, Vol. 6 (1972), p. 29.

[8.37] DE CANDIA, F., and VITORIA, V., "Elastic and Visoelastic Behavior of Ethylene-Propylene Copolymers," J. of Applied Polymer Science, Vol. 17 (1973), p. 3243.

[8.38] DISTEFANO, N., and TODESCHINI, R., "Modeling, Identification and Prediction of a Class of Nonlinear Viscoelastic Materials, Part I," Int. J. of Solids and Structures, Vol. 9 (1973), 805.

[8.39] DISTEFANO, N., and TODESCHINI, R., "Modeling, Identification and Prediction of a Class of Nonlinear Viscoelastic Materials, Part II," Int. J. of Solids and Structures, Vol. 9 (1973), p. 1431.

[8.40] HLAVACEK, B., SEYER, F. A., and STANISLAV, J., "Quantitative Analogies Between the Linear and the Nonlinear Viscoelastic Functions," Canadian Journal of Chemical Engineering, Vol. 51 (1973), p. 412.

[8.41] ZAKHARIEV, G., KHADZHIKOV, L., and MARINOV, P., "A Rheological Model for Polymers and Glass Reinforced Plastics," Polymer Mechanics, Vol. 7 (1974), p. 761.

[9.1] ONARAN, K. and FINDLEY, W. N., "Experimental Determination of Some Kernel Functions in the Multiple Integral Method for Non-Linear Creep of Polyvinyl Chloride", Trans. ASME, J. Appl. Mech., Vol. 38 E (1971), p. 30.

[9.2] NEIS, V. V. and SACKMAN, J. L., "An Experimental Study of a Non-linear Material with Memory", Trans. Soc. of Rheology, Vol. 11, issue 3 (1967), p. 307.

[9.3] NAKADA, O., "Theory of Non-linear Responses", J. of the Physical Society of Japan, Vol. 15 (1960), p. 2280.

[9.4] FINDLEY, W. N. and ONARAN, K., "Product Form of Kernel Functions for Non-linear Viscoelasticity of PVC under Constant Rate Stressing", Trans. Society of Rheology, Vol. 12 issue 2 (1968), p. 217.

[9.5] NOLTE, K. G. and FINDLEY, W. N., "A Linear Compressibility Assumption for the Multiple Integral Representation of Nonlinear Creep of Polyurethane", Trans. ASME, J. Appl. Mech., Vol. 37E (1970), p. 441.

[9.6] GOTTENBERG, W. G., BIRD, J. O. and AGRAWALL, G. L., "An Experimental Study of a Non-linear Viscoelastic Solid in Uniaxial Tension", Trans. ASME, J. of Appl. Mechanics, Vol. 36E (1969), p. 558.

[9.7] CHEUNG, J. B., "Nonlinear Viscoelastic Stress Analysis of Blood Vessels", Ph.D. thesis, University of Minnesota (1970).

[9.8] FINDLEY, W. N. and KHOSLA, G., "Application of the Superposition Principle and Theories of Mechanical Equation of State, Strain and Time Hardening to Creep of Plastic Under Changing Loads", J. Appl. Physics, Vol. 26 (1955), p. 821.

[9.9] FINDLEY, W. N. and LAI, J. S. Y., "A Modified Superposition Principle Applied to Creep of Non-linear Viscoelastic Material under Abrupt Changes in State of Combined Stress", Trans. Society of Rheology, Vol. 11 issue 3 (1967), p. 361.

[9.10] FINDLEY, W. N. and STANLEY, C. A., "Combined Stress Creep Experiments on Rigid Polyurethane Foam in the Non-linear Region with Application to Multiple Integral and Modified Superposition Theory", ASTM J. Mater., Vol. 4 (1968), p. 916.

[9.11] LAI, J. S. Y. and FINDLEY, W. N., "Stress Relaxation of Nonlinear Viscoelastic Material under Uniaxial Strain", Trans. Society of Rheology, Vol. 12, issue 2 (1968), p. 259.

[9.12] PIPKIN, A. C. and ROGERS, T. G., "A Nonlinear Integral Representation for Viscoelastic Behavior", J. of the Mechanics and Physics of Solids, Vol. 16 (1968), p. 59.

[9.13] TURNER, S., "The Strain Response of Plastics to Complex Stress Histories" Polymer Engineering and Science, Vol. 6 (1966), p. 306.

[9.14] ARUTYUNYAN, N. K., *Some Problems in the Theory of Creep*, translated from the Russian by H. E. Nowottny and Ed. by A. Graham, Pergamon Press, Oxford (1965).

[9.15] NOLTE, K. G. and FINDLEY, W. N., "Multiple Step Nonlinear Creep of Polyurethane Predicted from Constant Stress Creep by Three Integral Representations", Trans. Society of Rheology, Vol. 15, issue 1 (1971), p. 111.

[9.16] COLEMAN, B. D. and NOLL, W., "Foundations of Linear Viscoelasticity", Reviews of Modern Physics, Vol. 33 (1961), p. 239.

[9.17] STAFFORD, R. O., "On Mathematical Forms for the Material Functions in Nonlinear Viscoelasticity", J. Mech. Phys. Solids, Vol. 17 (1969), p. 339.

Additional Literature

[9.18] BENHAM, P. P., and HUTCHINSON, S. J., "A Comparison of Constant and Complex Creep Loading Programs for Several Thermoplastics," Polymer Engineering and Science, Vol. 11 (1971), p. 335.

[9.19] BROWN R. L., and SIDEBOTTOM, A. M., "A Comparison of Creep Theories for Multiaxial Loading of Polyethylene," Trans. Soc. of Rheology, Vol. 15 (1971), p. 3.

[9.20] LOCKETT, F. J., and TURNER, S., "Nonlinear Creep of Plastics," J, Mech. Phys. Solids, Vol. 19 (1971), p. 201.

[9.21] STOUFFER, D. C., and WINEMAN, A. S., "Constitutive Representation for Nonlinear, Aging, Environmental-dependent Viscoelastic Materials", Acta Mech., Vol. 13 (1972), p. 31.

[9.22] EWING, P. D., TURNER, S., and WILLIAMS, J. G., "Combined Tension-Torsion Creep of Polyethylene with Abrupt Changes of Stress," J. of Strain Analysis, Vol. 8 (1973), p. 83.

[10.1] ARUTYUNYAN, N. K., *Some Problems in the Theory of Creep*, translated from the Russian by H. E. Nowottny and edited by A. Graham, Pergamon Press, Oxford (1966).

[10.2] BYCHAWSKI, Z. and FOX, A., "Generalized Creep Function and the Problem of Inversion in the Theory of Nonlinear Viscoelasticity", Bull. L'Académie Polonaise des Sciences, Vol. 15 (1967), p. 389.

[10.3] LAI, J. S. Y. and FINDLEY, W. N., "Prediction of Uniaxial Stress Relaxation from Creep of Nonlinear Viscoelastic Material", Trans. Soc. Rheology, Vol. 12, Issue 2 (1968), p. 243.

[10.4] ABRAMOWITZ, M. and STEGUN, I., *Handbook of Mathematical Functions*, Dover, New York, 1965.

[10.5] LAI, J. S. Y. and FINDLEY, W. N., "Behavior of Nonlinear Viscoelastic Material Under Simultaneous Stress Relaxation in Tension and Creep in Torsion", Trans. ASME, J. Appl. Mech., Vol. 36E (1969), p. 22.

[10.6] NOLTE, K. G. and FINDLEY, W. N., "Approximation of Irregular Loading by Intervals of Constant Stress Rate to Predict Creep and Relaxation of Polyurethane by Three Integral Representations", Trans. Soc. Rheology, Vol. 18, Issue 1 (1974), p. 123.

Additional Literature

[10.7] MOLINARI, A., "Relation between Creep and Relaxation in Nonlinear Viscoelasticity," (in French), C. R. Ac. Sci. A, Vol. 277 (1973), p. 621.

[10.8] SERVAS, J. M., HUET, C., and MANDEL, J., "Correlation Between Flow and Relaxation Functions in Nonlinear Viscoelasticity," (in French), Cr. Ac. Sci. A, Vol. 277 (1973), p. 1003.

[11.1] LAI, J. S. Y. and FINDLEY, W. N., "Creep of Polyurethane Under Varying Temperature For Nonlinear Uniaxial Stress", Trans. Soc. of Rheology, Vol. 17, issue 1 (1973), p. 63.

[11.2] LAI, J. S. Y. and FINDLEY, W. N., "Elevated Temperature Creep of Polyurethane Under Nonlinear Torsional Stress with Step Changes in Torque"; Trans. Soc. of Rheology, Vol. 17, issue 1 (1973), p. 129.

[11.3] FINDLEY, W. N., "The Effect of Temperature and Combined Stresses On Creep of Plastics", Proceedings of Second Int. Reinforced Plastics Conf. Sponsored by the British Plastics Federation, London, England (1960), p. 16.1.

[11.4] KAUZMANN, W., "Flow of Solid Metals From the Standpoint of the Chemical Rate Theory", Trans. AIMME Inst. of Metals Div., Vol. 143 (1941), p. 57.

[11.5] BERNSTEIN, B., KEARSLY, E. A. and ZAPAS, L. J., "Thermodynamics of Perfect Elastic Fluids", J. of Research, N.B.S. 68B (1964), p. 103.

[11.6] LIANIS, G., "Integral Constitutive Equation of Nonlinear Thermoviscoelasticity", Purdue University Report. A & ES 65-1 (1965).

[11.7] MCGUIRT, C. W. and LIANIS, G., "Experimental Investigation of Non-Linear Non-Isothermal Viscoelasticity", Int. J. Engg. Sci., Vol. 7 (1969), p. 579.

[11.8] SCHAPERY, R. A., "A Theory of Nonlinear Thermoviscoelasticity Based on Irreversible Thermodynamics", Proc. 5th U. S. National Congress Appl. Mech., ASME (1966), p. 511.

[11.9] MARK, R. and FINDLEY, W. N., "Nonlinear Creep of Polyurethane under Combined Stresses and Elevated Temperature", Trans. Soc. of Rheology, Vol. 18, issue 4 (1974), p. 563.

[11.10] MARK, R. and FINDLEY, W. N., "Temperature-History Dependence in Combined Tension-Torsion Creep of Polyurethane under Varying Temperature", Trans. Soc. of Rheology, Vol. 19, issue 2 (1975), p. 201.

[11.11] SPENCER, R. S. and BOYER, R. F., "Thermal Expansion and Second-Order Transition Effects in High Polymers. III. Time Effects", J. Applied Physics, Vol. 17 (1946), p. 398.
[11.12] KOVÁCS, A. J., "La Construction Isotherme du Volume des Polymères Amorphes" J. Polymer Sci., Vol. 30 (1958), p. 131.
[11.13] FINDLEY, W. N. and REED, R. M., "Thermal Expansion Instability and Effect on Creep of a Highly Crosslinked Polyurethane", Poly. Engr. and Science, Vol. 14, (1974) p. 724.

Additional Literature

[11.14] MORGAN, C. J., and WARD, I. M., "The Temperature Dependence of Non-linear Creep and Recovery in Oriented Polypropylene," J. Mech. Phys. Solids, Vol. 22 (1971), p. 165.
[11.15] CROCHET, M. J., and NAGHDI, P. M., "A Class of Non-isothermal Viscoelastic Fluids," Int. J. Engr. Science, Vol. 10 (1972), p. 775.
[11.16] CROCHET, M. J., and NAGHDI, P. M., "On Thermal Effects in a Special Class of Viscoelastic Fluids," Univ. of Cal. Report No. AM-72-5, Nov. 1972.
[11.17] BIOT, M. A., "Nonlinear Thermoelasticity, Irreversible Thermodynamics and Elastic Instability," Indi. Math. J., Vol. 23 (1973), p. 309.
[11.18] COLEMAN, B. D., and DILL, E. H., "Thermodynamics and Stability of Motions of Materials with Memory," Arch. R. Mech., Vol. 51 (1973), p. 1
[11.19] CROCHET M. J., and NAGHDI, P. M., "On a Restricted Non-isothermal Theory of Simple Materials," Journal de Mécanique, Vol. 13 (1974), p. 97.
[11.20] PAVAN, A., RINK, M., BLUNDO, G., and DANUSSO, F., "Characterization of Thermomechanical Behavior of Polymers: Non-isothermal Creep Deflection Processes," Polymer, Vol. 15 (1974), p. 243.

12.1] FINDLEY, W. N. and POCZATEK, J. J., "Prediction of Creep-Deflection and Stress Distribution in Beams from Creep in Tension", Trans. ASME, J. of Appl. Mech., Vol. 22 (1955), p. 165.
[12.2] FINDLEY, W. N., POCZATEK, J. J. and MATHUR, P. N., "Prediction of Creep in Bending From Tension- and Compression-Creep Data when Creep Coefficients are Unequal", Trans. ASME, Vol. 80 (1958), p. 1294.
[12.3] HUANG, N. C. and LEE, E. H., "Nonlinear Viscoelasticity for Short-Time Ranges", Trans. ASME, J. of Appl. Mech., Vol. 33E (1966), p. 313.
[12.4] TING, E. C., "Stress Analysis for a Nonlinear Viscoelastic Cylinder with Ablating Inner Surface", Trans. ASME, J. of Appl. Mech., Vol. 37E (1970), p. 44.
[12.5] TING, E. C., "Stress Analysis for a Nonlinear Viscoelastic Compressible Cylinder," Tans. ASME, J. of Appl. Mech., Vol. 37E (1970), p. 1127.
[12.6] CHRISTENSEN, R. M., "On Obtaining Solutions in Nonlinear Viscoelasticity", Trans. ASME, J. of Appl. Mech., Vol. 35E (1968), p. 129.

[13.1] HETÉNYI, M., *Handbook of Experimental Stress Analysis*, Wiley, New York (1950).
[13.2] AVERY, D. H. and FINDLEY, W. N., "Quasi Static Mechanical Testing Techniques in Measurement of Mechanical Properties", *Techniques in Metals Research*, Ed. Bunshaw, R., Vol. V, part 1, Wiley, New York (1971).
[13.3] FINDLEY, W. N. and GJELSVIK, A., "A Biaxial Testing Machine for Plasticity, Creep, or Relaxation under Variable Principal-Stress Ratios", Proc. ASTM, Vol. 62 (1962), p. 1103.
[13.4] NOLTE, K. G. and FINDLEY, W. N., "Relationship Between the Creep of Solid and Foam Polyurethane Resulting from Combined Stresses", Trans. ASME, J. Basic Engineering, Vol. 92-D (1970), p. 105.

[13.5] Morrow, J. and Tuler, F. R., "Low-Cycle Fatigue Evaluation of Inconel 713C and Waspaloy", Paper 64-Met-15, AWS-ASME Metals Engineering Conference, Detroit, Mich. (1964).

[13.6] Gerard, G., "Compressive and Torsional Buckling of Thin-Walled Cylinders in the Yield Region", NACA TN-3726 (1956).

[13.7] Findley, W. N. and Reed, R. M., "Concerning Strain Gage Measurement of Creep of Plastics", Experimental Mechanics, Vol. 3 (1963), p. 29.

[13.8] Findley, W. N. and Reed, R. M., "An Extensometer for Circumferential Strains", ASTM Journal of Testing and Evaluation, Vol. 3 (1975), p. 300.

[13.9] Findley, W. N. and Reed, R. M., "A Constant Temperature and Humidity Room", Heating, Refrigeration and Air-Conditioning Journal, Vol. 12 (1970), p. 44.

[13.10] Hojo, H. and Findley, W. N., "Effect of Gas Diffusion on Creep Behavior of Polycarbonate", Polymer Engineering and Science, Vol. 13, No. 4 (1973), p. 255.

[13.11] Harrison, H. L. and Bollinger, J. G., *Introduction to Automatic Controls*, 2nd Ed., International, Scranton, Pa. (1969).

[13.12] Lai, J. S. Y. and Findley, W. N., "Stress Relaxation of Nonlinear Viscoelastic Material under Uniaxial Strain", Trans. Soc. Rheology, Vol. 12 issue 2 (1968), p. 259.

[A2.1] Green, A. E. and Rivlin, R. S., "The Mechanics of Nonlinear Materials with Memory, Part I", Archive for Rational Mechanics and Analysis, Vol. 1 (1957), p. 1.

[A2.2] Green, A. E. and Rivlin, R. S., "The Mechanics of Nonlinear Materials with Memory, Part III", Archive for Rational Mechanics and Analysis, Vol. 4 (1960), p. 387.

[A2.3] Green, A. E., Rivlin, R. S. and Spencer, A. J. M., "The Mechanics of Nonlinear Materials With Memory, Part II", Archive for Rational Mechanics and Analysis, Vol. 3 (1959), p. 82.

[A2.4] Noll, W., "A Mathematical Theory of the Mechanical Behavior of Continuous Media", Archive for Rational Mechanics and Analysis, Vol. 2 (1958), p. 197.

[A2.5] Pipkin, A. C., "Small Finite Deformations of Viscoelastic Solids", Reviews of Modern Physics, Vol. 36 (1964), p. 1034.

[A3.1] Sneddon, I. N., *Functional Analysis*, Encyclopedia of Physics, Ed. by S. Flügge, Vol. II, Math. Methods II, Springer-Verlag (1955).

[A3.2] Stackgold, I., *Boundary Value Problems of Mathematical Physics*, Vol. 1, McMillan Co., N.Y. (1967).

[A4.1] Erdélyi, A. et al., *Tables of Integral Transforms*, Vol. 1, McGraw-Hill, N.Y. (1954).

[A4.2] Scott, E. J., *Transform Calculus with an Introduction to Complex Variables*, Harper and Row, N.Y. (1955).

[A4.3] Churchill, R. C., *Operational Mathematics*, 2nd Ed., McGraw-Hill, N.Y. (1958).

[A4.4] Schapery, R. A., "Approximate Method of Transform Inversion for Viscoelastic Stress Analysis", Proc. Fourth U.S. Natl. Cong. App. Mech., ASME, N.Y. (1962), p. 1075.

[A5.1] Nolte, K. G. and Findley, W. N., "Multiple Step Nonlinear Creep of Polyurethane Predicted from Constant Stress Creep by Three Integral Representations", Trans. of Society of Rheology, Vol. 15, issue 1 (1971), p. 111.

[A6.1] MECHTLY, E. A., *The International System of Units – Physical Constants and Conversion Factors*, NASA SP-7012, 1969 revised edition, U.S. Government Printing Office.

[A6.2] PAGE, C. H. and VIGOUREUX, P., *The International System of Units (SI)*, NBS Special Publication 330, 1972 Edition, U.S. Government Printing Office.

[A6.3] HODGMAN, C. D., *Handbook of Chemistry and Physics*, The Chemical Rubber Pub. Co., Cleveland.

*Additional Literature for Chapter 2, "Historical Survey of Creep"**

1974 KREMPL, E., "Cyclic Creep: An Interpretive Literature Survey", WRC Bulletin 195, June.

YAMADA, H. and LI., C.-Y., "Stress Relaxation and Mechanical Equation of State in Austenitic Stainless Steel", Metallurgical Transactions, Vol. 4, pp. 2133–2136.

YAMADA, H. and LI, C.-Y., "Stress Relaxation and Mechanical Equation of State in BCC Metals in Monotonic Loading", Acta Metallurgica, Vol. 22, pp. 249–253.

1975 BHARGAVA, R., MOTEFF, J. and SWINDEMAN, R. W., "Correlation of the Microstructure with the Creep and Tensile Properties of A1S1 304 Stainless Steel", ASME Symposium on Structural Materials for Service at Elevated Temperatures in Nuclear Power Generation, Doc. MPC-1, pp. 31–54.

BODNER, S. R. and PARTOM, Y., "Constitutive Equations for Elastic Viscoplastic Strain Hardening Materials", Trans. ASME, J. Appl. Mech., Vol. 42, p. 385.

CRUME, S. V. and MEINECKE, E. A., Rheol. Acta, Vol. 14, p. 312.

GITTUS, J. H., *Creep Viscoelasticity and Creep Fracture in Solids*, Wiley, New York.

HART, E. W., LI, C.-Y., YAMADA, H. and WIRE, G. L., "Phenomenological Theory: A Guide to Constitutive Relations and Fundamental Deformation Properties", in *Constitutive Equations in Plasticity*, A. S. Argon (Ed.), MIT Press, pp. 149–197.

RICE, J. R., "Continuum Mechanics and Thermodynamics of Plasticity in Relation to Microscale Deformation Mechanics", in *Constitutive Equations in Plasticity*, Argon (Ed.), MIT Press.

SIKKA, V. K., BRINKMAN, C. R. and MCCOY, H. E., "Effect of Thermal Aging on Tensile and Creep Properties of Types 304 and 316 Stainless Steel", ASME Symposium on Structural Materials for Service at Elevated Temperatures in Nuclear Power Generation, Doc. MPC-1, pp. 316–350.

WOODFORD, D. A., "Measurement of the Mechanical State of a Low Alloy Steel at Elevated Temperature", Metallurgical Transactions, Vol. 6A, pp. 1693–1697.

1976 HART, E. W., "Constitutive Relations for the Nonelastic Deformation of Metals", Trans. ASME, J. Eng. Materials and Tech., Vol. 98, p. 193.

HUTCHINSON, J. W., "Bounds and Self-Consistent Estimates for Creep of Polycrystalline Materials", Proceedings of the Royal Society, London, Series A, Vol. 348, pp. 101–127.

MILLER, A. K., "An Inelastic Constitutive Model for Monotonic, Cyclic, and Creep Deformation: Part 1—Equations Development and Analytical Procedures, Part 2—Application to Type 304 Stainless Steel", Trans. ASME, J. Eng. Materials and Tech., Vol. 98, pp. 97–113.

PHILLIPS, A. and RICCIUTI, M., "Fundamental Experiments in Plasticity and Creep of Aluminum—Extension of Previous Results", International Journal of Solids Structure, Vol. 12, p. 159.

PONTER, A. R. S. and LECKIE, F. A., "Constitutive Relationships for the Time-Dependent Deformation of Metals", Trans. ASME, Journal of Engineering Materials and Technology, Vol. 98, pp. 47–51.

*These literature citation references were added to update the references subsequent to publication of the 1976 North-Holland Edition. Note that these references are not referred to in the text or in the index.

1977 FINDLEY, W. N. and REED, R. M., "Effect of Cross-Linking on Hydrostatic Creep of Epoxy", Polymer Engineering and Science, Vol. 17, pp. 837–841.
LIN, T. H., YU, C. L. and WENG, G. J., "Derivation of Polycrystal Creep Properties from the Creep Data of Single Crystals", Trans. ASME, Journal of Applied Mechanics, Vol. 44, pp. 73–78.
SWINDEMAN, R. W., BHARGAVA, R., SIKKA, V. K. and MOTEFF, J., "Substructures Developed During Creep and Cyclic Tests of Type 304 Stainless Steel (Heat 9T2796)", Oak Ridge National Laboratory, ORNL-5293, September.

1978 FINDLEY, W. N. and LAI, J. S., "Creep and Recovery of 2618 Aluminum Alloy Under Combined Stress with a Representation by a Viscous-Visco-Elastic Model", Trans. ASME, Journal of Applied Mechanics, Vol. 45, pp. 507–514.
GOODMAN, A. M., "Materials Data for High-Temperature Design", *Creep of Engineering Materials and Structures*, in G. Bernasconi and G. Piatti (Eds.), Applied Sciences Publishers, Ltd, London, pp. 289–339.
KREIG, R. D., SWEARENGEN, J. C. and ROHDE, R. W., "A Physically Based Internal Variable Model for Rate Dependent Plasticity", Inelastic Behavior of Pressure Vessel and Piping Components, PVP-PE-028 (ASME), p. 15.
MARK, R. and FINDLEY, W. N., "Thermal Expansion Instability and Creep in Amino-Cured Epoxy Resins", Polymer Engineering and Science, Vol. 18, pp. 6–15.
MARK, R. and FINDLEY, W. N., "Concerning a Creep Surface Derived from a Multiple Integral Representation for 304 Stainless Steel Under Combined Tension and Torsion", Trans. ASME, J. Applied Mechanics, Vol. 45, pp. 773–779.
MARK, R. and FINDLEY, W. N., "Nonlinear Variable Temperature Creep of Low Density Polyethylene", Journal of Rheology, Vol. 22, pp. 471–492.
MILLER, A. K., "A Unified Approach to Predicting Interactions Among Creep, Cyclic Plasticity, and Recovery", Nuclear Engineering and Design, Vol. 51, pp. 35–43.
PUGH, C. E. and ROBINSON, D. N., "Some Trends in Constitutive Equation Model Development for High-Temperature Behavior of Fast-Reactor Structural Alloys", Nuclear Engineering and Design, Vol. 48, pp. 269–276.
ROBINSON, D. N., "A Unified Creep-Plasticity Model for Structural Metals at High Temperature", ONRL-TM-5969, Oak Ridge National Laboratory, October.
SIDEBOTTOM, O. M., "Elevated-Temperature Creep and Relaxation of Torsion-Tension Members", Experimental Mechanics, Vol. 18, p. 121.

1979 FINDLEY, W. N., CHO, U. W. and DING, J. L., "Creep of Metals and Plastics Under Combined Stresses, A Review", Trans. ASME, J. Engineering Materials and Technology, Vol. 101, pp. 365–368.
FOX, ALFRED (Ed.), *Stress Relaxation Testing*, ASTM STP676, American Society for Testing and Materials.
HART, E. W., "Load Relaxation Testing and Material Constitutive Equations", in *Stress Relaxation Testing*, ASTM STP676, A. Fox (Ed.), pp. 5–20.
HENDERSON, J., "An Investigation of Multiaxial Creep Characteristics of Metals", Trans. ASME, Journal of Engineering Materials and Technology, Vol. 101, p. 356.
SWINDEMAN, R. W., "Correlation of Rupture Life, Creep Rate, and Microstructure for Type 304 Stainless Steel", Oak Ridge National Laboratory, ONRL-5523.
WENG, G. J., "A Physically Consistent Method for the Prediction of Creep Behavior of Metals", Trans. ASME, J. of Applied Mechanics, Vol. 46, pp. 800–804.

1980 CHO, U. W. and FINDLEY, W. N., "Creep and Creep Recovery of 304 Stainless Steel Under Combined Stress with a Representation by a Viscous-Viscoelastic Model", Trans. ASME, J. Applied Mechanics, Vol. 47, pp. 755–761.

KRAUS, H., *Creep Analysis*, Wiley, New York.
KUJAWSKI, D., KALLIANPUR, V. and KREMPL, E., "An Experimental Study of Uniaxial Creep, Cyclic Creep and Relaxation of A1S1 304 Stainless Steel at Room Temperature", Journal of Mechanics and Physics of Solids, Vol. 28, pp. 129–148.
LAI, J. S. and FINDLEY, W. N., "Creep of 2618 Aluminum Under Step Stress Changes Predicted by a Viscous-Viscoelastic Model", Trans. ASME, J. Applied Mechanics, Vol. 47, pp. 21–26.

1981 CHO, U. W. and FINDLEY, W. N., "Creep and Creep Recovery of 304 Stainless Steel at Low Stresses with Effects of Aging on Creep and Plastic Strains", Trans. ASME, J. Applied Mechanics, Vol. 48, pp. 785–790.
FINDLEY, W. N. and LAI, J. S., "Creep of 2618 Aluminum Under Side-Steps of Tension and Torsion and Stress Reversal Predicted by a Viscous-Viscoelastic Model", ASME, J. Applied Mechanics, Vol. 48, pp. 47–54.
KREMPL, E., "The Role of Aging in the Modelling of Elevated Temperature Deformation", in *Creep and Fracture of Engineering Materials and Structures*, B. Wilshire and D. R. J. Owen (Eds.), Pineridge Press, Swansea, U.K., pp. 201–211.
MURAKAMI, S., KUMAZAKI, H. and TAMAKI, K., "Experimental Study of Creep of Type 304 Stainless Steel Under Variable Stress", Preprint of JSME, No. 813-1, pp. 24–26 (in Japanese).
ONAT, E. T., "Representation of Inelastic Behavior", in *Creep and Fracture of Engineering Materials and Structures*, Wilshire and Owen (Eds.), Pineridge Press, Swansea, U.K., pp. 231–264.
REES, D. W. A., "Effects of Plastic Prestrain on the Creep of Aluminum Under Biaxial Stress", in *Creep and Fracture of Engineering Materials and Structures*, Wilshire and Owen (Eds.), Pineridge Press, Swansea, U.K., pp. 559–572.
WENG, G. J., "Self-Consistent Determination of Time-Dependent Behavior of Metals", ASME, J. of Applied Mechanics, Vol. 48, pp. 41–46.
WENG, G. J., "A Self-Consistent Scheme for the Relaxation Behavior of Metals", ASME, J. Applied Mechanics, Vol. 48, pp. 779–784.

1982 ALEXOPOULES, P., KEUSSEYAN, R. L., WIRE, G. L. and LI, C.-Y., "Experimental Investigation of Nonelastic Deformation Emphasizing Transient Phenomena by Using a State Variable Approach", in *Mechanical Testing for Deformation Model Development*, ASTM STP 765, R. W. Rohde and J. C. Swearengen (Eds.), pp. 148–184.
CHO, U. W. and FINDLEY, W. N., "Creep and Plastic Strains of 304 Stainless Steel at 593°C Under Step Stress Changes, Considering Aging", Trans. ASME, J. Applied Mechanics, Vol. 49, pp. 297–304.
FINDLEY, W. N., "A Compressive Creep Machine", Trans. ASTM, J. Testing and Evaluations, Vol. 10, pp. 179–180.
LAI, J. S. and FINDLEY, W. N., "Simultaneous Stress Relaxation in Tension and Creep in Torsion of 2618 Aluminum at Elevated Temperature", Trans. ASME, J. Applied Mechanics, Vol. 49, pp. 19–25.
MURAKAMI, S. and OHNO, N., "A Constitutive Equation of Creep Based on the Concept of a Creep-Hardening Surface", International Journal of Solids and Structures, Vol. 18, pp. 597–609.
OHASHI, Y. and OHNO, N., "Inelastic Stress-Responses of an Aluminum Alloy in Non-Proportional Deformations at Elevated Temperature", J. Mechanics and Physics of Solids, Vol. 30, pp. 287–304.
OHASHI, Y., OHNO, N. and KAWAI, M., "Evaluation of Creep Constitutive Equations for Type 304 Stainless Steel Under Repeated Multiaxial Loading", Trans. ASME, J. Engineering Materials and Technology, Vol. 104, pp. 159–164.
WENG, G. J., "A Unified, Self-Consistent Theory for the Plastic-Creep Deformation of Metals", Trans. ASME, J. Applied Mechanics, Vol. 104, pp. 728–734.

1983 CHO, U. W. and FINDLEY, W. N., "Creep and Plastic Strains Under Side Steps of Tension and Torsion for 304 Stainless Steel at 593°C", Trans. ASME, J. Applied Mechanics, Vol. 50, pp. 580–586.
CHO, U. W. and FINDLEY, W. N., "Creep and Plastic Strains Under Stress Reversal in Torsion with and without Simultaneous Tension for 304 Stainless Steel at 593°C", ASME, J. Appl. Mech., Vol. 50, pp. 587–592.
MEIJERS, P. and ROODE, F., "Experimental Verification of Constitutive Equations for Creep and Plasticity Based on Overlay Models", Trans. ASME, J. of Pressure Vessel Technology, Vol. 105, pp. 277–284.
OHASHI, Y., KAWAI, M. and SHIMIZU, H., "Effects of Prior Creep Subsequent Plasticity of Type 316 Stainless Steel at Elevated Temperature", Trans. ASME, J. Engineering Materials and Technology, Vol. 105, pp. 257–263.

1984 CHO, U. W. and FINDLEY, W. N., "Stress Relaxation in Tension Combined with Creep in Torsion of 304 Stainless Steel at 593°C", Proceedings Second International Conference on Creep and Fracture of Engineering Materials, Swansea, U.K., pp. 1334–1344.
CHO, U. W. and FINDLEY, W. N., "Creep and Creep Recovery of 2618-T61 Aluminum Under Variable Temperature", Trans. ASME, J. Appl. Mech., Vol. 106, pp. 816–820.
CHO, U. W. and FINDLEY, W. N., "Variable Temperature Creep and Creep Recovery of 304 Stainless Steel", Trans. ASME, J. Materials and Technology, Vol. 106, pp. 393–396.
DING, J. L. and FINDLEY, W. N., "48 Hour Multiaxial Creep and Creep Recovery of 2618 Aluminum Alloy at 200°C", Trans. ASME, J. Appl. Mech., Vol. 51, pp. 125–132.
DING, J. L. and FINDLEY, W. N., "Multiaxial Creep of 2618 Aluminum Under Proportional Loading Steps", Trans. ASME, J. Appl. Mech., Vol. 51, pp. 133–140.
DING, J. L. and FINDLEY, W. N., "Creep Experiments under Nonproportional Loadings with Stress Reversals for 2618 Aluminum", Trans. ASME, J. Engineering Materials and Technology, Vol. 106, pp. 397–404.
MROZ, Z. and TRAMPCZYNSKI, W. A., "On the Creep-Hardening Rule for Metals with a Memory of Maximal Prestress", International Journal of Solids and Structures, Vol. 20, pp. 467–486.

1985 DING, J. L. and FINDLEY, W. N., "Nonproportional Loading Steps in Multiaxial Creep of 2618 Aluminum", Trans. ASME, J. Applied Mechanics, Vol. 107, pp. 621–628.
OHASHI, Y., KAWAI, M. and KAITO, T., "Inelastic Behavior of Type 316 Stainless Steel Under Cyclic Stressings at Elevated Temperature", Trans. ASME, J. Engineering Materials and Technology, Vol. 107, pp. 101–109.
OHNO, N., MURAKAMI, S. and UENO, T., "A Constitutive Model of Creep Describing Creep Recovery and Material Softening Caused by Stress Reversals", ASME, J. Eng. Materials and Technology, Vol. 107, pp. 1–6.

1986 DING, J. L. and FINDLEY, W. N., "Simultaneous and Mixed Stress Relaxation in Tension and Creep in Torsion of 2618 Aluminum", Trans. ASME, J. Appl. Mech., Vol. 53, pp. 529–535.
FINDLEY, W. N., "Engineering Properties of Polymers", Invited Lecture ONR/NUSC Symposium Proc. on Dynamic Mechanical Properties of Elastomers, July 16–19, 1984, New London Laboratory, 7 Jan., Code 10.
MURAKAMI, S., OHNO, N. and TAGAMI, H., "Experimental Evaluation of Creep Constitutive Equations for Type 304 Stainless Steel under Non-Steady Multiaxial States of Stress", Trans. ASME, J. of Eng. Materials and Technology, Vol. 108, pp. 119–126.
OHASHI, Y., KAWAI, M. and MOMOSE, T., "Effects of Prior Plasticity on Subsequent Creep of Type 316 Stainless Steel at Elevated Temperatures", Trans. ASME, J. Engineering Materials and Technology, Vol. 108, pp. 69–73.

1987 DING, J. L. and FINDLEY, W. N., "On Using Stress Relaxation Tests to Characterize Material Behavior", Trans. ASME, J. App. Mech., Vol. 109, pp. 346–350.

FINDLEY, W. N., "26-Year Creep and Recovery of Poly(Vinyl Chloride) and Polyethylene", Polymer Engineering and Science, Vol. 27, pp. 582–585.

1988 ABDEL-KADER, M. S., EFTIS, J. and JONES, D. L., "Modelling the Viscoplastic Behavior of Inconel 718 at 1200° F", NASA Conference Publication 10010, pp. 37–68.

BASS, B. R., PUGH, C. E. and SWINDEMAN, R. W., "Applications of Elastic-Viscoplastic Constitutive Models in Dynamic Analyses of Crack Run-Arrest Events", NASA Conference Publication 10010, pp. 109–118.

DING, J. L. and LEE, S. R., "Development of Viscoplastic Constitutive Equation Through Biaxial Material Testing", Society for Experimental Mechanics, Experimental Mechanics, Vol. 28, No. 3, pp. 304–309.

DING, J. L. and LEE, S. R., "Constitutive Modelling of the Material Behavior at Elevated Temperatures and Related Material Testing", International Journal of Plasticity, Vol. 4, No. 2, pp. 149–161.

DING, J. L. and LEE, S. R., "Controlled-Strain Rate Tests at Very Low Strain Rates of 2618 Aluminum at 200° C", NASA Conference Publication 10010, pp. 225–238.

DING, J. L., LEE, S. R. and ORTIZ, R., "A Simple High Precision Biaxial Extensometer for High Temperature Creep Study", Trans. ASTM, J. Testing and Evaluation, Vol. 16, pp. 72–76.

EMRI, I., "Pressure Induced Ageing of Polymers", NASA Conference Publication 10010, pp. 77–88.

GU, R. J., "Creep Rupture Analysis of a Beam Resting on High Temperature Foundation", NASA Conference Publication 10010, pp. 137–186.

JAMES, G. H., IMBRIE, P. H., HILL, P. S., ALLEN, D. H. and HAISLER, W. E., "An Experimental Comparison of Several Current Viscoplastic Constitutive Models at Elevated Temperature", NASA Conference Publication 10010, pp. 253–290.

KIM, K. S., COOK, T. S. and MCKNIGHT, R. L., "Constitutive Response of Rene 80 under Thermal Mechanical Loads", NASA Conference Publication 10010, pp. 395–418.

KREMPL, E., LU, H. and YAO, D., "The Viscoplasticity Theory Based on Overstress Applied to the Modeling of a Nickel Base Superalloy at 815° C", NASA Conference Publication 10010, pp. 1–6.

KUMAR, S. and ARMENIADES, C., "Creep in Glossy Polymers and Composites: A Constitutive Equation", SPE Technical Papers, Vol. XXXIV, pp. 569–572.

LINDHOLM, U. S., CHAN, K. S., BODNER, S. R., WEBER, R. M. and WALKER, K. P., "Unified Constitutive Models for High-Temperature Structural Applications", NASA Conference Publication 10010, pp. 371–394.

ROBINSON, D. N. and ARNOLD, S. M., "Effects of State Recovery on Creep Buckling Under Variable Loading", NASA Conference Publication 10010, pp. 461–489.

SALEEB, A. F., CHANG, T. Y. and CHEN, J. Y., "On the Global/Local Time Incrementing for Viscoplastic Analysis", NASA Conference Publication 10010, pp. 307–316.

SHERWOOD, J. A. and STOUFFER, D. C., "A Constitutive Model with Damage for High Temperature Superalloys", NASA Conference Publication 10010, pp. 187–200.

SLAVIK, D. and SEHITOGLU, H., "A Unified Creep-Plasticity Model Suitable for Thermo-Mechanical Loading", NASA Conference Publication 10010, pp. 295–306.

SWINDEMAN, R. W., "Data Requirements to Model Creep in 9CR-IMO-V Steel", NASA Conference Publication 10010, pp. 217–224.

SUBJECT INDEX

AUTHOR INDEX

Italic numbers refer to pages in the Bibliography

A CATALOG OF SELECTED

DOVER BOOKS

IN ALL FIELDS OF INTEREST

A CATALOG OF SELECTED DOVER BOOKS IN ALL FIELDS OF INTEREST

DRAWINGS OF REMBRANDT, edited by Seymour Slive. Updated Lippmann, Hofstede de Groot edition, with definitive scholarly apparatus. All portraits, biblical sketches, landscapes, nudes. Oriental figures, classical studies, together with selection of work by followers. 550 illustrations. Total of 630pp. 9⅛ × 12¼.
21485-0, 21486-9 Pa., Two-vol. set $25.00

GHOST AND HORROR STORIES OF AMBROSE BIERCE, Ambrose Bierce. 24 tales vividly imagined, strangely prophetic, and decades ahead of their time in technical skill: "The Damned Thing," "An Inhabitant of Carcosa," "The Eyes of the Panther," "Moxon's Master," and 20 more. 199pp. 5⅜ × 8½. 20767-6 Pa. $3.95

ETHICAL WRITINGS OF MAIMONIDES, Maimonides. Most significant ethical works of great medieval sage, newly translated for utmost precision, readability. Laws Concerning Character Traits, Eight Chapters, more. 192pp. 5⅜ × 8½.
24522-5 Pa. $4.50

THE EXPLORATION OF THE COLORADO RIVER AND ITS CANYONS, J. W. Powell. Full text of Powell's 1,000-mile expedition down the fabled Colorado in 1869. Superb account of terrain, geology, vegetation, Indians, famine, mutiny, treacherous rapids, mighty canyons, during exploration of last unknown part of continental U.S. 400pp. 5⅜ × 8½. 20094-9 Pa. $6.95

HISTORY OF PHILOSOPHY, Julián Marías. Clearest one-volume history on the market. Every major philosopher and dozens of others, to Existentialism and later. 505pp. 5⅜ × 8½. 21739-6 Pa. $8.50

ALL ABOUT LIGHTNING, Martin A. Uman. Highly readable non-technical survey of nature and causes of lightning, thunderstorms, ball lightning, St. Elmo's Fire, much more. Illustrated. 192pp. 5⅜ × 8½. 25237-X Pa. $5.95

SAILING ALONE AROUND THE WORLD, Captain Joshua Slocum. First man to sail around the world, alone, in small boat. One of great feats of seamanship told in delightful manner. 67 illustrations. 294pp. 5⅜ × 8½. 20326-3 Pa. $4.95

LETTERS AND NOTES ON THE MANNERS, CUSTOMS AND CONDITIONS OF THE NORTH AMERICAN INDIANS, George Catlin. Classic account of life among Plains Indians: ceremonies, hunt, warfare, etc. 312 plates. 572pp. of text. 6⅛ × 9¼. 22118-0, 22119-9 Pa. Two-vol. set $15.90

ALASKA: The Harriman Expedition, 1899, John Burroughs, John Muir, et al. Informative, engrossing accounts of two-month, 9,000-mile expedition. Native peoples, wildlife, forests, geography, salmon industry, glaciers, more. Profusely illustrated. 240 black-and-white line drawings. 124 black-and-white photographs. 3 maps. Index. 576pp. 5⅜ × 8½. 25109-8 Pa. $11.95

THE BOOK OF BEASTS: Being a Translation from a Latin Bestiary of the Twelfth Century, T. H. White. Wonderful catalog real and fanciful beasts: manticore, griffin, phoenix, amphivius, jaculus, many more. White's witty erudite commentary on scientific, historical aspects. Fascinating glimpse of medieval mind. Illustrated. 296pp. 5⅜ × 8¼. (Available in U.S. only) 24609-4 Pa. $5.95

FRANK LLOYD WRIGHT: ARCHITECTURE AND NATURE With 160 Illustrations, Donald Hoffmann. Profusely illustrated study of influence of nature—especially prairie—on Wright's designs for Fallingwater, Robie House, Guggenheim Museum, other masterpieces. 96pp. 9¼ × 10¾. 25098-9 Pa. $7.95

FRANK LLOYD WRIGHT'S FALLINGWATER, Donald Hoffmann. Wright's famous waterfall house: planning and construction of organic idea. History of site, owners, Wright's personal involvement. Photographs of various stages of building. Preface by Edgar Kaufmann, Jr. 100 illustrations. 112pp. 9¼ × 10. 23671-4 Pa. $7.95

YEARS WITH FRANK LLOYD WRIGHT: Apprentice to Genius, Edgar Tafel. Insightful memoir by a former apprentice presents a revealing portrait of Wright the man, the inspired teacher, the greatest American architect. 372 black-and-white illustrations. Preface. Index. vi + 228pp. 8¼ × 11. 24801-1 Pa. $9.95

THE STORY OF KING ARTHUR AND HIS KNIGHTS, Howard Pyle. Enchanting version of King Arthur fable has delighted generations with imaginative narratives of exciting adventures and unforgettable illustrations by the author. 41 illustrations. xviii + 313pp. 6⅛ × 9¼. 21445-1 Pa. $6.50

THE GODS OF THE EGYPTIANS, E. A. Wallis Budge. Thorough coverage of numerous gods of ancient Egypt by foremost Egyptologist. Information on evolution of cults, rites and gods; the cult of Osiris; the Book of the Dead and its rites; the sacred animals and birds; Heaven and Hell; and more. 956pp. 6⅛ × 9¼. 22055-9, 22056-7 Pa., Two-vol. set $20.00

A THEOLOGICO-POLITICAL TREATISE, Benedict Spinoza. Also contains unfinished *Political Treatise.* Great classic on religious liberty, theory of government on common consent. R. Elwes translation. Total of 421pp. 5⅜ × 8½. 20249-6 Pa. $6.95

INCIDENTS OF TRAVEL IN CENTRAL AMERICA, CHIAPAS, AND YUCATAN, John L. Stephens. Almost single-handed discovery of Maya culture; exploration of ruined cities, monuments, temples; customs of Indians. 115 drawings. 892pp. 5⅜ × 8½. 22404-X, 22405-8 Pa., Two-vol. set $15.90

LOS CAPRICHOS, Francisco Goya. 80 plates of wild, grotesque monsters and caricatures. Prado manuscript included. 183pp. 6⅜ × 9⅝. 22384-1 Pa. $4.95

AUTOBIOGRAPHY: The Story of My Experiments with Truth, Mohandas K. Gandhi. Not hagiography, but Gandhi in his own words. Boyhood, legal studies, purification, the growth of the Satyagraha (nonviolent protest) movement. Critical, inspiring work of the man who freed India. 480pp. 5⅜ × 8½. (Available in U.S. only) 24593-4 Pa. $6.95

HOW TO WRITE, Gertrude Stein. Gertrude Stein claimed anyone could understand her unconventional writing—here are clues to help. Fascinating improvisations, language experiments, explanations illuminate Stein's craft and the art of writing. Total of 414pp. 4⅜ × 6⅜. 23144-5 Pa. $5.95

ADVENTURES AT SEA IN THE GREAT AGE OF SAIL: Five Firsthand Narratives, edited by Elliot Snow. Rare true accounts of exploration, whaling, shipwreck, fierce natives, trade, shipboard life, more. 33 illustrations. Introduction. 353pp. 5⅜ × 8½. 25177-2 Pa. $7.95

THE HERBAL OR GENERAL HISTORY OF PLANTS, John Gerard. Classic descriptions of about 2,850 plants—with over 2,700 illustrations—includes Latin and English names, physical descriptions, varieties, time and place of growth, more. 2,706 illustrations. xlv + 1,678pp. 8½ × 12¼. 23147-X Cloth. $75.00

DOROTHY AND THE WIZARD IN OZ, L. Frank Baum. Dorothy and the Wizard visit the center of the Earth, where people are vegetables, glass houses grow and Oz characters reappear. Classic sequel to *Wizard of Oz*. 256pp. 5⅜ × 8. 24714-7 Pa. $4.95

SONGS OF EXPERIENCE: Facsimile Reproduction with 26 Plates in Full Color, William Blake. This facsimile of Blake's original "Illuminated Book" reproduces 26 full-color plates from a rare 1826 edition. Includes "The Tyger," "London," "Holy Thursday," and other immortal poems. 26 color plates. Printed text of poems. 48pp. 5¼ × 7. 24636-1 Pa. $3.50

SONGS OF INNOCENCE, William Blake. The first and most popular of Blake's famous "Illuminated Books," in a facsimile edition reproducing all 31 brightly colored plates. Additional printed text of each poem. 64pp. 5¼ × 7. 22764-2 Pa. $3.50

PRECIOUS STONES, Max Bauer. Classic, thorough study of diamonds, rubies, emeralds, garnets, etc.: physical character, occurrence, properties, use, similar topics. 20 plates, 8 in color. 94 figures. 659pp. 6⅛ × 9¼. 21910-0, 21911-9 Pa., Two-vol. set $15.90

ENCYCLOPEDIA OF VICTORIAN NEEDLEWORK, S. F. A. Caulfeild and Blanche Saward. Full, precise descriptions of stitches, techniques for dozens of needlecrafts—most exhaustive reference of its kind. Over 800 figures. Total of 679pp. 8⅛ × 11. Two volumes. Vol. 1 22800-2 Pa. $11.95
Vol. 2 22801-0 Pa. $11.95

THE MARVELOUS LAND OF OZ, L. Frank Baum. Second Oz book, the Scarecrow and Tin Woodman are back with hero named Tip, Oz magic. 136 illustrations. 287pp. 5⅜ × 8½. 20692-0 Pa. $5.95

WILD FOWL DECOYS, Joel Barber. Basic book on the subject, by foremost authority and collector. Reveals history of decoy making and rigging, place in American culture, different kinds of decoys, how to make them, and how to use them. 140 plates. 156pp. 7⅞ × 10¾. 20011-6 Pa. $8.95

HISTORY OF LACE, Mrs. Bury Palliser. Definitive, profusely illustrated chronicle of lace from earliest times to late 19th century. Laces of Italy, Greece, England, France, Belgium, etc. Landmark of needlework scholarship. 266 illustrations. 672pp. 6⅛ × 9¼. 24742-2 Pa. $14.95

ILLUSTRATED GUIDE TO SHAKER FURNITURE, Robert Meader. All furniture and appurtenances, with much on unknown local styles. 235 photos. 146pp. 9 × 12. 22819-3 Pa. $7.95

WHALE SHIPS AND WHALING: A Pictorial Survey, George Francis Dow. Over 200 vintage engravings, drawings, photographs of barks, brigs, cutters, other vessels. Also harpoons, lances, whaling guns, many other artifacts. Comprehensive text by foremost authority. 207 black-and-white illustrations. 288pp. 6 × 9. 24808-9 Pa. $8.95

THE BERTRAMS, Anthony Trollope. Powerful portrayal of blind self-will and thwarted ambition includes one of Trollope's most heartrending love stories. 497pp. 5⅜ × 8½. 25119-5 Pa. $8.95

ADVENTURES WITH A HAND LENS, Richard Headstrom. Clearly written guide to observing and studying flowers and grasses, fish scales, moth and insect wings, egg cases, buds, feathers, seeds, leaf scars, moss, molds, ferns, common crystals, etc.—all with an ordinary, inexpensive magnifying glass. 209 exact line drawings aid in your discoveries. 220pp. 5⅜ × 8½. 23330-8 Pa. $3.95

RODIN ON ART AND ARTISTS, Auguste Rodin. Great sculptor's candid, wide-ranging comments on meaning of art; great artists; relation of sculpture to poetry, painting, music; philosophy of life, more. 76 superb black-and-white illustrations of Rodin's sculpture, drawings and prints. 119pp. 8⅝ × 11¼. 24487-3 Pa. $6.95

FIFTY CLASSIC FRENCH FILMS, 1912-1982: A Pictorial Record, Anthony Slide. Memorable stills from Grand Illusion, Beauty and the Beast, Hiroshima, Mon Amour, many more. Credits, plot synopses, reviews, etc. 160pp. 8¼ × 11. 25256-6 Pa. $11.95

THE PRINCIPLES OF PSYCHOLOGY, William James. Famous long course complete, unabridged. Stream of thought, time perception, memory, experimental methods; great work decades ahead of its time. 94 figures. 1,391pp. 5⅜ × 8½. 20381-6, 20382-4 Pa., Two-vol. set $19.90

BODIES IN A BOOKSHOP, R. T. Campbell. Challenging mystery of blackmail and murder with ingenious plot and superbly drawn characters. In the best tradition of British suspense fiction. 192pp. 5⅜ × 8½. 24720-1 Pa. $3.95

CALLAS: PORTRAIT OF A PRIMA DONNA, George Jellinek. Renowned commentator on the musical scene chronicles incredible career and life of the most controversial, fascinating, influential operatic personality of our time. 64 black-and-white photographs. 416pp. 5⅜ × 8¼. 25047-4 Pa. $7.95

GEOMETRY, RELATIVITY AND THE FOURTH DIMENSION, Rudolph Rucker. Exposition of fourth dimension, concepts of relativity as Flatland characters continue adventures. Popular, easily followed yet accurate, profound. 141 illustrations. 133pp. 5⅜ × 8½. 23400-2 Pa. $3.95

HOUSEHOLD STORIES BY THE BROTHERS GRIMM, with pictures by Walter Crane. 53 classic stories—Rumpelstiltskin, Rapunzel, Hansel and Gretel, the Fisherman and his Wife, Snow White, Tom Thumb, Sleeping Beauty, Cinderella, and so much more—lavishly illustrated with original 19th century drawings. 114 illustrations. x + 269pp. 5⅜ × 8½. 21080-4 Pa. $4.50

SUNDIALS, Albert Waugh. Far and away the best, most thorough coverage of ideas, mathematics concerned, types, construction, adjusting anywhere. Over 100 illustrations. 230pp. 5⅜ × 8½. 22947-5 Pa. $4.50

PICTURE HISTORY OF THE NORMANDIE: With 190 Illustrations, Frank O. Braynard. Full story of legendary French ocean liner: Art Deco interiors, design innovations, furnishings, celebrities, maiden voyage, tragic fire, much more. Extensive text. 144pp. 8⅜ × 11¾. 25257-4 Pa. $9.95

THE FIRST AMERICAN COOKBOOK: A Facsimile of "American Cookery," 1796, Amelia Simmons. Facsimile of the first American-written cookbook published in the United States contains authentic recipes for colonial favorites—pumpkin pudding, winter squash pudding, spruce beer, Indian slapjacks, and more. Introductory Essay and Glossary of colonial cooking terms. 80pp. 5⅜ × 8½. 24710-4 Pa. $3.50

101 PUZZLES IN THOUGHT AND LOGIC, C. R. Wylie, Jr. Solve murders and robberies, find out which fishermen are liars, how a blind man could possibly identify a color—purely by your own reasoning! 107pp. 5⅜ × 8½. 20367-0 Pa. $2.50

THE BOOK OF WORLD-FAMOUS MUSIC—CLASSICAL, POPULAR AND FOLK, James J. Fuld. Revised and enlarged republication of landmark work in musico-bibliography. Full information about nearly 1,000 songs and compositions including first lines of music and lyrics. New supplement. Index. 800pp. 5⅜ × 8¼. 24857-7 Pa. $14.95

ANTHROPOLOGY AND MODERN LIFE, Franz Boas. Great anthropologist's classic treatise on race and culture. Introduction by Ruth Bunzel. Only inexpensive paperback edition. 255pp. 5⅜ × 8½. 25245-0 Pa. $5.95

THE TALE OF PETER RABBIT, Beatrix Potter. The inimitable Peter's terrifying adventure in Mr. McGregor's garden, with all 27 wonderful, full-color Potter illustrations. 55pp. 4¼ × 5½. (Available in U.S. only) 22827-4 Pa. $1.75

THREE PROPHETIC SCIENCE FICTION NOVELS, H. G. Wells. *When the Sleeper Wakes, A Story of the Days to Come* and *The Time Machine* (full version). 335pp. 5⅜ × 8½. (Available in U.S. only) 20605-X Pa. $5.95

APICIUS COOKERY AND DINING IN IMPERIAL ROME, edited and translated by Joseph Dommers Vehling. Oldest known cookbook in existence offers readers a clear picture of what foods Romans ate, how they prepared them, etc. 49 illustrations. 301pp. 6⅛ × 9¼. 23563-7 Pa. $6.50

SHAKESPEARE LEXICON AND QUOTATION DICTIONARY, Alexander Schmidt. Full definitions, locations, shades of meaning of every word in plays and poems. More than 50,000 exact quotations. 1,485pp. 6½ × 9¼. 22726-X, 22727-8 Pa., Two-vol. set $27.90

THE WORLD'S GREAT SPEECHES, edited by Lewis Copeland and Lawrence W. Lamm. Vast collection of 278 speeches from Greeks to 1970. Powerful and effective models; unique look at history. 842pp. 5⅜ × 8½. 20468-5 Pa. $11.95

PLANTS OF THE BIBLE, Harold N. Moldenke and Alma L. Moldenke. Standard reference to all 230 plants mentioned in Scriptures. Latin name, biblical reference, uses, modern identity, much more. Unsurpassed encyclopedic resource for scholars, botanists, nature lovers, students of Bible. Bibliography. Indexes. 123 black-and-white illustrations. 384pp. 6 × 9. 25069-5 Pa. $8.95

FAMOUS AMERICAN WOMEN: A Biographical Dictionary from Colonial Times to the Present, Robert McHenry, ed. From Pocahontas to Rosa Parks, 1,035 distinguished American women documented in separate biographical entries. Accurate, up-to-date data, numerous categories, spans 400 years. Indices. 493pp. 6½ × 9¼. 24523-3 Pa. $9.95

THE FABULOUS INTERIORS OF THE GREAT OCEAN LINERS IN HISTORIC PHOTOGRAPHS, William H. Miller, Jr. Some 200 superb photographs capture exquisite interiors of world's great "floating palaces"—1890's to 1980's: *Titanic, Ile de France, Queen Elizabeth, United States, Europa,* more. Approx. 200 black-and-white photographs. Captions. Text. Introduction. 160pp. 8⅜ × 11¾. 24756-2 Pa. $9.95

THE GREAT LUXURY LINERS, 1927–1954: A Photographic Record, William H. Miller, Jr. Nostalgic tribute to heyday of ocean liners. 186 photos of Ile de France, Normandie, Leviathan, Queen Elizabeth, United States, many others. Interior and exterior views. Introduction. Captions. 160pp. 9 × 12. 24056-8 Pa. $9.95

A NATURAL HISTORY OF THE DUCKS, John Charles Phillips. Great landmark of ornithology offers complete detailed coverage of nearly 200 species and subspecies of ducks: gadwall, sheldrake, merganser, pintail, many more. 74 full-color plates, 102 black-and-white. Bibliography. Total of 1,920pp. 8⅝ × 11¼. 25141-1, 25142-X Cloth. Two-vol. set $100.00

THE SEAWEED HANDBOOK: An Illustrated Guide to Seaweeds from North Carolina to Canada, Thomas F. Lee. Concise reference covers 78 species. Scientific and common names, habitat, distribution, more. Finding keys for easy identification. 224pp. 5⅜ × 8½. 25215-9 Pa. $5.95

THE TEN BOOKS OF ARCHITECTURE: The 1755 Leoni Edition, Leon Battista Alberti. Rare classic helped introduce the glories of ancient architecture to the Renaissance. 68 black-and-white plates. 336pp. 8⅜ × 11¼. 25239-6 Pa. $14.95

MISS MACKENZIE, Anthony Trollope. Minor masterpieces by Victorian master unmasks many truths about life in 19th-century England. First inexpensive edition in years. 392pp. 5⅜ × 8½. 25201-9 Pa. $7.95

THE RIME OF THE ANCIENT MARINER, Gustave Doré, Samuel Taylor Coleridge. Dramatic engravings considered by many to be his greatest work. The terrifying space of the open sea, the storms and whirlpools of an unknown ocean, the ice of Antarctica, more—all rendered in a powerful, chilling manner. Full text. 38 plates. 77pp. 9¼ × 12. 22305-1 Pa. $4.95

THE EXPEDITIONS OF ZEBULON MONTGOMERY PIKE, Zebulon Montgomery Pike. Fascinating first-hand accounts (1805–6) of exploration of Mississippi River, Indian wars, capture by Spanish dragoons, much more. 1,088pp. 5⅜ × 8½. 25254-X, 25255-8 Pa. Two-vol. set $23.90

A CONCISE HISTORY OF PHOTOGRAPHY: Third Revised Edition, Helmut Gernsheim. Best one-volume history—camera obscura, photochemistry, daguerreotypes, evolution of cameras, film, more. Also artistic aspects—landscape, portraits, fine art, etc. 281 black-and-white photographs. 26 in color. 176pp. 8⅜ × 11¼. 25128-4 Pa. $12.95

THE DORÉ BIBLE ILLUSTRATIONS, Gustave Doré. 241 detailed plates from the Bible: the Creation scenes, Adam and Eve, Flood, Babylon, battle sequences, life of Jesus, etc. Each plate is accompanied by the verses from the King James version of the Bible. 241pp. 9 × 12. 23004-X Pa. $8.95

HUGGER-MUGGER IN THE LOUVRE, Elliot Paul. Second Homer Evans mystery-comedy. Theft at the Louvre involves sleuth in hilarious, madcap caper. "A knockout."—Books. 336pp. 5⅜ × 8½. 25185-3 Pa. $5.95

FLATLAND, E. A. Abbott. Intriguing and enormously popular science-fiction classic explores the complexities of trying to survive as a two-dimensional being in a three-dimensional world. Amusingly illustrated by the author. 16 illustrations. 103pp. 5⅜ × 8½. 20001-9 Pa. $2.25

THE HISTORY OF THE LEWIS AND CLARK EXPEDITION, Meriwether Lewis and William Clark, edited by Elliott Coues. Classic edition of Lewis and Clark's day-by-day journals that later became the basis for U.S. claims to Oregon and the West. Accurate and invaluable geographical, botanical, biological, meteorological and anthropological material. Total of 1,508pp. 5⅜ × 8½. 21268-8, 21269-6, 21270-X Pa. Three-vol. set $25.50

LANGUAGE, TRUTH AND LOGIC, Alfred J. Ayer. Famous, clear introduction to Vienna, Cambridge schools of Logical Positivism. Role of philosophy, elimination of metaphysics, nature of analysis, etc. 160pp. 5⅜ × 8½. (Available in U.S. and Canada only) 20010-8 Pa. $2.95

MATHEMATICS FOR THE NONMATHEMATICIAN, Morris Kline. Detailed, college-level treatment of mathematics in cultural and historical context, with numerous exercises. For liberal arts students. Preface. Recommended Reading Lists. Tables. Index. Numerous black-and-white figures. xvi + 641pp. 5⅜ × 8½. 24823-2 Pa. $11.95

28 SCIENCE FICTION STORIES, H. G. Wells. Novels, *Star Begotten* and *Men Like Gods*, plus 26 short stories: "Empire of the Ants," "A Story of the Stone Age," "The Stolen Bacillus," "In the Abyss," etc. 915pp. 5⅜ × 8½. (Available in U.S. only) 20265-8 Cloth. $10.95

HANDBOOK OF PICTORIAL SYMBOLS, Rudolph Modley. 3,250 signs and symbols, many systems in full; official or heavy commercial use. Arranged by subject. Most in Pictorial Archive series. 143pp. 8⅛ × 11. 23357-X Pa. $5.95

INCIDENTS OF TRAVEL IN YUCATAN, John L. Stephens. Classic (1843) exploration of jungles of Yucatan, looking for evidences of Maya civilization. Travel adventures, Mexican and Indian culture, etc. Total of 669pp. 5⅜ × 8½. 20926-1, 20927-X Pa., Two-vol. set $9.90

SIR HARRY HOTSPUR OF HUMBLETHWAITE, Anthony Trollope. Incisive, unconventional psychological study of a conflict between a wealthy baronet, his idealistic daughter, and their scapegrace cousin. The 1870 novel in its first inexpensive edition in years. 250pp. 5⅜ × 8½. 24953-0 Pa. $5.95

LASERS AND HOLOGRAPHY, Winston E. Kock. Sound introduction to burgeoning field, expanded (1981) for second edition. Wave patterns, coherence, lasers, diffraction, zone plates, properties of holograms, recent advances. 84 illustrations. 160pp. 5⅜ × 8¼. (Except in United Kingdom) 24041-X Pa. $3.50

INTRODUCTION TO ARTIFICIAL INTELLIGENCE: SECOND, ENLARGED EDITION, Philip C. Jackson, Jr. Comprehensive survey of artificial intelligence—the study of how machines (computers) can be made to act intelligently. Includes introductory and advanced material. Extensive notes updating the main text. 132 black-and-white illustrations. 512pp. 5⅜ × 8½. 24864-X Pa. $8.95

HISTORY OF INDIAN AND INDONESIAN ART, Ananda K. Coomaraswamy. Over 400 illustrations illuminate classic study of Indian art from earliest Harappa finds to early 20th century. Provides philosophical, religious and social insights. 304pp. 6⅜ × 9⅝. 25005-9 Pa. $8.95

THE GOLEM, Gustav Meyrink. Most famous supernatural novel in modern European literature, set in Ghetto of Old Prague around 1890. Compelling story of mystical experiences, strange transformations, profound terror. 13 black-and-white illustrations. 224pp. 5⅜ × 8½. (Available in U.S. only) 25025-3 Pa. $5.95

ARMADALE, Wilkie Collins. Third great mystery novel by the author of *The Woman in White* and *The Moonstone*. Original magazine version with 40 illustrations. 597pp. 5⅜ × 8½. 23429-0 Pa. $9.95

PICTORIAL ENCYCLOPEDIA OF HISTORIC ARCHITECTURAL PLANS, DETAILS AND ELEMENTS: With 1,880 Line Drawings of Arches, Domes, Doorways, Facades, Gables, Windows, etc., John Theodore Haneman. Sourcebook of inspiration for architects, designers, others. Bibliography. Captions. 141pp. 9 × 12. 24605-1 Pa. $6.95

BENCHLEY LOST AND FOUND, Robert Benchley. Finest humor from early 30's, about pet peeves, child psychologists, post office and others. Mostly unavailable elsewhere. 73 illustrations by Peter Arno and others. 183pp. 5⅜ × 8½. 22410-4 Pa. $3.95

ERTÉ GRAPHICS, Erté. Collection of striking color graphics: *Seasons, Alphabet, Numerals, Aces* and *Precious Stones*. 50 plates, including 4 on covers. 48pp. 9⅜ × 12¼. 23580-7 Pa. $6.95

THE JOURNAL OF HENRY D. THOREAU, edited by Bradford Torrey, F. H. Allen. Complete reprinting of 14 volumes, 1837-61, over two million words; the sourcebooks for *Walden*, etc. Definitive. All original sketches, plus 75 photographs. 1,804pp. 8½ × 12¼. 20312-3, 20313-1 Cloth., Two-vol. set $80.00

CASTLES: THEIR CONSTRUCTION AND HISTORY, Sidney Toy. Traces castle development from ancient roots. Nearly 200 photographs and drawings illustrate moats, keeps, baileys, many other features. Caernarvon, Dover Castles, Hadrian's Wall, Tower of London, dozens more. 256pp. 5⅜ × 8¼. 24898-4 Pa. $5.95